Microsoft® Visual C# 2008
Step by Step

John Sharp

图书在版编目(CIP)数据

Microsoft Visual C# 2008 从入门到精通:英文/(美)夏普(Master,C.)著. – 上海:上海世界图书出版公司,2009.1

ISBN 978 – 7 – 5062 – 9172 – 9

Ⅰ. M… Ⅱ. 夏… Ⅲ. C 语言 – 程序设计 – 英文 Ⅳ. TP312

中国版本图书馆 CIP 数据核字(2008)第 180648 号

Microsoft Visual C# 2008 从入门到精通

[美]约翰·夏普(Content Master) 著

上海世界图书出版公司 出版发行

上海市尚文路 185 号 B 楼

邮政编码 200010

(公司电话:021 – 63783016 转发行部)

上海竟成印务有限公司印刷

如发现印装质量问题,请与印刷厂联系

(质检科电话:021 – 56422678)

各地新华书店经销

开本:787×960 1/16 印张:44 字数:1400 000

2009 年 1 月第 1 版 2009 年 1 月第 1 次印刷

ISBN 978 – 7 – 5062 – 9172 – 9/T · 187

图字:09 – 2008 – 627 号

定价:188.00 元

http://www.wpcsh.com.cn

http://www.mspress.com.cn

Contents at a Glance

Table of Contents

What do you think of this book? We want to hear from you!

Microsoft is interested in hearing your feedback so we can continually improve our books and learning resources for you. To participate in a brief online survey, please visit:

www.microsoft.com/learning/booksurvey

v

Acknowledgments

An old Latin proverb says "Tempora mutantur, nos et mutantur in illis," which roughly translates into English as "Times change, and we change with them." This proverb has a quaint, sedate feel and was obviously penned before the Romans had heard of Microsoft, Windows, the .NET Framework, and C#; otherwise, they would have written something more like "Times change, and we run like mad trying to keep up!" When I look back over the last seven or eight years, I am absolutely flabbergasted to see how much the .NET Framework, and the C# language in particular, has evolved. I am also very thankful, because it keeps me in gainful employment, performing biannual updates on this book. I am not complaining because the .NET Framework is a superb platform for building applications and services, and I thank the visionaries in the various product groups at Microsoft who have dedicated several millennia of person-years of effort in its development. In my opinion, C# is the greatest vehicle for taking full advantage of the .NET Framework. I have thoroughly enjoyed watching its development and learning the new features that each new release provides. This book is my attempt to convey my enthusiasm for the language to other programmers who are just starting along the C# path of discovery.

As with all projects of this type, writing a book is a group effort. The team I have had the pleasure of working with at Microsoft Press is second to none. In particular, I would like to single out Lynn Finnel who has kept the faith in me over several editions of this book, Christina Palaia and Jennifer Harris for their thorough editing of my manuscripts, and Stephen Sagman who has worked like a Trojan keeping us all in order and on schedule. I must pay special thanks to Kurt Meyer for his sterling efforts in reviewing my work, correcting my mistakes, and suggesting modifications, and of course to Jon Jagger who coauthored the first edition of this book with me back in 2001.

My long-suffering family have been wonderful, as they always are. Diana is now familiar with terms such as "DLINQ" and "lambda expression" and throws them into conversation with effortless aplomb. (For example, "Will you ever stop talking about DLINQ and lambda expressions?") James is still convinced that I spend my life playing computer games rather than working. Francesca has developed a frowning nod that says, "I have no idea what you are talking about, but I will nod anyway in the hope that you might stop." And Ginger, my arch-competitor for the chair in my study, has tried her best to completely distract me and delay my efforts in the ways that only a cat can.

As ever, "Up the Gills!"

—John Sharp

Introduction

Microsoft Visual C# is a powerful but simple language aimed primarily at developers creating applications by using the Microsoft .NET Framework. It inherits many of the best features of C++ and Microsoft Visual Basic but few of the inconsistencies and anachronisms, resulting in a cleaner and more logical language. With the advent of C# 2.0 in 2005, several important new features were added to the language, including generics, iterators, and anonymous methods. C# 3.0, available as part of Microsoft Visual Studio 2008, adds further features, such as extension methods, lambda expressions, and, most famously of all, the Language Integrated Query facility, or LINQ. The development environment provided by Visual Studio 2008 makes these powerful features easy to use, and the many new wizards and enhancements included in Visual Studio 2008 can greatly improve your productivity as a developer.

Who This Book Is For

The aim of this book is to teach you the fundamentals of programming with C# by using Visual Studio 2008 and the .NET Framework version 3.5. You will learn the features of the C# language, and then use them to build applications running on the Microsoft Windows operating system. By the time you complete this book, you will have a thorough understanding of C# and will have used it to build Windows Presentation Foundation (WPF) applications, access Microsoft SQL Server databases, develop ASP.NET Web applications, and build and consume a Windows Communication Foundation service.

Finding Your Best Starting Point in This Book

This book is designed to help you build skills in a number of essential areas. You can use this book if you are new to programming or if you are switching from another programming language such as C, C++, Sun Microsystems Java, or Visual Basic. Use the following table to find your best starting point.

If you are	Follow these steps
New to object-oriented programming	1. Install the practice files as described in the next section, "Installing and Using the Practice Files." 2. Work through the chapters in Parts I, II, and III sequentially. 3. Complete Parts IV, V, and VI as your level of experience and interest dictates.
Familiar with procedural programming languages such as C, but new to C#	1. Install the practice files as described in the next section, "Installing and Using the Practice Files." Skim the first five chapters to get an overview of C# and Visual Studio 2008, and then concentrate on Chapters 6 through 21. 2. Complete Parts IV, V, and VI as your level of experience and interest dictates.
Migrating from an object-oriented language such as C++ or Java	1. Install the practice files as described in the next section, "Installing and Using the Practice Files." 2. Skim the first seven chapters to get an overview of C# and Visual Studio 2008, and then concentrate on Chapters 8 through 21. 3. For information about building Windows-based applications and using a database, read Parts IV and V. 4. For information about building Web applications and Web services, read Part VI.
Switching from Visual Basic 6	1. Install the practice files as described in the next section, "Installing and Using the Practice Files." 2. Work through the chapters in Parts I, II, and III sequentially. 3. For information about building Windows-based applications, read Part IV. 4. For information about accessing a database, read Part V. 5. For information about creating Web applications and Web services, read Part VI. 6. Read the Quick Reference sections at the end of the chapters for information about specific C# and Visual Studio 2008 constructs.
Referencing the book after working through the exercises	1. Use the index or the table of contents to find information about particular subjects. 2. Read the Quick Reference sections at the end of each chapter to find a brief review of the syntax and techniques presented in the chapter.

Conventions and Features in This Book

This book presents information using conventions designed to make the information readable and easy to follow. Before you start, read the following list, which explains conventions you'll see throughout the book and points out helpful features that you might want to use.

Conventions

- Each exercise is a series of tasks. Each task is presented as a series of numbered steps (1, 2, and so on). A round bullet (•) indicates an exercise that has only one step.

- Notes labeled "tip" provide additional information or alternative methods for completing a step successfully.

- Notes labeled "important" alert you to information you need to check before continuing.

- Text that you type appears in bold.

- A plus sign (+) between two key names means that you must press those keys at the same time. For example, "Press Alt+Tab" means that you hold down the Alt key while you press the Tab key.

Other Features

- Sidebars throughout the book provide more in-depth information about the exercise. The sidebars might contain background information, design tips, or features related to the information being discussed.

- Each chapter ends with a Quick Reference section. The Quick Reference section contains quick reminders of how to perform the tasks you learned in the chapter.

System Requirements

You'll need the following hardware and software to complete the practice exercises in this book:

- Windows Vista Home Premium Edition, Windows Vista Business Edition, or Windows Vista Ultimate Edition. The exercises will also run using Microsoft Windows XP Professional Edition with Service Pack 2

> **Important** If you are using Windows XP, some of the dialog boxes and screen shots described in this book might look a little different from those that you see. This is because of differences in the user interface in the Windows Vista operating system and the way in which Windows Vista manages security.

- Microsoft Visual Studio 2008 Standard Edition, Visual Studio 2008 Enterprise Edition, or Microsoft Visual C# 2008 Express Edition and Microsoft Visual Web Developer 2008 Express Edition

- Microsoft SQL Server 2005 Express Edition, Service Pack 2

- 1.6-GHz Pentium III+ processor, or faster

- 1 GB of available, physical RAM

- Video (800 × 600 or higher resolution) monitor with at least 256 colors

- CD-ROM or DVD-ROM drive

- Microsoft mouse or compatible pointing device

You will also need to have Administrator access to your computer to configure SQL Server 2005 Express Edition and to perform the exercises.

Code Samples

The companion CD inside this book contains the code samples that you'll use as you perform the exercises. By using the code samples, you won't waste time creating files that aren't relevant to the exercise. The files and the step-by-step instructions in the lessons also let you learn by doing, which is an easy and effective way to acquire and remember new skills.

Installing the Code Samples

Follow these steps to install the code samples and required software on your computer so that you can use them with the exercises.

1. Remove the companion CD from the package inside this book and insert it into your CD-ROM drive.

> **Note** An end-user license agreement should open automatically. If this agreement does not appear, open My Computer on the desktop or Start menu, double-click the icon for your CD-ROM drive, and then double-click StartCD.exe.

2. Review the end-user license agreement. If you accept the terms, select the accept option, and then click *Next*.

   A menu will appear with options related to the book.

3. Click *Install Code Samples*.

4. Follow the instructions that appear.

   The code samples are installed to the following location on your computer:

   Documents\Microsoft Press\Visual CSharp Step By Step

Using the Code Samples

Each chapter in this book explains when and how to use any code samples for that chapter. When it's time to use a code sample, the book will list the instructions for how to open the files.

> **Important** The code samples have been tested by using an account that is a member of the local Administrators group. It is recommended that you perform the exercises by using an account that has Administrator rights.

For those of you who like to know all the details, here's a list of the code sample Visual Studio 2008 projects and solutions, grouped by the folders where you can find them.

Project	Description
Chapter 1	
TextHello	This project gets you started. It steps through the creation of a simple program that displays a text-based greeting.
WPFHello	This project displays the greeting in a window by using Windows Presentation Foundation.
Chapter 2	
PrimitiveDataTypes	This project demonstrates how to declare variables by using each of the primitive types, how to assign values to these variables, and how to display their values in a window.
MathsOperators	This program introduces the arithmetic operators (+ − * / %).

Project	Description
Chapter 3	
Methods	In this project, you'll reexamine the code in the previous project and investigate how it uses methods to structure the code.
DailyRate	This project walks you through writing your own methods, running the methods, and stepping through the method calls by using the Visual Studio 2008 debugger.
Chapter 4	
Selection	This project shows how to use a cascading *if* statement to implement complex logic, such as comparing the equivalence of two dates.
SwitchStatement	This simple program uses a *switch* statement to convert characters into their XML representations.
Chapter 5	
WhileStatement	This project uses a *while* statement to read the contents of a source file one line at a time and display each line in a text box on a form.
DoStatement	This project uses a *do* statement to convert a decimal number to its octal representation.
Chapter 6	
MathsOperators	This project reexamines the MathsOperators project from Chapter 2, "Working with Variables, Operators, and Expressions," and causes various unhandled exceptions to make the program fail. The *try* and *catch* keywords then make the application more robust so that it no longer fails.
Chapter 7	
Classes	This project covers the basics of defining your own classes, complete with public constructors, methods, and private fields. It also shows how to create class instances by using the *new* keyword and how to define static methods and fields.
Chapter 8	
Parameters	This program investigates the difference between value parameters and reference parameters. It demonstrates how to use the *ref* and *out* keywords.
Chapter 9	
StructsAndEnums	This project defines a *struct* type to represent a calendar date.

Project	Description
Chapter 10	
Cards	This project uses the *ArrayList* collection class to group together playing cards in a hand.
Chapter 11	
ParamsArrays	This project demonstrates how to use the *params* keyword to create a single method that can accept any number of *int* arguments.
Chapter 12	
Vehicles	This project creates a simple hierarchy of vehicle classes by using inheritance. It also demonstrates how to define a virtual method.
ExtensionMethod	This project shows how to create an extension method for the *int* type, providing a method that converts an integer value from base 10 to a different number base.
Chapter 13	
Tokenizer	This project uses a hierarchy of interfaces and classes to simulate both reading a C# source file and classifying its contents into various kinds of tokens (identifiers, keywords, operators, and so on). As an example of use, it also derives classes from the key interfaces to display the tokens in a rich text box in color syntax.
Chapter 14	
UsingStatement	This project revisits a small piece of code from Chapter 5, "Using Compound Assignment and Iteration Statements," and reveals that it is not exception-safe. It shows you how to make the code exception-safe with a *using* statement.
Chapter 15	
WindowProperties	This project presents a simple Windows application that uses several properties to display the size of its main window. The display updates automatically as the user resizes the window.
AutomaticProperties	This project shows how to create automatic properties for a class and use them to initialize instances of the class.
Chapter 16	
Indexers	This project uses two indexers: one to look up a person's phone number when given a name, and the other to look up a person's name when given a phone number.
Chapter 17	
Delegates	This project displays the time in digital format by using delegate callbacks. The code is then simplified by using events.

Project	Description
Chapter 18	
BinaryTree	This solution shows you how to use generics to build a typesafe structure that can contain elements of any type.
BuildTree	This project demonstrates how to use generics to implement a typesafe method that can take parameters of any type.
Chapter 19	
BinaryTree	This project shows you how to implement the generic IEnumerator<T> interface to create an enumerator for the generic BinaryTree class.
IteratorBinaryTree	This solution uses an iterator to generate an enumerator for the generic BinaryTree class.
Chapter 20	
QueryBinaryTree	This project shows how to use LINQ queries to retrieve data from a binary tree object.
Chapter 21	
Operators	This project builds three structs, called Hour, Minute, and Second, that contain user-defined operators. The code is then simplified by using a conversion operator.
Chapter 22	
BellRingers	This project is a Windows Presentation Foundation application demonstrating how to define styles and use basic WPF controls.
Chapter 23	
BellRingers	This project is an extension of the application created in Chapter 22, "Introducing Windows Presentation Foundation," but with drop-down and pop-up menus added to the user interface.
Chapter 24	
CustomerDetails	This project demonstrates how to implement business rules for validating user input in a WPF application using customer information as an example.
Chapter 25	
ReportOrders	This project shows how to access a database by using ADO.NET code. The application retrieves information from the Orders table in the Northwind database.
DLINQOrders	This project shows how to use DLINQ to access a database and retrieve information from the Orders table in the Northwind database.

Project	Description
Chapter 26	
Suppliers	This project demonstrates how to use data binding with a WPF application to display and format data retrieved from a database in controls on a WPF form. The application also enables the user to modify information in the Products table in the Northwind database.
Chapter 27	
Litware	This project creates a simple Microsoft ASP.NET Web site that enables the user to input information about employees working for a fictitious software development company.
Chapter 28	
Litware	This project is an extended version of the Litware project from the previous chapter and shows how to validate user input in an ASP. NET Web application.
Chapter 29	
Northwind	This project shows how to use Forms-based security for authenticating the user. The application also demonstrates how to use ADO.NET from an ASP.NET Web form, showing how to query and update a database in a scalable manner, and how to create applications that span multiple Web forms.
Chapter 30	
NorthwindServices	This project implements a Windows Communication Foundation Web service, providing remote access across the Internet to data in the *Products* table in the Northwind database.

Uninstalling the Code Samples

Follow these steps to remove the code samples from your computer.

1. In *Control Panel*, open *Add or Remove Programs*.

2. From the list of Currently Installed Programs, select Microsoft Visual C# 2008 Step by Step.

3. Click *Remove*.

4. Follow the instructions that appear to remove the code samples.

Support for This Book

Every effort has been made to ensure the accuracy of this book and the contents of the companion CD. As corrections or changes are collected, they will be added to a Microsoft Knowledge Base article.

Microsoft Press provides support for books and companion CDs at the following Web site:

http://www.microsoft.com/learning/support/books/

Questions and Comments

If you have comments, questions, or ideas regarding the book or the companion CD, or questions that are not answered by visiting the site above, please send them to Microsoft Press via e-mail to

mspinput@microsoft.com

Or via postal mail to

Microsoft Press
Attn: *Microsoft Visual C# 2008 Step by Step* Series Editor
One Microsoft Way
Redmond, WA 98052-6399

Please note that Microsoft software product support is not offered through the above addresses.

Part I
Introducing Microsoft Visual C# and Microsoft Visual Studio 2008

Chapter 1
Welcome to C#

After completing this chapter, you will be able to:

- Use the Microsoft Visual Studio 2008 programming environment.

- Create a C# console application.

- Explain the purpose of namespaces.

- Create a simple graphical C# application.

Microsoft Visual C# is Microsoft's powerful component-oriented language. C# plays an important role in the architecture of the Microsoft .NET Framework, and some people have drawn comparisons to the role that C played in the development of UNIX. If you already know a language such as C, C++, or Java, you'll find the syntax of C# reassuringly familiar. If you are used to programming in other languages, you should soon be able to pick up the syntax and feel of C#; you just need to learn to put the braces and semicolons in the right place. Hopefully, this is just the book to help you!

In Part I, you'll learn the fundamentals of C#. You'll discover how to declare variables and how to use arithmetic operators such as the plus sign (+) and minus sign (–) to manipulate the values in variables. You'll see how to write methods and pass arguments to methods. You'll also learn how to use selection statements such as *if* and iteration statements such as *while*. Finally, you'll understand how C# uses exceptions to handle errors in a graceful, easy-to-use manner. These topics form the core of C#, and from this solid foundation, you'll progress to more advanced features in Part II through Part VI.

Beginning Programming with the Visual Studio 2008 Environment

Visual Studio 2008 is a tool-rich programming environment containing all the functionality you need to create large or small C# projects. You can even create projects that seamlessly combine modules compiled using different programming languages. In the first exercise, you start the Visual Studio 2008 programming environment and learn how to create a console application.

 Note A console application is an application that runs in a command prompt window, rather than providing a graphical user interface.

Create a console application in Visual Studio 2008

- If you are using Visual Studio 2008 Standard Edition or Visual Studio 2008 Professional Edition, perform the following operations to start Visual Studio 2008:

  1. On the Microsoft Windows task bar, click the *Start* button, point to *All Programs*, and then point to the *Microsoft Visual Studio 2008* program group.

  2. In the Microsoft Visual Studio 2008 program group, click *Microsoft Visual Studio 2008*.

  Visual Studio 2008 starts, like this:

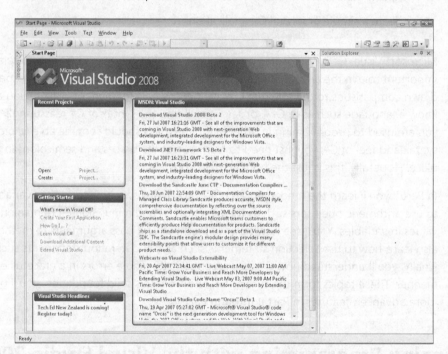

Note If this is the first time you have run Visual Studio 2008, you might see a dialog box prompting you to choose your default development environment settings. Visual Studio 2008 can tailor itself according to your preferred development language. The various dialog boxes and tools in the integrated development environment (IDE) will have their default selections set for the language you choose. Select *Visual C# Development Settings* from the list, and then click the *Start Visual Studio* button. After a short delay, the Visual Studio 2008 IDE appears.

- If you are using Visual C# 2008 Express Edition, on the Microsoft Windows task bar, click the *Start* button, point to *All Programs*, and then click *Microsoft Visual C# 2008 Express Edition*.

Visual C# 2008 Express Edition starts, like this:

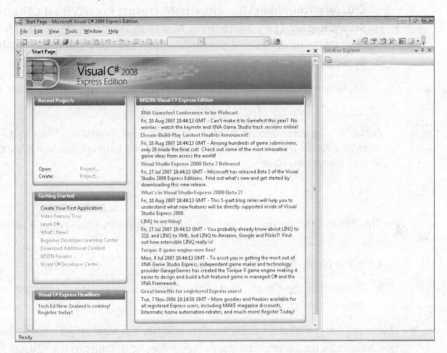

> **Note** To avoid repetition, throughout this book, I simply state, "Start Visual Studio" when you need to open Visual Studio 2008 Standard Edition, Visual Studio 2008 Professional Edition, or Visual C# 2008 Express Edition. Additionally, unless explicitly stated, all references to Visual Studio 2008 apply to Visual Studio 2008 Standard Edition, Visual Studio 2008 Professional Edition, and Visual C# 2008 Express Edition.

- If you are using Visual Studio 2008 Standard Edition or Visual Studio 2008 Professional Edition, perform the following tasks to create a new console application.

    1. On the *File* menu, point to *New*, and then click *Project*.

        The *New Project* dialog box opens. This dialog box lists the templates that you can use as a starting point for building an application. The dialog box categorizes templates according to the programming language you are using and the type of application.

    2. In the *Project types* pane, click *Visual C#*. In the *Templates* pane, click the *Console Application* icon.

3. In the *Location* field, if you are using the Windows Vista operating system, type **C:\Users*YourName*\Documents\Microsoft Press\Visual CSharp Step By Step\Chapter 1**. If you are using Microsoft Windows XP or Windows Server 2003, type **C:\Documents and Settings*YourName*\My Documents\Microsoft Press\Visual CSharp Step by Step\Chapter 1**.

Replace the text *YourName* in these paths with your Windows user name.

Note To save space throughout the rest of this book, I will simply refer to the path "C:\Users*YourName*\Documents" or "C:\Documents and Settings*YourName*\My Documents" as your Documents folder.

Tip If the folder you specify does not exist, Visual Studio 2008 creates it for you.

4. In the *Name* field, type **TextHello**.

5. Ensure that the *Create directory for solution* check box is selected, and then click *OK*.

- If you are using Visual C# 2008 Express Edition, the *New Project* dialog box won't allow you to specify the location of your project files; it defaults to the C:\Users*YourName*\AppData\Local\Temporary Projects folder. Change it by using the following procedure:

   1. On the *Tools* menu, click *Options*.

   2. In the *Options* dialog box, turn on the *Show All Settings* check box, and then click *Projects and Solutions* in the tree view in the left pane.

   3. In the right pane, in the *Visual Studio projects location* text box, specify the *Microsoft Press\Visual CSharp Step By Step\Chapter 1* folder under your Documents folder.

   4. Click *OK*.

- If you are using Visual C# 2008 Express Edition, perform the following tasks to create a new console application.

   1. On the *File* menu, click *New Project*.

   2. In the *New Project* dialog box, click the *Console Application* icon.

   3. In the *Name* field, type **TextHello**.

   4. Click *OK*.

Visual Studio creates the project using the Console Application template and displays the starter code for the project, like this:

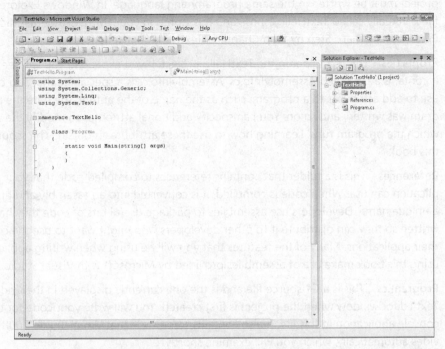

The *menu bar* at the top of the screen provides access to the features you'll use in the programming environment. You can use the keyboard or the mouse to access the menus and commands exactly as you can in all Windows-based programs. The *toolbar* is located beneath the menu bar and provides button shortcuts to run the most frequently used commands. The *Code and Text Editor* window occupying the main part of the IDE displays the contents of source files. In a multi-file project, when you edit more than one file, each source file has its own tab labeled with the name of the source file. You can click the tab to bring the named source file to the foreground in the *Code and Text Editor* window. The *Solution Explorer* displays the names of the files associated with the project, among other items. You can also double-click a file name in the *Solution Explorer* to bring that source file to the foreground in the *Code and Text Editor* window.

Before writing the code, examine the files listed in the *Solution Explorer*, which Visual Studio 2008 has created as part of your project:

■ **Solution 'TextHello'** This is the top-level solution file, of which there is one per application. If you use Windows Explorer to look at your Documents\Microsoft Press\Visual CSharp Step by Step\Chapter 1\TextHello folder, you'll see that the actual name of this file is TextHello.sln. Each solution file contains references to one or more project files.

- **TextHello**   This is the C# project file. Each project file references one or more files containing the source code and other items for the project. All the source code in a single project must be written in the same programming language. In Windows Explorer, this file is actually called TextHello.csproj, and it is stored in your \My Documents\Microsoft Press\Visual CSharp Step by Step\Chapter 1\TextHello\TextHello folder.

- **Properties**   This is a folder in the TextHello project. If you expand it, you will see that it contains a file called AssemblyInfo.cs. AssemblyInfo.cs is a special file that you can use to add attributes to a program, such as the name of the author, the date the program was written, and so on. You can specify additional attributes to modify the way in which the program runs. Learning how to use these attributes is outside the scope of this book.

- **References**   This is a folder that contains references to compiled code that your application can use. When code is compiled, it is converted into an assembly and given a unique name. Developers use assemblies to package useful bits of code they have written so they can distribute it to other developers who might want to use the code in their applications. Many of the features that you will be using when writing applications using this book make use of assemblies provided by Microsoft with Visual Studio 2008.

- **Program.cs**   This is a C# source file and is the one currently displayed in the Code and Text Editor window when the project is first created. You will write your code for the console application in this file. It also contains some code that Visual Studio 2008 provides automatically, which you will examine shortly.

Writing Your First Program

The Program.cs file defines a class called *Program* that contains a method called *Main*. All methods must be defined inside a class. You will learn more about classes in Chapter 7, "Creating and Managing Classes and Objects." The *Main* method is special—it designates the program's entry point. It must be a static method. (You will look at methods in detail in Chapter 3, "Writing Methods and Applying Scope," and I discuss static methods in Chapter 7.)

 Important   C# is a case-sensitive language. You must spell *Main* with a capital *M*.

In the following exercises, you'll write the code to display the message Hello World in the console; you'll build and run your Hello World console application; and you'll learn how namespaces are used to partition code elements.

Write the code by using IntelliSense

1. In the *Code and Text Editor* window displaying the Program.cs file, place the cursor in the *Main* method immediately after the opening brace, {, and then press Enter to create a new line. On the new line, type the word **Console**, which is the name of a built-in class. As you type the letter *C* at the start of the word *Console*, an IntelliSense list appears. This list contains all of the C# keywords and data types that are valid in this context. You can either continue typing or scroll through the list and double-click the Console item with the mouse. Alternatively, after you have typed *Con*, the IntelliSense list will automatically home in on the *Console* item and you can press the Tab or Enter key to select it.

Main should look like this:

```
static void Main(string[] args)
{
    Console
}
```

 Note *Console* is a built-in class that contains the methods for displaying messages on the screen and getting input from the keyboard.

2. Type a period immediately after *Console*. Another IntelliSense list appears, displaying the methods, properties, and fields of the *Console* class.

3. Scroll down through the list, select *WriteLine*, and then press Enter. Alternatively, you can continue typing the characters *W, r, i, t, e, L* until *WriteLine* is selected, and then press Enter.

The IntelliSense list closes, and the word *WriteLine* is added to the source file. *Main* should now look like this:

```
static void Main(string[] args)
{
    Console.WriteLine
}
```

4. Type an opening parenthesis , (. Another IntelliSense tip appears.

This tip displays the parameters that the *WriteLine* method can take. In fact, *WriteLine* is an *overloaded method*, meaning that the *Console* class contains more than one method named *WriteLine*—it actually provides 19 different versions of this method. Each version of the *WriteLine* method can be used to output different types of data. (Chapter 3 describes overloaded methods in more detail.) *Main* should now look like this:

```
static void Main(string[] args)
{
    Console.WriteLine(
}
```

Tip You can click the up and down arrows in the tip to scroll through the different overloads of *WriteLine*.

5. Type a closing parenthesis,) followed by a semicolon, ;.

Main should now look like this:

```
static void Main(string[] args)
{
    Console.WriteLine();
}
```

6. Move the cursor, and type the string **"Hello World"**, including the quotation marks, between the left and right parentheses following the *WriteLine* method.

Main should now look like this:

```
static void Main(string[] args)
{
    Console.WriteLine("Hello World");
}
```

Tip Get into the habit of typing matched character pairs, such as (and) and { and }, before filling in their contents. It's easy to forget the closing character if you wait until after you've entered the contents.

IntelliSense Icons

When you type a period after the name of a class, IntelliSense displays the name of every member of that class. To the left of each member name is an icon that depicts the type of member. Common icons and their types include the following:

Icon	Meaning
	method (discussed in Chapter 3)
	property (discussed in Chapter 15)
	class (discussed in Chapter 7)
	struct (discussed in Chapter 9)
	enum (discussed in Chapter 9)

Icon	Meaning
⊷O	interface (discussed in Chapter 13)
(delegate icon)	delegate (discussed in Chapter 17)
(extension method icon)	extension method (discussed in Chapter 12)

You will also see other IntelliSense icons appear as you type code in different contexts.

 Note You will frequently see lines of code containing two forward slashes followed by ordinary text. These are comments. They are ignored by the compiler but are very useful for developers because they help document what a program is actually doing. For example:

```
Console.ReadLine(); // Wait for the user to press the Enter key
```

The compiler will skip all text from the two slashes to the end of the line. You can also add multiline comments that start with a forward slash followed by an asterisk (/*). The compiler will skip everything until it finds an asterisk followed by a forward slash sequence (*/), which could be many lines lower down. You are actively encouraged to document your code with as many meaningful comments as necessary.

Build and run the console application

1. On the *Build* menu, click *Build Solution*.

   This action compiles the C# code, resulting in a program that you can run. The *Output* window appears below the *Code and Text Editor* window.

   Tip If the *Output* window does not appear, on the *View* menu, click *Output* to display it.

   In the *Output* window, you should see messages similar to the following indicating how the program is being compiled.

   ```
   ------ Build started: Project: TextHello, Configuration: Debug Any CPU ----
   C:\Windows\Microsoft.NET\Framework\v3.5\Csc.exe /config /nowarn:1701;1702 …
   Compile complete -- 0 errors, 0 warnings
   TextHello -> C:\Documents and Settings\John\My Documents\Microsoft Press\…
   ========== Build: 1 succeeded or up-to-date, 0 failed, 0 skipped ========
   ```

   If you have made some mistakes, they will appear in the *Error List* window. The following image shows what happens if you forget to type the closing quotation marks after the text Hello World in the *WriteLine* statement. Notice that a single mistake can sometimes cause multiple compiler errors.

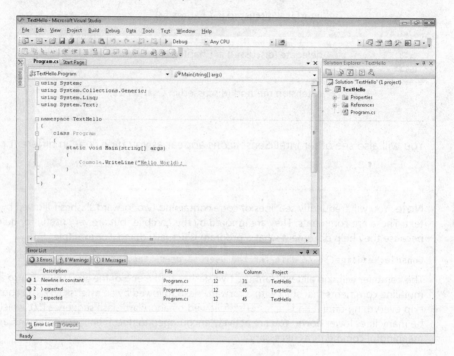

Tip You can double-click an item in the *Error List* window, and the cursor will be placed on the line that caused the error. You should also notice that Visual Studio displays a wavy red line under any lines of code that will not compile when you enter them.

If you have followed the previous instructions carefully, there should be no errors or warnings, and the program should build successfully.

Tip There is no need to save the file explicitly before building because the *Build Solution* command automatically saves the file. If you are using Visual Studio 2008 Standard Edition or Visual Studio 2008 Professional Edition, the project is saved in the location specified when you created it. If you are using Visual C# 2008 Express Edition, the project is saved in a temporary location and is copied to the folder you specified in the *Options* dialog box only when you explicitly save the project by using the *Save All* command on the *File* menu or when you close Visual C# 2008 Express Edition.

An asterisk after the file name in the tab above the *Code and Text Editor* window indicates that the file has been changed since it was last saved.

2. On the *Debug* menu, click *Start Without Debugging*.

A command window opens, and the program runs. The message Hello World appears, and then the program waits for you to press any key, as shown in the following graphic:

> **Note** The prompt "Press any key to continue . . ." is generated by Visual Studio; you did not write any code to do this. If you run the program by using the *Start Debugging* command on the *Debug* menu, the application runs, but the command window closes immediately without waiting for you to press a key.

3. Ensure that the command window displaying the program's output has the focus, and then press Enter.

The command window closes, and you return to the Visual Studio 2008 programming environment.

4. In *Solution Explorer*, click the TextHello project (not the solution), and then click the *Show All Files* toolbar button on the *Solution Explorer* toolbar—this is the second button from the left on the toolbar in the Solution Explorer window.

Entries named *bin* and *obj* appear above the Program.cs file. These entries correspond directly to folders named *bin* and *obj* in the project folder (Microsoft Press\Visual CSharp Step by Step\Chapter 1\TextHello\TextHello). Visual Studio creates these folders when you build your application, and they contain the executable version of the program together with some other files used to build and debug the application.

5. In *Solution Explorer*, click the plus sign (**+**) to the left of the *bin* entry.

Another folder named *Debug* appears.

> **Note** You may also see a folder called *Release*.

6. In *Solution Explorer*, click the plus sign (**+**) to the left of the *Debug* folder.

Four more items named TextHello.exe, TextHello.pdb, TextHello.vshost.exe, and TextHello.vshost.exe.manifest appear, like this:

Show All Files

 Note If you are using Visual C# 2008 Express Edition, you might not see all of these files.

The file TextHello.exe is the compiled program, and it is this file that runs when you click *Start Without Debugging* on the *Debug* menu. The other files contain information that is used by Visual Studio 2008 if you run your program in *Debug* mode (when you click *Start Debugging* on the *Debug* menu).

Using Namespaces

The example you have seen so far is a very small program. However, small programs can soon grow into much bigger programs. As a program grows, two issues arise. First, it is harder to understand and maintain big programs than it is to understand and maintain smaller programs. Second, more code usually means more names, more methods, and more classes. As the number of names increases, so does the likelihood of the project build failing because two or more names clash (especially when a program also uses third-party libraries written by developers who have also used a variety of names).

In the past, programmers tried to solve the name-clashing problem by prefixing names with some sort of qualifier (or set of qualifiers). This solution is not a good one because it's not scalable; names become longer, and you spend less time writing software and more time typing (there is a difference) and reading and rereading incomprehensibly long names.

Namespaces help solve this problem by creating a named container for other identifiers, such as classes. Two classes with the same name will not be confused with each other if they live in different namespaces. You can create a class named *Greeting* inside the namespace named *TextHello*, like this:

```
namespace TextHello
{
    class Greeting
    {
        ...
    }
}
```

You can then refer to the *Greeting* class as *TextHello.Greeting* in your programs. If another developer also creates a *Greeting* class in a different namespace, such as *NewNamespace*, and installs it on your computer, your programs will still work as expected because they are using the *TextHello.Greeting* class. If you want to refer to the other developer's *Greeting* class, you must specify it as *NewNamespace.Greeting*.

It is good practice to define all your classes in namespaces, and the Visual Studio 2008 environment follows this recommendation by using the name of your project as the top-level namespace. The .NET Framework software development kit (SDK) also adheres to this recommendation; every class in the .NET Framework lives inside a namespace. For example, the *Console* class lives inside the *System* namespace. This means that its full name is actually *System.Console*.

Of course, if you had to write the full name of a class every time you used it, the situation would be no better than prefixing qualifiers or even just naming the class with some globally unique name such *SystemConsole* and not bothering with a namespace. Fortunately, you can solve this problem with a *using* directive in your programs. If you return to the TextHello program in Visual Studio 2008 and look at the file Program.cs in the *Code and Text Editor* window, you will notice the following statements at the top of the file:

```
using System;
using System.Collections.Generic;
using System.Linq;
using System.Text;
```

A *using* statement brings a namespace into scope. In subsequent code in the same file, you no longer have to explicitly qualify objects with the namespace to which they belong. The four namespaces shown contain classes that are used so often that Visual Studio 2008 automatically adds these *using* statements every time you create a new project. You can add further *using* directives to the top of a source file.

The following exercise demonstrates the concept of namespaces in more depth.

Try longhand names

1. In the *Code and Text Editor* window displaying the Program.cs file, comment out the first *using* directive at the top of the file, like this:

```
//using System;
```

2. On the *Build* menu, click *Build Solution*.

   The build fails, and the *Error List* window displays the following error message:

```
The name 'Console' does not exist in the current context.
```

3. In the *Error List* window, double-click the error message.

   The identifier that caused the error is selected in the Program.cs source file.

4. In the *Code and Text Editor* window, edit the *Main* method to use the fully qualified name *System.Console*.

   Main should look like this:

```
static void Main(string[] args)
{
    System.Console.WriteLine("Hello World");
}
```

 Note   When you type *System.* the names of all the items in the *System* namespace are displayed by IntelliSense.

5. On the *Build* menu, click *Build Solution*.

   The build should succeed this time. If it doesn't, make sure that *Main* is exactly as it appears in the preceding code, and then try building again.

6. Run the application to make sure it still works by clicking *Start Without Debugging* on the *Debug* menu.

Namespaces and Assemblies

A *using* statement simply brings the items in a namespace into scope and frees you from having to fully qualify the names of classes in your code. Classes are compiled into *assemblies*. An assembly is a file that usually has the *.dll* file name extension, although strictly speaking, executable programs with the *.exe* file name extension are also assemblies.

An assembly can contain many classes. The classes that the .NET Framework class library comprises, such as *System.Console,* are provided in assemblies that are installed on your computer together with Visual Studio. You will find that the .NET Framework class library contains many thousands of classes. If they were all held in the same assembly, the assembly would be huge and difficult to maintain. (If Microsoft updated a single method in a single class, it would have to distribute the entire class library to all developers!)

For this reason, the .NET Framework class library is split into a number of assemblies, partitioned by the functional area to which the classes they contain relate. For example, there is a "core" assembly that contains all the common classes, such as *System.Console*, and there are further assemblies that contain classes for manipulating databases, accessing Web services, building graphical user interfaces, and so on. If you want to make use of a class in an assembly, you must add to your project a reference to that assembly. You can then add *using* statements to your code that bring the items in namespaces in that assembly into scope.

You should note that there is not necessarily a 1:1 equivalence between an assembly and a namespace; a single assembly can contain classes for multiple namespaces, and a single namespace can span multiple assemblies. This all sounds very confusing at first, but you will soon get used to it.

When you use Visual Studio to create an application, the template you select automatically includes references to the appropriate assemblies. For example, in *Solution Explorer* for the TextHello project, click the plus sign (+) to the left of the *References* folder. You will see that a Console application automatically includes references to assemblies called *System*, *System.Core*, *System.Data*, and *System.Xml*. You can add references for additional assemblies to a project by right-clicking the *References* folder and clicking *Add Reference*—you will practice performing this task in later exercises.

Creating a Graphical Application

So far, you have used Visual Studio 2008 to create and run a basic Console application. The Visual Studio 2008 programming environment also contains everything you need to create graphical Windows-based applications. You can design the form-based user interface of a Windows-based application interactively. Visual Studio 2008 then generates the program statements to implement the user interface you've designed.

Visual Studio 2008 provides you with two views of a graphical application: the *design view* and the *code view*. You use the *Code and Text Editor* window to modify and maintain the code and logic for a graphical application, and you use the *Design View* window to lay out your user interface. You can switch between the two views whenever you want.

In the following set of exercises, you'll learn how to create a graphical application by using Visual Studio 2008. This program will display a simple form containing a text box where you can enter your name and a button that displays a personalized greeting in a message box when you click the button.

> **Note** Visual Studio 2008 provides two templates for building graphical applications—the Windows Forms Application template and the WPF Application template. Windows Forms is a technology that first appeared with the .NET Framework version 1.0. WPF, or Windows Presentation Foundation, is an enhanced technology that first appeared with the .NET Framework version 3.0. It provides many additional features and capabilities over Windows Forms, and you should consider using it in preference to Windows Forms for all new development.

Create a graphical application in Visual Studio 2008

- If you are using Visual Studio 2008 Standard Edition or Visual Studio 2008 Professional Edition, perform the following operations to create a new graphical application:

  1. On the *File* menu, point to *New*, and then click *Project*.

     The *New Project* dialog box opens.

  2. In the *Project Types* pane, click *Visual C#*.

  3. In the *Templates* pane, click the *WPF Application* icon.

  4. Ensure that the *Location* field refers to your *Documents\Microsoft Press\Visual CSharp Step by Step\Chapter 1* folder.

  5. In the *Name* field, type **WPFHello**.

  6. In the *Solution* field, ensure that *Create new solution* is selected.

     This action creates a new solution for holding the project. The alternative, *Add to Solution*, adds the project to the TextHello solution.

  7. Click *OK*.

- If you are using Visual C# 2008 Express Edition, perform the following tasks to create a new graphical application.

  1. On the *File* menu, click *New Project*.

  2. If the *New Project* message box appears, click *Save* to save your changes to the TextHello project. In the *Save Project* dialog box, verify that the *Location* field is set to *Microsoft Press\Visual CSharp Step By Step\Chapter 1* under your Documents folder, and then click *Save*.

3. In the *New Project* dialog box, click the *WPF Application* icon.

4. In the *Name* field, type **WPFHello**.

5. Click *OK*.

Visual Studio 2008 closes your current application and creates the new WPF application. It displays an empty WPF form in the *Design View* window, together with another window containing an XAML description of the form, as shown in the following graphic:

> **Tip** Close the *Output* and *Error List* windows to provide more space for displaying the *Design View* window.

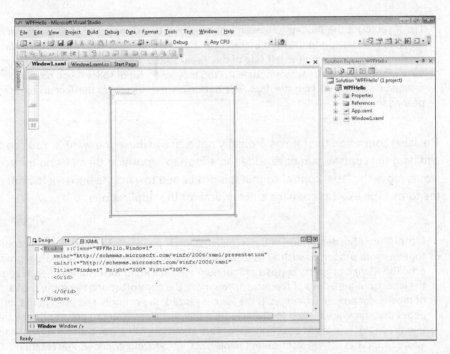

XAML stands for Extensible Application Markup Language and is an XML-like language used by WPF applications to define the layout of a form and its contents. If you have knowledge of XML, XAML should look familiar. You can actually define a WPF form completely by writing an XAML description if you don't like using the Design View window of Visual Studio or if you don't have access to Visual Studio; Microsoft provides an XAML editor called XAMLPad that is installed with the Windows SDK.

In the following exercise, you'll use the Design View window to add three controls to the Windows form and examine some of the C# code automatically generated by Visual Studio 2008 to implement these controls.

Create the user interface

1. Click the *Toolbox* tab that appears to the left of the form in the Design View window.

   The Toolbox appears, partially obscuring the form, and displaying the various components and controls that you can place on a Windows form. The Common section displays a list of controls that are used by most WPF applications. The Controls section displays a more extensive list of controls.

2. In the Common section, click Label, and then click the visible part of the form.

   A label control is added to the form (you will move it to its correct location in a moment), and the *Toolbox* disappears from view.

   Tip If you want the *Toolbox* to remain visible but not to hide any part of the form, click the *Auto Hide* button to the right in the *Toolbox* title bar (it looks like a pin). The *Toolbox* appears permanently on the left side of the Visual Studio 2008 window, and the *Design View* window shrinks to accommodate it. (You may lose a lot of space if you have a low-resolution screen.) Clicking the *Auto Hide* button once more causes the *Toolbox* to disappear again.

3. The label control on the form is probably not exactly where you want it. You can click and drag the controls you have added to a form to reposition them. Using this technique, move the label control so that it is positioned toward the upper-left corner of the form. (The exact placement is not critical for this application.)

   Note The XAML description of the form in the lower pane now includes the label control, together with properties such as its location on the form, governed by the *Margin* property. The *Margin* property consists of four numbers indicating the distance of each edge of the label from the edges of the form. If you move the control around the form, the value of the *Margin* property changes. If the form is resized, the controls anchored to the form's edges that move are resized to preserve their margin values. You can prevent this by setting the *Margin* values to zero. You learn more about the *Margin* and also the *Height* and *Width* properties of WPF controls in Chapter 22, "Introducing Windows Presentation Foundation."

4. On the *View* menu, click *Properties Window*.

   The *Properties* window appears on the lower-right side of the screen, under *Solution Explorer* (if it was not already displayed). The *Properties* window provides another way for you to modify the properties for items on a form, as well as other items in a project. It is context sensitive in that it displays the properties for the currently selected item. If you click the title bar of the form displayed in the *Design View* window, you can see that the *Properties* window displays the properties for the form itself. If you click the label control, the window displays the properties for the label instead. If you click anywhere else on the form, the *Properties* window displays the properties for a mysterious item called a *grid*. A grid acts as a container for items on a WPF form, and you can use the grid, among other things, to indicate how items on the form should be aligned and grouped together.

5. Click the label control on the form. In the *Properties* window, locate the *Text* section.

   By using the properties in this section, you can specify the font and font size for the label but not the actual text that the label displays.

6. Change the *FontSize* property to **20**, and then click the title bar of the form.

   The size of the text in the label changes, although the label is no longer big enough to display the text. Change the *FontSize* property back to **12**.

> **Note** The text displayed in the label might not resize itself immediately in the *Design View* window. It will correct itself when you build and run the application, or if you close and open the form in the *Design View* window.

7. Scroll the XAML description of the form in the lower pane to the right, and examine the properties of the label control.

   The label control consists of a `<Label>` tag containing property values, followed by the text for the label itself ("Label"), followed by a closing `</Label>` tag.

8. Change the text Label (just before the closing tag) to **Please enter your name**, as shown in the following image.

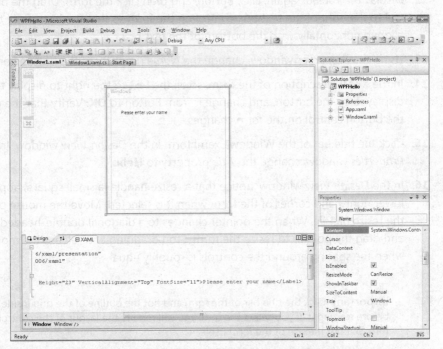

Notice that the text displayed in the label on the form changes, although the label is still too small to display it correctly.

9. Click the form in the *Design View* window, and then display the *Toolbox* again.

> **Note** If you don't click the form in the *Design View* window, the *Toolbox* displays the message "There are no usable controls in this group."

10. In the *Toolbox*, click *TextBox*, and then click the form. A text box control is added to the form. Move the text box control so that it is directly underneath the label control.

> **Tip** When you drag a control on a form, alignment indicators appear automatically when the control becomes aligned vertically or horizontally with other controls. This gives you a quick visual cue for making sure that controls are lined up neatly.

11. While the text box control is selected, in the *Properties* window, change the value of the *Name* property displayed at the top of the window to **userName**.

> **Note** You will learn more about naming conventions for controls and variables in Chapter 2, "Working with Variables, Operators, and Expressions."

12. Display the *Toolbox* again, click *Button*, and then click the form. Drag the button control to the right of the text box control on the form so that the bottom of the button is aligned horizontally with the bottom of the text box.

13. Using the *Properties* window, change the *Name* property of the button control to **ok**.

14. In the XAML description of the form, scroll the text to the right to display the caption displayed by the button, and change it from Button to **OK**. Verify that the caption of the button control on the form changes.

15. Click the title bar of the Window1.xaml form in the *Design View* window. In the *Properties* window, change the *Title* property to **Hello**.

16. In the *Design View* window, notice that a resize handle (a small square) appears on the lower right-hand corner of the form when it is selected. Move the mouse pointer over the resize handle. When the pointer changes to a diagonal double-headed arrow, click and drag the pointer to resize the form. Stop dragging and release the mouse button when the spacing around the controls is roughly equal.

> **Important** Click the title bar of the form and not the outline of the grid inside the form before resizing it. If you select the grid, you will modify the layout of the controls on the form but not the size of the form itself.

Note If you make the form narrower, the *OK* button remains a fixed distance from the right-hand edge of the form, determined by its *Margin* property. If you make the form too narrow, the *OK* button will overwrite the text box control. The right-hand margin of the label is also fixed, and the text for the label will start to disappear when the label shrinks as the form becomes narrower.

The form should now look similar to this:

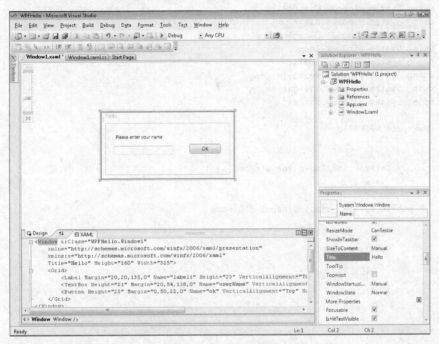

17. On the *Build* menu, click *Build Solution*, and verify that the project builds successfully.

18. On the *Debug* menu, click *Start Without Debugging*.

   The application should run and display your form. You can type your name in the text box and click *OK*, but nothing happens yet. You need to add some code to process the *Click* event for the *OK* button, which is what you will do next.

19. Click the *Close* button (the *X* in the upper-right corner of the form) to close the form and return to Visual Studio.

You have managed to create a graphical application without writing a single line of C# code. It does not do much yet (you will have to write some code soon), but Visual Studio actually generates a lot of code for you that handles routine tasks that all graphical applications must perform, such as starting up and displaying a form. Before adding your own code to the application, it helps to have an understanding of what Visual Studio has generated for you.

In *Solution Explorer*, click the plus sign (**+**) beside the file Window1.xaml. The file Window1.xaml.cs appears. Double-click the file Window1.xaml.cs. The code for the form is displayed in the *Code and Text Editor* window. It looks like this:

```
using System;
using System.Collections.Generic;
using System.Linq;
using System.Text;
using System.Windows;
using System.Windows.Controls;
using System.Windows.Data;
using System.Windows.Documents;
using System.Windows.Input;
using System.Windows.Media;
using System.Windows.Media.Imaging;
using System.Windows.Navigation;
using System.Windows.Shapes;

namespace WPFHello
{
    /// <summary>
    /// Interaction logic for Window1.xaml
    /// </summary>

    public partial class Window1 : Window
    {

        public Window1()
        {
            InitializeComponent();
        }

    }
}
```

Apart from a good number of *using* statements bringing into scope some namespaces that most WPF applications use, the file contains the definition of a class called *Window1* but not much else. There is a little bit of code for the *Window1* class known as a constructor that calls a method called *InitializeComponent*, but that is all. (A *constructor* is a special method with the same name as the class. It is executed when an instance of the class is created and can contain code to initialize the instance. You will learn about constructors in Chapter 7.) In fact, the application contains a lot more code, but most of it is generated automatically based on the XAML description of the form, and it is hidden from you. This hidden code performs operations such as creating and displaying the form, and creating and positioning the various controls on the form.

The purpose of the code that you *can* see in this class is so that you can add your own methods to handle the logic for your application, such as what happens when the user clicks the *OK* button.

Tip You can also display the C# code file for a WPF form by right-clicking anywhere in the *Design View* window and then clicking *View Code*.

At this point you might well be wondering where the *Main* method is and how the form gets displayed when the application runs; remember that *Main* defines the point at which the program starts. In *Solution Explorer*, you should notice another source file called App.xaml. If you double-click this file, the *Design View* window displays the message "Intentionally Left Blank," but the file has an XAML description. One property in the XAML code is called *StartupUri*, and it refers to the Window1.xaml file as shown here:

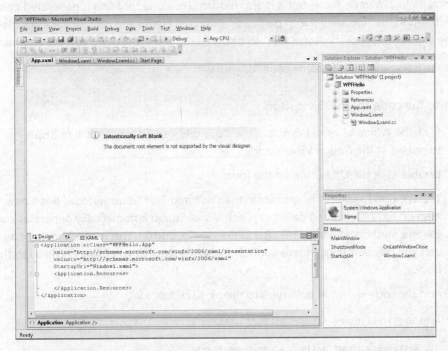

If you click the plus sign (+) adjacent to App.xaml in *Solution Explorer*, you will see that there is also an Application.xaml.cs file. If you double-click this file, you will find it contains the following code:

```
using System;
using System.Collections.Generic;
using System.Configuration;
using System.Data;
using System.Linq;
using System.Windows;
```

```
namespace WPFHello
{
    /// <summary>
    /// Interaction logic for App.xaml
    /// </summary>

    public partial class App : Application
    {

    }
}
```

Once again, there are a number of *using* statements, but not a lot else, not even a *Main* method. In fact, *Main* is there, but it is also hidden. The code for *Main* is generated based on the settings in the App.xaml file; in particular, *Main* will create and display the form specified by the *StartupUri* property. If you want to display a different form, you edit the App.xaml file.

The time has come to write some code for yourself!

Write the code for the OK button

1. Click the *Window1.xaml* tab above the *Code and Text Editor* window to display Window1 in the *Design View* window.

2. Double-click the *OK* button on the form.

   The Window1.xaml.cs file appears in the *Code and Text Editor* window, but a new method has been added called *ok_Click*. Visual Studio automatically generates code to call this method whenever the user clicks the *OK* button. This is an example of an event, and you will learn much more about how events work as you progress through this book.

3. Add the code shown in bold type to the *ok_Click* method:

   ```
   void ok_Click(object sender, RoutedEventArgs e)
   {
       MessageBox.Show("Hello " + userName.Text);
   }
   ```

   This is the code that will run when the user clicks the *OK* button. Do not worry too much about the syntax of this code just yet (just make sure you copy it exactly as shown) because you will learn all about methods in Chapter 3. The interesting part is the *MessageBox.Show* statement. This statement displays a message box containing the text "Hello" with whatever name the user typed into the username text box on the appended form.

4. Click the *Window1.xaml* tab above the *Code and Text Editor* window to display Window1 in the *Design View* window again.

5. In the lower pane displaying the XAML description of the form, examine the *Button* element, but be careful not to change anything. Notice that it contains an element called *Click* that refers to the *ok_Click* method:

```
<Button Height="23" … Click="ok_Click">OK</Button>
```

6. On the *Debug* menu, click *Start Without Debugging*.

7. When the form appears, type your name in the text box, and then click *OK*. A message box appears, welcoming you by name.

8. Click *OK* in the message box.

   The message box closes.

9. Close the form.

   - If you want to continue to the next chapter

     Keep Visual Studio 2008 running, and turn to Chapter 2.

   - If you want to exit Visual Studio 2008 now

     On the *File* menu, click *Exit*. If you see a *Save* dialog box, click *Yes* (if you are using Visual Studio 2008) or *Save* (if you are using Visual C# 2008 Express Edition) and save the project.

Chapter 1 Quick Reference

To	Do this	Key combination
Create a new console application using Visual Studio 2008 Standard or Professional Edition	On the *File* menu, point to *New*, and then click *Project* to open the *New Project* dialog box. For the project type, select *Visual C#*. For the template, select *Console Application*. Select a directory for the project files in the *Location* box. Choose a name for the project. Click *OK*.	
Create a new console application using Visual C# 2008 Express Edition	On the *Tools* menu, click *Options*. In the *Options* dialog box, click *Projects and Solutions*. In the *Visual Studio projects location* box, specify a directory for the project files. On the *File* menu, click *New Project* to open the *New Project* dialog box. For the template, select *Console Application*. Choose a name for the project. Click *OK*.	
Create a new graphical application using Visual Studio 2008 Standard or Professional Edition	On the *File* menu, point to *New*, and then click *Project* to open the *New Project* dialog box. For the project type, select Visual C#. For the template, select *WPF Application*. Select a directory for the project files in the *Location* box. Choose a name for the project. Click *OK*.	
Create a new graphical application using Visual C# 2008 Express Edition	On the *Tools* menu, click *Options*. In the *Options* dialog box, click *Projects and Solutions*. In the *Visual Studio projects location* box, specify a directory for the project files. On the *File* menu, click *New Project* to open the *New Project* dialog box. For the template, select *WPF Application*. Choose a name for the project. Click *OK*.	
Build the application	On the *Build* menu, click *Build Solution*.	F6
Run the application	On the *Debug* menu, click *Start Without Debugging*.	Ctrl+F5

Chapter 2
Working with Variables, Operators, and Expressions

After completing this chapter, you will be able to:

- Understand statements, identifiers, and keywords.

- Use variables to store information.

- Work with primitive data types.

- Use arithmetic operators such as the plus sign (+) and the minus sign (–).

- Increment and decrement variables.

In Chapter 1, "Welcome to C#," you learned how to use the Microsoft Visual Studio 2008 programming environment to build and run a Console program and a Windows Presentation Foundation (WPF) application. In this chapter, you are introduced to the elements of Microsoft Visual C# syntax and semantics, including statements, keywords, and identifiers. You'll study the primitive types that are built into the C# language and the characteristics of the values that each type holds. You'll also see how to declare and use local variables (variables that exist only in a method or other small section of code), learn about the arithmetic operators that C# provides, find out how to use operators to manipulate values, and learn how to control expressions containing two or more operators.

Understanding Statements

A *statement* is a command that performs an action. You combine statements to create methods. You'll learn more about methods in Chapter 3, "Writing Methods and Applying Scope," but for now, think of a method as a named sequence of statements. *Main*, which was introduced in the previous chapter, is an example of a method. Statements in C# follow a well-defined set of rules describing their format and construction. These rules are collectively known as *syntax*. (In contrast, the specification of what statements *do* is collectively known as *semantics*.) One of the simplest and most important C# syntax rules states that you must terminate all statements with a semicolon. For example, without its terminating semicolon, the following statement won't compile:

```
Console.WriteLine("Hello World");
```

> **Tip** C# is a "free format" language, which means that white space, such as a space character or a newline, is not significant except as a separator. In other words, you are free to lay out your statements in any style you choose. However, you should adopt a simple, consistent layout style and keep to it to make your programs easier to read and understand.

The trick to programming well in any language is learning the syntax and semantics of the language and then using the language in a natural and idiomatic way. This approach makes your programs more easily maintainable. In the chapters throughout this book, you'll see examples of the most important C# statements.

Using Identifiers

Identifiers are the names you use to identify the elements in your programs, such as namespaces, classes, methods, and variables (you will learn about variables shortly). In C#, you must adhere to the following syntax rules when choosing identifiers:

- You can use only letters (uppercase and lowercase), digits, and underscore characters.

- An identifier must start with a letter (an underscore is considered a letter).

For example, *result*, *_score*, *footballTeam*, and *plan9* are all valid identifiers, whereas *result%*, *footballTeam$*, and *9plan* are not.

> **Important** C# is a case-sensitive language: *footballTeam* and *FootballTeam* are not the same identifier.

Identifying Keywords

The C# language reserves 77 identifiers for its own use, and you cannot reuse these identifiers for your own purposes. These identifiers are called *keywords*, and each has a particular meaning. Examples of keywords are *class*, *namespace*, and *using*. You'll learn the meaning of most of the C# keywords as you proceed through this book. The keywords are listed in the following table.

abstract	do	in	protected	true
as	double	int	public	try
base	else	interface	readonly	typeof
bool	enum	internal	ref	uint
break	event	is	return	ulong
byte	explicit	lock	sbyte	unchecked
case	extern	long	sealed	unsafe
catch	false	namespace	short	ushort
char	finally	new	sizeof	using
checked	fixed	null	stackalloc	virtual
class	float	object	static	void
const	for	operator	string	volatile
continue	foreach	out	struct	while
decimal	goto	override	switch	
default	if	params	this	
delegate	implicit	private	throw	

> **Tip** In the Visual Studio 2008 *Code and Text Editor* window, keywords are colored blue when you type them.

C# also uses the following identifiers. These identifiers are not reserved by C#, which means that you can use these names as identifiers for your own methods, variables, and classes, but you should really avoid doing so if at all possible.

from	join	select	yield
get	let	set	
group	orderby	value	
into	partial	where	

Using Variables

A *variable* is a storage location that holds a value. You can think of a variable as a box in the computer's memory holding temporary information. You must give each variable in a program an unambiguous name that uniquely identifies it in the context in which it is used. You use a variable's name to refer to the value it holds. For example, if you want to store the value of the cost of an item in a store, you might create a variable simply called *cost* and store the item's cost in this variable. Later on, if you refer to the *cost* variable, the value retrieved will be the item's cost that you stored there earlier.

Naming Variables

You should adopt a naming convention for variables that helps you avoid confusion concerning the variables you have defined. The following list contains some general recommendations:

- Don't use underscores in identifiers.

- Don't create identifiers that differ only by case. For example, do not create one variable named *myVariable* and another named *MyVariable* for use at the same time because it is too easy to get them confused.

> **Note** Using identifiers that differ only by case can limit the ability to reuse classes in applications developed using other languages that are not case sensitive, such as Microsoft Visual Basic.

- Start the name with a lowercase letter.

- In a multiword identifier, start the second and each subsequent word with an upper-case letter. (This is called camelCase notation.)

- Don't use Hungarian notation. (Microsoft Visual C++ developers reading this book are probably familiar with Hungarian notation. If you don't know what Hungarian notation is, don't worry about it!)

> **Important** You should treat the first two of the preceding recommendations as compulsory because they relate to Common Language Specification (CLS) compliance. If you want to write programs that can interoperate with other languages, such as Microsoft Visual Basic .NET, you must comply with these recommendations.

For example, *score*, *footballTeam*, *_score*, and *FootballTeam* are all valid variable names, but only the first two are recommended.

Declaring Variables

Variables hold values. C# has many different types of values that it can store and process—integers, floating-point numbers, and strings of characters, to name three. When you declare a variable, you must specify the type of data it will hold.

You declare the type and name of a variable in a declaration statement. For example, the following statement declares that the variable named *age* holds *int* (integer) values. As always, the statement must be terminated with a semicolon.

```
int age;
```

The variable type *int* is the name of one of the *primitive* C# types, integer, which is a whole number. (You'll learn about several primitive data types later in this chapter.)

> **Note** Microsoft Visual Basic programmers should note that C# does not allow implicit variable declarations. You must explicitly declare all variables before you use them.

After you've declared your variable, you can assign it a value. The following statement assigns *age* the value 42. Again, you'll see that the semicolon is required.

```
age = 42;
```

The equal sign (=) is the *assignment* operator, which assigns the value on its right to the variable on its left. After this assignment, the *age* variable can be used in your code to refer to the value it holds. The next statement writes the value of the *age* variable, 42, to the console:

```
Console.WriteLine(age);
```

> **Tip** If you leave the mouse pointer over a variable in the Visual Studio 2008 *Code and Text Editor* window, a ScreenTip appears, telling you the type of the variable.

Working with Primitive Data Types

C# has a number of built-in types called *primitive data types*. The following table lists the most commonly used primitive data types in C# and the range of values that you can store in each.

Data type	Description	Size (bits)	Range [1]	Sample usage
int	Whole numbers	32	-2^{31} through $2^{31} - 1$	`int count;` `count = 42;`
long	Whole numbers (bigger range)	64	-2^{63} through $2^{63} - 1$	`long wait;` `wait = 42L;`
float	Floating-point numbers	32	$\pm 1.5 \times 10^{45}$ through $\pm 3.4 \times 10^{38}$	`float away;` `away = 0.42F;`
double	Double-precision (more accurate) floating-point numbers	64	$\pm 5.0 \times 10^{-324}$ through $\pm 1.7 \times 10^{308}$	`double trouble;` `trouble = 0.42;`
decimal	Monetary values	128	28 significant figures	`decimal coin;` `coin = 0.42M;`

Data type	Description	Size (bits)	Range [1]	Sample usage

string	Sequence of characters	16 bits per character	Not applicable	`string vest;`
				`vest =`
				`"fortytwo";`

char	Single character	16	0 through $2^{16} - 1$	`char grill;` `grill = 'x';`
bool	Boolean	8	True or false	`bool teeth;` `teeth = false;`

Unassigned Local Variables

When you declare a variable, it contains a random value until you assign a value to it. This behavior was a rich source of bugs in C and C++ programs that created a variable and accidentally used it as a source of information before giving it a value. C# does not allow you to use an unassigned variable. You must assign a value to a variable before you can use it; otherwise, your program might not compile. This requirement is called the *Definite Assignment Rule*. For example, the following statements will generate a compile-time error because *age* is unassigned:

```
int age;
Console.WriteLine(age); // compile-time error
```

Displaying Primitive Data Type Values

In the following exercise, you'll use a C# program named *PrimitiveDataTypes* to demonstrate how several primitive data types work.

Display primitive data type values

1. Start Visual Studio 2008 if it is not already running.

2. If you are using Visual Studio 2008 Standard Edition or Visual Studio 2008 Professional Edition, on the *File* menu, point to *Open*, and then click *Project/Solution*.

   If you are using Visual C# 2008 Express Edition, on the *File* menu, click *Open Project*.

   The *Open Project* dialog box appears.

3. Move to the \Microsoft Press\Visual CSharp Step by Step\Chapter 2\PrimitiveDataTypes folder in your Documents folder. Select the PrimitiveDataTypes solution file, and then click *Open*.

   The solution loads, and *Solution Explorer* displays the PrimitiveDataTypes project.

> **Note** Solution file names have the .sln suffix, such as PrimitiveDataTypes.sln. A solution can contain one or more projects. Project files have the .csproj suffix. If you open a project rather than a solution, Visual Studio 2008 will automatically create a new solution file for it. If you build the solution, Visual Studio 2008 automatically saves any new or updated files, so you will be prompted to provide a name and location for the new solution file.

4. On the *Debug* menu, click *Start Without Debugging*.

   The following application window appears:

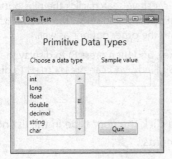

5. In the *Choose a data type* list, click the string type.

   The value "forty two" appears in the *Sample value* box.

6. Click the int type in the list.

   The value to do appears in the *Sample value* box, indicating that the statements to display an int value still need to be written.

7. Click each data type in the list. Confirm that the code for the double and bool types also must be completed.

8. Click *Quit* to close the window and stop the program.

   Control returns to the Visual Studio 2008 programming environment.

Use primitive data types in code

1. In *Solution Explorer*, double-click *Window1.xaml*.

   The WPF form for the application appears in the *Design View* window.

2. Right-click anywhere in the *Design View* window displaying the Window1.xaml form, and then click *View Code*.

   The *Code and Text Editor* window opens, displaying the Window1.xaml.cs file.

> **Note** Remember that you can also use *Solution Explorer* to access the code; click the plus sign, +, to the left of the Window1.xaml file, and then double-click *Window1.xaml.cs*.

3. In the *Code and Text Editor* window, find the *showFloatValue* method.

> **Tip** To locate an item in your project, on the *Edit* menu, point to *Find and Replace*, and then click *Quick Find*. A dialog box opens, asking what you want to search for. Type the name of the item you're looking for, and then click *Find Next*. By default, the search is not case-sensitive. If you want to perform a case-sensitive search, click the plus button, +, next to the *Find Options* label to display additional options, and select the *Match Case* check box. If you have time, you can experiment with the other options as well.
>
> You can also press Ctrl+F (press the Control key, and then press F) to display the *Quick Find* dialog box rather than using the *Edit* menu. Similarly, you can press Ctrl+H to display the *Quick Replace* dialog box.

The *showFloatValue* method runs when you click the *float* type in the list box. This method contains the following three statements:

```
float variable;
variable=0.42F;
value.Text = "0.42F";
```

The first statement declares a variable named *variable* of type *float*.

The second statement assigns *variable* the value 0.42F. (The *F* is a type suffix specifying that 0.42 should be treated as a *float* value. If you forget the *F*, the value 0.42 will be treated as a *double*, and your program will not compile because you cannot assign a value of one type to a variable of a different type without writing additional code—C# is very strict in this respect.)

The third statement displays the value of this variable in the *value* text box on the form. This statement requires a little bit of your attention. The way in which you display an item in a text box is to set its *Text* property. Notice that you access the property of an object by using the same "dot" notation that you saw for running a method. (Remember *Console.WriteLine* from Chapter 1?) The data that you put in the *Text* property must be a string (a sequence of characters enclosed in double quotation marks), and not a number. If you try to assign a number to the *Text* property, your program will not compile. In this program, the statement simply displays the text "0.42F" in the text box. In a real-world application, you would add statements that convert the value of the variable *variable* into a string and then put this into the *Text* property, but you need to know a little bit more about C# and the Microsoft .NET Framework before you can do that. (Chapter 11, "Understanding Parameter Arrays," and Chapter 21, "Operator Overloading," cover data type conversions.)

4. In the *Code and Text Editor* window, locate the *showIntValue* method. It looks like this:

```
private void showIntValue()
{
    value.Text = "to do";
}
```

The *showIntValue* method is called when you click the *int* type in the list box.

> **Tip** Another way to find a method in the *Code and Text Editor* window is to click the *Members* drop-down list that appears above the window, to the right. This window displays a list of all the methods (and other items) in the class displayed in the *Code and Text Editor* window. You can click the name of a member, and you will be taken directly to it in the *Code and Text Editor* window.

5. Type the following two statements at the start of the *showIntValue* method, on a new line after the opening brace, as shown in bold type in the following code:

```
private void showIntValue()
{
    int variable;
    variable = 42;
}
```

6. In the original statement in this method, change the string *"to do"* to *"42"*.

The method should now look exactly like this:

```
private void showIntValue()
{
    int variable;
    variable = 42;
    value.Text = "42";
}
```

> **Note** If you have previous programming experience, you might be tempted to change the third statement to
>
> ```
> value.Text = variable;
> ```
>
> This looks like it should display the value of *variable* in the *value* text box on the form. However, C# performs strict type checking; text boxes can display only *string* values, and *variable* is an *int*, so this statement will not compile. You will see how to convert between numeric and string values later in this chapter.

7. On the *Debug* menu, click *Start Without Debugging*.

The form appears again.

8. Select the *int* type in the *Choose a data type* list. Confirm that the value 42 is displayed in the *Sample value* text box.

9. Click *Quit* to close the window and stop the program.

10. In the *Code and Text Editor* window, find the *showDoubleValue* method.

11. Edit the *showDoubleValue* method exactly as shown in bold type in the following code:

```
private void showDoubleValue()
{
    double variable;
    variable = 0.42;
    value.Text = "0.42";
}
```

12. In the *Code and Text Editor* window, locate the *showBoolValue* method.

13. Edit the *showBoolValue* method exactly as follows:

```
private void showBoolValue()
{
    bool variable;
    variable = false;
    value.Text = "false";
}
```

14. On the *Debug* menu, click *Start Without Debugging*.

15. In the *Choose a data type* list, select the *int*, *double*, and *bool* types. In each case, verify that the correct value is displayed in the *Sample value* text box.

16. Click *Quit* to stop the program.

Using Arithmetic Operators

C# supports the regular arithmetic operations you learned in your childhood: the plus sign (+) for addition, the minus sign (–) for subtraction, the asterisk (*) for multiplication, and the forward slash (/) for division. The symbols +, –, *, and / are called *operators* because they "operate" on values to create new values. In the following example, the variable *moneyPaid-ToConsultant* ends up holding the product of 750 (the daily rate) and 20 (the number of days the consultant was employed):

```
long moneyPaidToConsultant;
moneyPaidToConsultant = 750 * 20;
```

> **Note** The values that an operator operates on are called *operands*. In the expression 750 * 20, the * is the operator, and 750 and 20 are the operands.

Operators and Types

Not all operators are applicable to all data types. The operators that you can use on a value depend on the value's type. For example, you can use all the arithmetic operators on values of type *char*, *int*, *long*, *float*, *double*, or *decimal*. However, with the exception of the plus operator, +, you can't use the arithmetic operators on values of type *string* or *bool*. So the following statement is not allowed because the *string* type does not support the minus operator (subtracting one string from another would be meaningless):

```
// compile-time error
Console.WriteLine("Gillingham" - "Forest Green Rovers");
```

You can use the + operator to concatenate string values. You need to be careful because this can have results you might not expect. For example, the following statement writes "431" (not "44") to the console:

```
Console.WriteLine("43" + "1");
```

> **Tip** The .NET Framework provides a method called *Int32.Parse* that you can use to convert a string value to an integer if you need to perform arithmetic computations on values held as strings.

You should also be aware that the type of the result of an arithmetic operation depends on the type of the operands used. For example, the value of the expression 5.0/2.0 is 2.5; the type of both operands is *double* (in C#, literal numbers with decimal points are always *double*, not *float*, to maintain as much accuracy as possible), so the type of the result is also *double*. However, the value of the expression 5/2 is 2. In this case, the type of both operands is *int*, so the type of the result is also *int*. C# always rounds values down in circumstances like this. The situation gets a little more complicated if you mix the types of the operands. For example, the expression 5/2.0 consists of an *int* and a *double*. The C# compiler detects the mismatch and generates code that converts the *int* into a *double* before performing the operation. The result of the operation is therefore a *double* (2.5). However, although this works, it is considered poor practice to mix types in this way.

Numeric Types and Infinite Values

There are one or two other features of numbers in C# that you should be aware of. For example, the result of dividing any number by zero is infinity, which is outside the range of the *int*, *long*, and *decimal* types, and consequently evaluating an expression such as 5/0 results in an error. However, the *double* and *float* types actually have a special value that can represent infinity, and the value of the expression 5.0/0.0 is *Infinity*. The one exception to this rule is the value of the expression 0.0/0.0. Usually, if you divide zero by anything, the result is zero, but if you divide anything by zero the result is infinity. The expression 0.0/0.0 results in a paradox—the value must be zero and infinity at the same time. C# has another special value for this situation called *NaN*, which stands for "not a number." So if you evaluate 0.0/0.0, the result is *NaN*. *NaN* and *Infinity* propagate through expressions. If you evaluate 10 + *NaN*, the result is *NaN*, and if you evaluate 10 + *Infinity*, the result is *Infinity*. The one exception to this rule is the expression *Infinity* * 0, which results in 0, whereas the result of the expression *NaN* * 0 is *NaN*.

C# also supports one less-familiar arithmetic operator: the remainder, or modulus, operator, which is represented by the percent sign (%). The result of *x* % *y* is the remainder after dividing *x* by *y*. For example, 9 % 2 is 1 because 9 divided by 2 is 4, remainder 1.

Note If you are familiar with C or C++, you will know that you can't use the remainder operator on *float* or *double* values in these languages. However, C# relaxes this rule. The remainder operator is valid with all numeric types, and the result is not necessarily an integer. For example, the result of the expression 7.0 % 2.4 is 2.2.

Examining Arithmetic Operators

The following exercise demonstrates how to use the arithmetic operators on *int* values using a previously written C# program called MathsOperators.

Work with arithmetic operators

1. Open the MathsOperators project, located in the \Microsoft Press\Visual CSharp Step by Step\Chapter 2\MathsOperators folder in your Documents folder.

2. On the *Debug* menu, click *Start Without Debugging*.

   A form appears on the screen.

3. Type **54** in the *left operand* text box.

4. Type **13** in the *right operand* text box.

   You can now apply any of the operators to the values in the text boxes.

5. Click the – *Subtraction* button, and then click *Calculate*.

   The text in the *Expression* text box changes to 54 – 13, and the value 41 appears in the *Result* box, as shown in the following image:

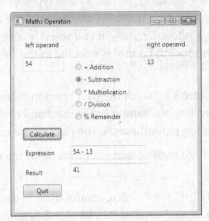

6. Click the / *Division* button, and then click *Calculate*.

   The text in the *Expression* text box changes to 54/13, and the value 4 appears in the *Result* text box. In real life, 54/13 is 4.153846 recurring, but this is not real life; this is C# performing integer division, and when you divide one integer by another integer, the answer you get back is an integer, as explained earlier.

7. Click the % *Remainder* button, and then click *Calculate*.

   The text in the *Expression* text box changes to 54 % 13, and the value 2 appears in the *Result* text box. This is because the remainder after dividing 54 by 13 is 2. (54 – ((54/13) * 13)) is 2 if you do the arithmetic rounding down to an integer at each stage—my old math master at school would be horrified to be told that (54/13) * 13 does not equal 54!

8. Test the other combinations of numbers and operators. When you have finished, click *Quit* to return to the Visual Studio 2008 programming environment.

Now take a look at the MathsOperators program code.

Examine the MathsOperators program code

1. Display the Window1.xaml form in the *Design View* window (double-click the file *Window1.xaml* in *Solution Explorer*).

2. On the *View* menu, point to *Other Windows*, and then click *Document Outline*.

   The *Document Outline* window appears, showing the names and types of the controls on the form. If you click each of the controls on the form, the name of the control is highlighted in the *Document Outline* window. Similarly, if you select a control in the *Document Outline* window, the corresponding control is selected in the *Design View* window.

3. On the form, click the two *TextBox* controls in which the user types numbers. In the *Document Outline* window, verify that they are named *lhsOperand* and *rhsOperand*. (You can see the name of a control in the parentheses to the right of the control.)

   When the form runs, the *Text* property of each of these controls holds the values that the user enters.

4. Toward the bottom of the form, verify that the TextBox control used to display the expression being evaluated is named expression and that the TextBox control used to display the result of the calculation is named result.

5. Close the Document Outline window.

6. Display the code for the Window1.xaml.cs file in the Code and Text Editor window.

7. In the Code and Text Editor window, locate the subtractValues method. It looks like this:

```
private void subtractValues()
{
    int lhs = int.Parse(lhsOperand.Text);
    int rhs = int.Parse(rhsOperand.Text);
    int outcome;
    outcome = lhs - rhs;
    expression.Text = lhsOperand.Text + " - " + rhsOperand.Text;
    result.Text = outcome.ToString();
}
```

The first statement in this method declares an *int* variable called *lhs* and initializes it with the integer corresponding to the value typed by the user in the *lhsOperand* text box. Remember that the *Text* property of a text box control contains a string, so you must convert this string to an integer before you can assign it to an *int* variable. The *int* data type provides the *int.Parse* method, which does precisely this.

The second statement declares an *int* variable called *rhs* and initializes it to the value in the *rhsOperand* text box after converting it to an *int*.

The third statement declares an *int* variable called *outcome*.

The fourth statement subtracts the value of the *rhs* variable from the value of the *lhs* variable and assigns the result to *outcome*.

The fifth statement concatenates three strings indicating the calculation being performed (using the plus operator, +) and assigns the result to the *expression.Text* property. This causes the string to appear in the *expression* text box on the form.

The sixth statement displays the result of the calculation by assigning it to the *Text* property of the *result* text box. Remember that the *Text* property is a string and that the result of the calculation is an *int*, so you must convert the string to an *int* before assigning it to the *Text* property. This is what the *ToString* method of the *int* type does.

The *ToString* Method

Every class in the .NET Framework has a *ToString* method. The purpose of *ToString* is to convert an object to its string representation. In the preceding example, the *ToString* method of the integer object, *outcome*, is used to convert the integer value of *outcome* to the equivalent string value. This conversion is necessary because the value is displayed in the *Text* property of the *result* text box—the *Text* property can contain only strings. When you create your own classes, you can define your own implementation of the *ToString* method to specify how your class should be represented as a string. You learn more about creating your own classes in Chapter 7, "Creating and Managing Classes and Objects."

Controlling Precedence

Precedence governs the order in which an expression's operators are evaluated. Consider the following expression, which uses the + and * operators:

```
2 + 3 * 4
```

This expression is potentially ambiguous; do you perform the addition first or the multiplication? In other words, does 3 bind to the + operator on its left or to the * operator on its right? The order of the operations matters because it changes the result:

- If you perform the addition first, followed by the multiplication, the result of the addition (2 + 3) forms the left operand of the * operator, and the result of the whole expression is 5 * 4, which is 20.

- If you perform the multiplication first, followed by the addition, the result of the multiplication (3 * 4) forms the right operand of the + operator, and the result of the whole expression is 2 + 12, which is 14.

In C#, the multiplicative operators (*, /, and %) have precedence over the additive operators (+ and −), so in expressions such as 2 + 3 * 4, the multiplication is performed first, followed by the addition. The answer to 2 + 3 * 4 is therefore 14. As each new operator is discussed in later chapters, its precedence will be explained.

You can use parentheses to override precedence and force operands to bind to operators in a different way. For example, in the following expression, the parentheses force the 2 and the 3 to bind to the + operator (making 5), and the result of this addition forms the left operand of the * operator to produce the value 20:

```
(2 + 3) * 4
```

> **Note** The term *parentheses* or *round brackets* refers to ( ). The term *braces* or *curly brackets* refers to { }. The term *square brackets* refers to [ ].

Using Associativity to Evaluate Expressions

Operator precedence is only half the story. What happens when an expression contains different operators that have the same precedence? This is where associativity becomes important. *Associativity* is the direction (left or right) in which the operands of an operator are evaluated. Consider the following expression that uses the / and * operators:

```
4 / 2 * 6
```

This expression is still potentially ambiguous. Do you perform the division first, or the multiplication? The precedence of both operators is the same (they are both multiplicative), but the order in which the expression is evaluated is important because you get one of two possible results:

- If you perform the division first, the result of the division (4/2) forms the left operand of the * operator, and the result of the whole expression is (4/2) * 6, or 12.

- If you perform the multiplication first, the result of the multiplication (2 * 6) forms the right operand of the / operator, and the result of the whole expression is 4/(2 * 6), or 4/12.

In this case, the associativity of the operators determines how the expression is evaluated. The * and / operators are both left-associative, which means that the operands are evaluated from left to right. In this case, 4/2 will be evaluated before multiplying by 6, giving the result 12. As each new operator is discussed in subsequent chapters, its associativity is also covered.

Associativity and the Assignment Operator

In C#, the equal sign (=) is an operator. All operators return a value based on their operands. The assignment operator (=) is no different. It takes two operands; the operand on its right side is evaluated and then stored in the operand on its left side. The value of the assignment operator is the value that was assigned to the left operand. For example, in the following assignment statement, the value returned by the assignment operator is 10, which is also the value assigned to the variable *myInt*:

```
int myInt;
myInt = 10; //value of assignment expression is 10
```

At this point, you are probably thinking that this is all very nice and esoteric, but so what? Well, because the assignment operator returns a value, you can use this same value with another occurrence of the assignment statement, like this:

```
int myInt;
int myInt2;
myInt2 = myInt = 10;
```

The value assigned to the variable *myInt2* is the value that was assigned to *myInt*. The assignment statement assigns the same value to both variables. This technique is very useful if you want to initialize several variables to the same value. It makes it very clear to anyone reading your code that all the variables must have the same value:

```
myInt5 = myInt4 = myInt3 = myInt2 = myInt = 10;
```

From this discussion, you can probably deduce that the assignment operator associates from right to left. The rightmost assignment occurs first, and the value assigned propagates through the variables from right to left. If any of the variables previously had a value, it is overwritten by the value being assigned.

Incrementing and Decrementing Variables

If you want to add 1 to a variable, you can use the + operator:

```
count = count + 1;
```

However, adding 1 to a variable is so common that C# provides its own operator just for this purpose: the ++ operator. To increment the variable *count* by 1, you can write the following statement:

```
count++;
```

Similarly, C# provides the −− operator that you can use to subtract 1 from a variable, like this:

```
count--;
```

> **Note** The ++ and −− operators are *unary* operators, meaning that they take only a single operand. They share the same precedence and left associativity as the ! unary operator, which is discussed in Chapter 4, "Using Decision Statements."

Prefix and Postfix

The increment, ++, and decrement, −−, operators are unusual in that you can place them either before or after the variable. Placing the operator symbol before the variable is called the prefix form of the operator, and using the operator symbol after the variable is called the postfix form. Here are examples:

```
count++; // postfix increment
++count; // prefix increment
count--; // postfix decrement
--count; // prefix decrement
```

Whether you use the prefix or postfix form of the ++ or −− operator makes no difference to the variable being incremented or decremented. For example, if you write *count++*, the value of *count* increases by 1, and if you write *++count*, the value of *count* also increases by 1. Knowing this, you're probably wondering why there are two ways to write the same thing. To understand the answer, you must remember that ++ and −− are operators and that all operators are used to evaluate an expression that has a value. The value returned by *count++* is the value of *count* before the increment takes place, whereas the value returned by *++count* is the value of *count* after the increment takes place. Here is an example:

```
int x;
x = 42;
Console.WriteLine(x++); // x is now 43, 42 written out
x = 42;
Console.WriteLine(++x); // x is now 43, 43 written out
```

The way to remember which operand does what is to look at the order of the elements (the operand and the operator) in a prefix or postfix expression. In the expression *x++*, the variable *x* occurs first, so its value is used as the value of the expression before *x* is incremented. In the expression *++x*, the operator occurs first, so its operation is performed before the value of *x* is evaluated as the result.

These operators are most commonly used in *while* and *do* statements, which are presented in Chapter 5, "Using Compound Assignment and Iteration Statements." If you are using the increment and decrement operators in isolation, stick to the postfix form and be consistent.

Declaring Implicitly Typed Local Variables

Earlier in this chapter, you saw that you declare a variable by specifying a data type and an identifier, like this:

```
int myInt;
```

It was also mentioned that you should assign a value to a variable before you attempt to use it. You can declare and initialize a variable in the same statement, like this:

```
int myInt = 99;
```

or even like this, assuming that *myOtherInt* is an initialized integer variable:

```
int myInt = myOtherInt * 99;
```

Now, remember that the value you assign to a variable must be of the same type as the variable. For example, you can assign an *int* value only to an *int* variable. The C# compiler can quickly work out the type of an expression used to initialize a variable and tell you if it does not match the type of the variable. You can also ask the C# compiler to infer the type of a variable from an expression and use this type when declaring the variable by using the *var* keyword in place of the type, like this:

```
var myVariable = 99;
var myOtherVariable = "Hello";
```

Variables *myVariable* and *myOtherVariable* are referred to as *implicitly typed* variables. The *var* keyword causes the compiler to deduce the type of the variables from the types of the expressions used to initialize them. In these examples, *myVariable* is an *int*, and *myOtherVariable* is a *string*. It is important to understand that this is a convenience for declaring variables only and that after a variable has been declared, you can assign only values of the inferred type to it—you cannot assign *float*, *double*, or *string* values to *myVariable* at a later point in your program, for example. You should also understand that you can use the *var* keyword only when you supply an expression to initialize a variable. The following declaration is illegal and will cause a compilation error:

```
var yetAnotherVariable; // Error - compiler cannot infer type
```

> **Important** If you have programmed with Visual Basic in the past, you may be familiar with the *Variant* type, which you can use to store any type of value in a variable. I emphasize here and now that you should forget everything you ever learned when programming with Visual Basic about *Variant* variables. Although the keywords look similar, *var* and *Variant* mean totally different things. When you declare a variable in C# using the *var* keyword, the type of values that you assign to the variable *cannot change* from that used to initialize the variable.

If you are a purist, you are probably gritting your teeth at this point and wondering why on earth the designers of a neat language such as C# should allow a feature such as *var* to creep in. After all, it sounds like an excuse for extreme laziness on the part of programmers and can make it more difficult to understand what a program is doing or track down bugs (and it can even easily introduce new bugs into your code). However, trust me that *var* has a very valid place in C#, as you will see when you work through many of the following chapters. However, for the time being, we will stick to using explicitly typed variables except for when implicit typing becomes a necessity.

- If you want to continue to the next chapter

  Keep Visual Studio 2008 running, and turn to Chapter 3.

- If you want to exit Visual Studio 2008 now

  On the *File* menu, click *Exit*. If you see a *Save* dialog box, click *Yes* (if you are using Visual Studio 2008) or click *Save* (if you are using Visual C# 2008 Express Edition) and save the project.

Chapter 2 Quick Reference

To	Do this
Declare a variable	Write the name of the data type, followed by the name of the variable, followed by a semicolon. For example: `int outcome;`
Change the value of a variable	Write the name of the variable on the left, followed by the assignment operator, followed by the expression calculating the new value, followed by a semicolon. For example: `outcome = 42;`
Convert a *string* to an *int*	Call the *System.Int32.Parse* method. For example: `System.Int32.Parse("42");`
Override precedence	Use parentheses in the expression to force the order of evaluation. For example: `(3 + 4) * 5`
Initialize several variables to the same value	Use an assignment statement that initializes all the variables. For example: `myInt4 = myInt3 = myInt2 = myInt = 10;`
Increment or decrement a variable	Use the ++ or -- operator. For example: `count++;`

(Footnotes)

1 The value of 2^{16} is 65,536; the value of 2^{31} is 2,147,483,648; and the value of 2^{63} is 9,223,372,036,854,775,808.

Chapter 3
Writing Methods and Applying Scope

After completing this chapter, you will be able to:

- Declare and call methods.

- Pass information to a method.

- Return information from a method.

- Define local and class scope.

- Use the integrated debugger to step in and out of methods as they run.

In Chapter 2, "Working with Variables, Operators, and Expressions," you learned how to declare variables, how to create expressions using operators, and how precedence and associativity control how expressions containing multiple operators are evaluated. In this chapter, you'll learn about methods. You'll also learn how to use arguments and parameters to pass information to a method and how to return information from a method by using return statements. Finally, you'll see how to step in and out of methods by using the Microsoft Visual Studio 2008 integrated debugger. This information is useful when you need to trace the execution of your methods if they do not work quite as you expected.

Declaring Methods

A *method* is a named sequence of statements. If you have previously programmed using languages such as C or Microsoft Visual Basic, you know that a method is very similar to a function or a subroutine. A method has a name and a body. The method name should be a meaningful identifier that indicates the overall purpose of the method (*CalculateIncomeTax*, for example). The method body contains the actual statements to be run when the method is called. Additionally, methods can be given some data for processing and can return information, which is usually the result of the processing. Methods are a fundamental and powerful mechanism.

Specifying the Method Declaration Syntax

The syntax of a Microsoft Visual C# method is as follows:

```
returnType methodName ( parameterList )
{
    // method body statements go here
}
```

- The returnType is the name of a type and specifies the kind of information the method returns as a result of its processing. This can be any type, such as int or string. If you're writing a method that does not return a value, you must use the keyword void in place of the return type.

- The methodName is the name used to call the method. Method names follow the same identifier rules as variable names. For example, addValues is a valid method name, whereas add$Values is not. For now, you should follow the camelCase convention for method names—for example, displayCustomer.

- The parameterList is optional and describes the types and names of the information that you can pass into the method for it to process. You write the parameters between the opening and closing parentheses as though you're declaring variables, with the name of the type followed by the name of the parameter. If the method you're writing has two or more parameters, you must separate them with commas.

- The method body statements are the lines of code that are run when the method is called. They are enclosed between opening and closing braces { }.

> **Important** C, C++, and Microsoft Visual Basic programmers should note that C# does not support global methods. You must write all your methods inside a class, or your code will not compile.

Here's the definition of a method called *addValues* that returns an *int* result and has two *int* parameters called *leftHandSide* and *rightHandSide*:

```
int addValues(int leftHandSide, int rightHandSide)
{
    // ...
    // method body statements go here
    // ...
}
```

> **Note** You must explicitly specify the types of any parameters and the return type of a method. You cannot use the *var* keyword.

Here's the definition of a method called *showResult* that does not return a value and has a single *int* parameter called *answer*:

```
void showResult(int answer)
{
    // ...
}
```

Notice the use of the keyword *void* to indicate that the method does not return anything.

> **Important** Visual Basic programmers should notice that C# does not use different keywords to distinguish between a method that returns a value (a function) and a method that does not return a value (a procedure or subroutine). You must always specify either a return type or *void*.

Writing *return* Statements

If you want a method to return information (in other words, its return type is not *void*), you must write a *return* statement inside the method. You do this by using the keyword *return* followed by an expression that calculates the returned value, and a semicolon. The type of expression must be the same as the type specified by the method. In other words, if a method returns an *int*, the *return* statement must return an *int*; otherwise, your program will not compile. Here is an example:

```
int addValues(int leftHandSide, int rightHandSide)
{
    // ...
    return leftHandSide + rightHandSide;
}
```

The *return* statement is usually positioned at the end of your method because it causes the method to finish. Any statements that occur after the *return* statement are not executed (although the compiler warns you about this problem if you place statements after the *return* statement).

If you don't want your method to return information (in other words, its return type is *void*), you can use a variation of the *return* statement to cause an immediate exit from the method. You write the keyword *return* immediately followed by a semicolon. For example:

```
void showResult(int answer)
{
    // display the answer
    ...
    return;
}
```

If your method does not return anything, you can also omit the *return* statement because the method finishes automatically when execution arrives at the closing brace at the end of the method. Although this practice is common, it is not always considered good style.

In the following exercise, you will examine another version of the MathsOperators application from Chapter 2. This version has been improved by the careful use of some small methods.

Examine method definitions

1. Start Visual Studio 2008 if it is not already running.

2. Open the *Methods* project in the \Microsoft Press\Visual CSharp Step by Step\ Chapter 3\Methods folder in your Documents folder.

3. On the *Debug* menu, click *Start Without Debugging*.

   Visual Studio 2008 builds and runs the application.

4. Refamiliarize yourself with the application and how it works, and then click *Quit*.

5. Display the code for Window1.xaml.cs in the *Code and Text Editor* window.

6. In the *Code and Text Editor* window, locate the *addValues* method.

   The method looks like this:

   ```
   private int addValues(int leftHandSide, int rightHandSide)
   {
       expression.Text = leftHandSide.ToString() + " + " + rightHandSide.ToString();
       return leftHandSide + rightHandSide;
   }
   ```

   The *addValues* method contains two statements. The first statement displays the calculation being performed in the *expression* text box on the form. The values of the parameters *leftHandSide* and *rightHandSide* are converted to strings (using the *ToString* method you met in Chapter 2) and concatenated together with a string representation of the plus operator (+) in the middle.

   The second statement uses the + operator to add the values of the *leftHandSide* and *rightHandSide int* variables together and returns the result of this operation. Remember that adding two *int* values together creates another *int* value, so the return type of the *addValues* method is *int*.

   If you look at the methods *subtractValues*, *multiplyValues*, *divideValues*, and *remainderValues*, you will see that they follow a similar pattern.

7. In the *Code and Text Editor* window, locate the *showResult* method.

The *showResult* method looks like this:

```
private void showResult(int answer)
{
    result.Text = answer.ToString();
}
```

This method contains one statement that displays a string representation of the *answer* parameter in the *result* text box.

> **Tip** There is no minimum length for a method. If a method helps to avoid repetition and makes your program easier to understand, the method is useful regardless of how small it is.
>
> There is also no maximum length for a method, but usually you want to keep your method code small enough to get the job done. If your method is more than one screen in length, consider breaking it into smaller methods for readability.

Calling Methods

Methods exist to be called! You call a method by name to ask it to perform its task. If the method requires information (as specified by its parameters), you must supply the information requested. If the method returns information (as specified by its return type), you should arrange to capture this information somehow.

Specifying the Method Call Syntax

The syntax of a C# method call is as follows:

```
result = methodName ( argumentList )
```

- The *methodName* must exactly match the name of the method you're calling. Remember, C# is a case-sensitive language.

- The *result* = clause is optional. If specified, the variable identified by result contains the value returned by the method. If the method is void (it does not return a value), you must omit the *result* = clause of the statement.

- The *argumentList* supplies the optional information that the method accepts. You must supply an argument for each parameter, and the value of each argument must be compatible with the type of its corresponding parameter. If the method you're calling has two or more parameters, you must separate the arguments with commas.

> **Important** You must include the parentheses in every method call, even when calling a method that has no arguments.

To clarify these points, take a look at the *addValues* method again:

```
int addValues(int leftHandSide, int rightHandSide)
{
    // ...
}
```

The *addValues* method has two *int* parameters, so you must call it with two comma-separated *int* arguments:

```
addValues(39, 3);     // okay
```

You can also replace the literal values 39 and 3 with the names of *int* variables. The values in those variables are then passed to the method as its arguments, like this:

```
int arg1 = 99;
int arg2 = 1;
addValues(arg1, arg2);
```

If you try to call *addValues* in some other way, you will probably not succeed for the reasons described in the following examples:

```
addValues;              // compile-time error, no parentheses
addValues();            // compile-time error, not enough arguments
addValues(39);          // compile-time error, not enough arguments
addValues("39", "3");   // compile-time error, wrong types
```

The *addValues* method returns an *int* value. This *int* value can be used wherever an *int* value can be used. Consider these examples:

```
int result = addValues(39, 3);     // on right-hand side of an assignment
showResult(addValues(39, 3));      // as argument to another method call
```

The following exercise continues looking at the Methods application. This time you will examine some method calls.

Examine method calls

1. Return to the Methods project. (This project is already open in Visual Studio 2008 if you're continuing from the previous exercise. If you are not, open it from the \Microsoft Press\Visual CSharp Step by Step\Chapter 3\Methods folder in your Documents folder.)

2. Display the code for Window1.xaml.cs in the *Code and Text Editor* window.

3. Locate the *calculateClick* method, and look at the first two statements of this method after the *try* statement and opening brace. (We cover the purpose of *try* statements in Chapter 6, "Managing Errors and Exceptions.")

The statements are as follows:

```
int leftHandSide = System.Int32.Parse(lhsOperand.Text);
int rightHandSide = System.Int32.Parse(rhsOperand.Text);
```

These two statements declare two *int* variables called *leftHandSide* and *rightHandSide*. However, the interesting parts are the way in which the variables are initialized. In both cases, the *Parse* method of the *System.Int32* class is called (*System* is a namespace, and *Int32* is the name of the class in this namespace). You have seen this method before; it takes a single *string* parameter and converts it to an *int* value. These two lines of code take whatever the user has typed into the *lhsOperand* and *rhsOperand* text box controls on the form and converts them to *int* values.

4. Look at the fourth statement in the *calculateClick* method (after the *if* statement and another opening brace):

```
calculatedValue = addValues(leftHandSide, rightHandSide);
```

This statement calls the *addValues* method, passing the values of the *leftHandSide* and *rightHandSide* variables as its arguments. The value returned by the *addValues* method is stored in the *calculatedValue* variable.

5. Look at the next statement:

```
showResult(calculatedValue);
```

This statement calls the *showResult* method, passing the value in the *calculatedValue* variable as its argument. The *showResult* method does not return a value.

6. In the *Code and Text Editor* window, find the *showResult* method you looked at earlier.

The only statement of this method is this:

```
result.Text = answer.ToString();
```

Notice that the *ToString* method call uses parentheses even though there are no arguments.

> **Tip** You can call methods belonging to other objects by prefixing the method with the name of the object. In the preceding example, the expression *answer.ToString()* calls the method named *ToString* belonging to the object called *answer*.

Applying Scope

In some of the examples, you can see that you can create variables inside a method. These variables come into existence at the point where they are defined, and subsequent statements in the same method can then use these variables; a variable can be used only after it has been created. When the method has finished, these variables disappear.

If a variable can be used at a particular location in a program, the variable is said to be in *scope* at that location. To put it another way, the scope of a variable is simply the region of the program in which that variable is usable. Scope applies to methods as well as variables. The scope of an identifier (of a variable or method) is linked to the location of the declaration that introduces the identifier in the program, as you'll now learn.

Defining Local Scope

The opening and closing braces that form the body of a method define a scope. Any variables you declare inside the body of a method are scoped to that method; they disappear when the method ends and can be accessed only by code running in that method. These variables are called *local variables* because they are local to the method in which they are declared; they are not in scope in any other method. This arrangement means that you cannot use local variables to share information between methods. Consider this example:

```
class Example
{
    void firstMethod()
    {
        int myVar;
        ...
    }
    void anotherMethod()
    {
        myVar = 42; // error - variable not in scope
        ...
    }
}
```

This code would fail to compile because *anotherMethod* is trying to use the variable *myVar*, which is not in scope. The variable *myVar* is available only to statements in *firstMethod* and that occur after the line of code that declares *myVar*.

Defining Class Scope

The opening and closing braces that form the body of a class also create a scope. Any variables you declare inside the body of a class (but not inside a method) are scoped to that class. The proper C# name for the variables defined by a class is a *field*. In contrast with local variables, you can use fields to share information between methods. Here is an example:

```
class Example
{
    void firstMethod()
    {
        myField = 42; // ok
        ...
    }

    void anotherMethod()
    {
        myField++; // ok
        ...
    }

    int myField = 0;
}
```

The variable *myField* is defined in the class but outside the methods *firstMethod* and *anotherMethod*. Therefore, *myField* has class scope and is available for use by all methods in the class.

There is one other point to notice about this example. In a method, you must declare a variable before you can use it. Fields are a little different. A method can use a field before the statement that defines the field—the compiler sorts out the details for you!

Overloading Methods

If two identifiers have the same name and are declared in the same scope, they are said to be *overloaded*. Often an overloaded identifier is a bug that gets trapped as a compile-time error. For example, if you declare two local variables with the same name in the same method, you get a compile-time error. Similarly, if you declare two fields with the same name in the same class or two identical methods in the same class, you also get a compile-time error. This fact may seem hardly worth mentioning, given that everything so far has turned out to be a compile-time error. However, there is a way that you can overload an identifier, and that way is both useful and important.

Consider the *WriteLine* method of the *Console* class. You have already used this method for outputting a string to the screen. However, when you type *WriteLine* in the *Code and Text Editor* window when writing C# code, you will notice that IntelliSense gives you 19 different options! Each version of the *WriteLine* method takes a different set of parameters; one version takes no parameters and simply outputs a blank line, another version takes a *bool* parameter and outputs a string representation of its value (*true* or *false*), yet another implementation takes a *decimal* parameter and outputs it as a string, and so on. At compile time, the compiler looks at the types of the arguments you are passing in and then calls the version of the method that has a matching set of parameters. Here is an example:

```
static void Main()
{
    Console.WriteLine("The answer is ");
    Console.WriteLine(42);
}
```

Overloading is primarily useful when you need to perform the same operation on different data types. You can overload a method when the different implementations have different sets of parameters; that is, when they have the same name but a different number of parameters, or when the types of the parameters differ. This capability is allowed so that, when you call a method, you can supply a comma-separated list of arguments, and the number and type of the arguments are used by the compiler to select one of the overloaded methods. However, note that although you can overload the parameters of a method, you can't overload the return type of a method. In other words, you can't declare two methods with the same name that differ only in their return type. (The compiler is clever, but not that clever.)

Writing Methods

In the following exercises, you'll create a method that calculates how much a consultant would charge for a given number of consultancy days at a fixed daily rate. You will start by developing the logic for the application and then use the Generate Method Stub Wizard to help you write the methods that are used by this logic. Next, you'll run these methods in a Console application to get a feel for the program. Finally, you'll use the Visual Studio 2008 debugger to step in and out of the method calls as they run.

Develop the logic for the application

1. Using Visual Studio 2008, open the DailyRate project in the \Microsoft Press\Visual CSharp Step by Step\Chapter 3\DailyRate folder in your Documents folder.

2. In the *Solution Explorer*, double-click the file *Program.cs* to display the code for the program in the *Code and Text Editor* window.

3. Add the following statements to the body of the *run* method, between the opening and closing braces:

   ```
   double dailyRate = readDouble("Enter your daily rate: ");
   int noOfDays = readInt("Enter the number of days: ");
   writeFee(calculateFee(dailyRate, noOfDays));
   ```

   The *run* method is called by the *Main* method when the application starts. (The way in which it is called requires an understanding of classes, which we look at in Chapter 7, "Creating and Managing Classes and Objects.")

The block of code you have just added to the *run* method calls the *readDouble* method (which you will write shortly) to ask the user for the daily rate for the consultant. The next statement calls the *readInt* method (which you will also write) to obtain the number of days. Finally, the *writeFee* method (to be written) is called to display the results on the screen. Notice that the value passed to *writeFee* is the value returned by the *calculateFee* method (the last one you will need to write), which takes the daily rate and the number of days and calculates the total fee payable.

> **Note** You have not yet written the *readDouble*, *readInt*, *writeFee*, or *calculateFee* method, so IntelliSense does not display these methods when you type this code. Do not try to build the application yet, because it will fail.

Write the methods using the Generate Method Stub Wizard

1. In the *Code and Text Editor* window, right-click the *readDouble* method call in the *run* method.

   A shortcut menu appears that contains useful commands for generating and editing code, as shown here:

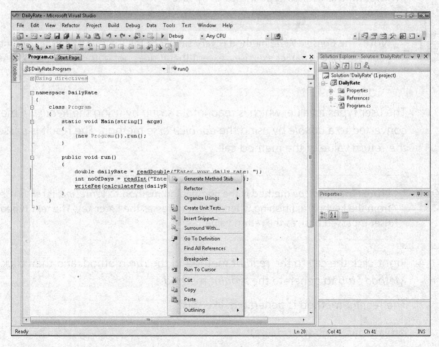

2. On the shortcut menu, click *Generate Method Stub*.

The Generate Method Stub Wizard examines the call to the *readDouble* method, ascertains the type of its parameters and return value, and generates a method with a default implementation, like this:

```
private double readDouble(string p)
{
    throw new NotImplementedException();
}
```

The new method is created with the *private* qualifier, which is described in Chapter 7. The body of the method currently just throws a *NotImplementedException*. (Exceptions are described in Chapter 6.) You will replace the body with your own code in the next step.

3. Delete the *throw new NotImplementedException();* statement from the *readDouble* method, and replace it with the following lines of code:

```
Console.Write(p);
string line = Console.ReadLine();
return double.Parse(line);
```

This block of code outputs the string in variable *p* to the screen. This variable is the string parameter passed in when the method is called, and it contains a message prompting the user to type in the daily rate.

> **Note** The *Console.Write* method is very similar to the *Console.WriteLine* statement that you have used in earlier exercises, except that it does not output a newline character after the message.

The user types a value, which is read into a *string* by using the *ReadLine* method and converted to a *double* by using the *double.Parse* method. The result is passed back as the return value of the method call.

> **Note** The *ReadLine* method is the companion method to *WriteLine*; it reads user input from the keyboard, finishing when the user presses the Enter key. The text typed by the user is passed back as the return value.

4. Right-click the call to the *readInt* method in the *run* method, and then click *Generate Method Stub* to generate the *readInt* method.

The *readInt* method is generated, like this:

```
private int readInt(string p)
{
    throw new NotImplementedException();
}
```

5. Replace the *throw new NotImplementedException();* statement in the body of the *readInt* method with the following code:

```
Console.Write(p);
string line = Console.ReadLine();
return int.Parse(line);
```

This block of code is similar to the code for the *readDouble* method. The only difference is that the method returns an *int* value, so the *string* typed by the user is converted to a number by using the *int.Parse* method.

6. Right-click the call to the *calculateFee* method in the *run* method, and then click *Generate Method Stub*.

The *calculateFee* method is generated, like this:

```
private object calculateFee(double dailyRate, int noOfDays)
{
    throw new NotImplementedException();
}
```

Notice that the Generate Method Stub Wizard uses the name of the arguments passed in to generate names for the parameters. (You can of course change the parameter names if they are not suitable.) What is more intriguing is the type returned by the method, which is *object*. The Generate Method Stub Wizard is unable to determine exactly which type of value should be returned by the method from the context in which it is called. The *object* type just means a "thing," and you should change it to the type you require when you add the code to the method. You will learn more about the *object* type in Chapter 7.

7. Change the definition of the *calculateFee* method so that it returns a *double*, as shown in bold type here:

```
private double calculateFee(double dailyRate, int noOfDays)
{
    throw new NotImplementedException();
}
```

8. Replace the body of the *calculateFee* method with the following statement, which calculates the fee payable by multiplying the two parameters together and then returns it:

```
return dailyRate * noOfDays;
```

9. Right-click the call to the *writeFee* method in the *run* method, and then click *Generate Method Stub*.

Note that the Generate Method Stub Wizard uses the definition of the *calculateFee* method to work out that its parameter should be a *double*. Also, the method call does not use a return value, so the type of the method is *void*:

```
private void writeFee(double p)
{
    ...
}
```

> **Tip** If you feel sufficiently comfortable with the syntax, you can also write methods by typing them directly into the *Code and Text Editor* window. You do not always have to use the *Generate Method Stub* menu option.

10. Type the following statements inside the *writeFee* method:

```
Console.WriteLine("The consultant's fee is: {0}", p * 1.1);
```

> **Note** This version of the *WriteLine* method demonstrates the use of a format string. The text *{0}* in the string used as the first argument to the *WriteLine* method is a placeholder that is replaced with the value of the expression following the string (*p * 1.1*) when it is evaluated at run time. Using this technique is preferable to alternatives, such as converting the value of the expression *p * 1.1* to a string and using the + operator to concatenate it to the message.

11. On the *Build* menu, click *Build Solution*.

Refactoring Code

A very useful feature of Visual Studio 2008 is the ability to refactor code.

Occasionally, you will find yourself writing the same (or similar) code in more than one place in an application. When this occurs, highlight the block of code you have just typed, and on the *Refactor* menu, click *Extract Method*. The *Extract Method* dialog box appears, prompting you for the name of a new method to create containing this code. Type a name, and click *OK*. The new method is created containing your code, and the code you typed is replaced with a call to this method. *Extract Method* is also intelligent enough to work out whether the method should take any parameters and return a value.

Test the program

1. On the *Debug* menu, click *Start Without Debugging*.

   Visual Studio 2008 builds the program and then runs it. A console window appears.

2. At the *Enter your daily rate* prompt, type **525**, and then press Enter.

3. At the *Enter the number of days* prompt, type **17**, and then press Enter.

   The program writes the following message to the console window:

   ```
   The consultant's fee is: 9817.5
   ```

4. Press the Enter key to close the application and return to the Visual Studio 2008 programming environment.

In the final exercise, you'll use the Visual Studio 2008 debugger to run your program in slow motion. You'll see when each method is called (this action is referred to as *stepping into the method*) and then see how each *return* statement transfers control back to the caller (also known as *stepping out of the method*). While you are stepping in and out of methods, you'll use the tools on the *Debug* toolbar. However, the same commands are also available on the *Debug* menu when an application is running in Debug mode.

Step through the methods using the Visual Studio 2008 debugger

1. In the *Code and Text Editor* window, find the *run* method.

2. Move the mouse to the first statement in the *run* method:

   ```
   double dailyRate = readDouble("Enter your daily rate: ");
   ```

3. Right-click anywhere on this line, and on the shortcut menu, click *Run To Cursor*.

   The program starts and runs until it reaches the first statement in the *run* method, and then it pauses. A yellow arrow in the left margin of the *Code and Text Editor* window indicates the current statement, which is also highlighted with a yellow background.

4. On the *View* menu, point to *Toolbars*, and then make sure the *Debug* toolbar is selected.

   If it was not already visible, the *Debug* toolbar opens. It may appear docked with the other toolbars. If you cannot see the toolbar, try using the *Toolbars* command on the *View* menu to hide it, and notice which buttons disappear. Then display the toolbar again. The *Debug* toolbar looks like this (the toolbar differs slightly between Visual Studio 2008 and Microsoft Visual C# 2008 Express Edition):

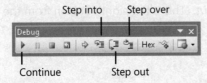

> **Tip** To make the *Debug* toolbar appear in its own window, use the handle at the left end of the toolbar to drag it over the *Code and Text Editor* window.

5. On the *Debug* toolbar, click the *Step Into* button. (This is the sixth button from the left.)

   This action causes the debugger to step into the method being called. The yellow cursor jumps to the opening brace at the start of the *readDouble* method.

6. Click *Step Into* again. The cursor advances to the first statement:

```
Console.Write(p);
```

> **Tip** You can also press F11 rather than repeatedly clicking *Step Into* on the *Debug* toolbar.

7. On the *Debug* toolbar, click *Step Over*. (This is the seventh button from the left.)

   This action causes the method to execute the next statement without debugging it (stepping into it). The yellow cursor moves to the second statement of the method, and the program displays the *Enter your daily rate* prompt in a Console window before returning to Visual Studio 2008. (The Console window might be hidden behind Visual Studio.)

> **Tip** You can also press F10 rather than clicking *Step Over* on the *Debug* toolbar.

8. On the *Debug* toolbar, click *Step Over*.

   This time, the yellow cursor disappears and the Console window gets the focus because the program is executing the *Console.ReadLine* method and is waiting for you to type something.

9. Type **525** in the Console window, and then press Enter.

   Control returns to Visual Studio 2008. The yellow cursor appears on the third line of the method.

10. Without clicking, move the mouse over the reference to the *line* variable on either the second or the third line of the method (it doesn't matter which).

    A ScreenTip appears, displaying the current value of the *line* variable ("525"). You can use this feature to make sure that a variable has been set to an expected value while stepping through methods.

11. On the *Debug* toolbar, click *Step Out*. (This is the eighth button from the left.)

This action causes the current method to continue running uninterrupted to its end. The *readDouble* method finishes, and the yellow cursor is placed back at the first statement of the *run* method.

 Tip You can also press Shift+F11 rather than clicking *Step Out* on the *Debug* toolbar.

12. On the *Debug* toolbar, click *Step Into*.

The yellow cursor moves to the second statement in the *run* method:

```
int noOfDays = readInt("Enter the number of days: ");
```

13. On the *Debug* toolbar, click *Step Over*.

This time you have chosen to run the method without stepping through it. The Console window appears again, prompting you for the number of days.

14. In the Console window, type **17**, and then press Enter.

Control returns to Visual Studio 2008. The yellow cursor moves to the third statement of the *run* method:

```
writeFee(calculateFee(dailyRate, noOfDays));
```

15. On the *Debug* toolbar, click *Step Into*.

The yellow cursor jumps to the opening brace at the start of the *calculateFee* method. This method is called first, before *writeFee*, because the value returned by this method is used as the parameter to *writeFee*.

16. On the *Debug* toolbar, click *Step Out*.

The yellow cursor jumps back to the third statement of the *run* method.

17. On the *Debug* toolbar, click *Step Into*.

This time, the yellow cursor jumps to the opening brace at the start of the *writeFee* method.

18. Place the mouse over the *p* variable in the method definition.

The value of *p*, 8925.0, is displayed in a ScreenTip.

19. On the *Debug* toolbar, click *Step Out*.

The message *The consultant's fee is: 9817.5* is displayed in the Console window. (You may need to bring the Console window to the foreground to display it if it is hidden behind Visual Studio 2008.) The yellow cursor returns to the third statement in the *run* method.

20. On the *Debug* toolbar, click *Continue* (this is the first button on the toolbar) to cause the program to continue running without stopping at each statement.

Tip You can also press F5 to continue execution in the debugger.

The application completes and finishes running.

Congratulations! You've successfully written and called methods and used the Visual Studio 2008 debugger to step in and out of methods as they run.

- If you want to continue to the next chapter

  Keep Visual Studio 2008 running, and turn to Chapter 4, "Using Decision Statements."

- If you want to exit Visual Studio 2008 now

  On the *File* menu, click *Exit*. If you see a *Save* dialog box, click *Yes* (if you are using Visual Studio 2008) or *Save* (if you are using Visual C# 2008 Express Edition) and save the project.

Chapter 3 Quick Reference

To	Do this

Declare a method	Write the method inside a class. For example:
	
	```int addValues(int leftHandSide, int rightHandSide)```
	```{```
	``` ...```
	```}```

Return a value from inside a method	Write a *return* statement inside the method. For example: ```return leftHandSide + rightHandSide;```
Return from a method before the end of the method	Write a *return* statement inside the method. For example: ```return;```
Call a method	Write the name of the method, together with any arguments between parentheses. For example: ```addValues(39, 3);```
Use the Generate Method Stub Wizard	Right-click a call to the method, and then click *Generate Method Stub* on the shortcut menu.
Display the *Debug* toolbar	On the *View* menu, point to *Toolbars*, and then click *Debug*.
Step into a method	On the *Debug* toolbar, click *Step Into*. or On the *Debug* menu, click *Step Into*.
Step out of a method	On the *Debug* toolbar, click *Step Out*. or On the *Debug* menu, click *Step Out*.

Chapter 4
Using Decision Statements

After completing this chapter, you will be able to:

- Declare Boolean variables.

- Use Boolean operators to create expressions whose outcome is either true or false.

- Write if statements to make decisions based on the result of a Boolean expression.

- Write switch statements to make more complex decisions.

In Chapter 3, "Writing Methods and Applying Scope," you learned how to group related statements into methods. You also learned how to use parameters to pass information to a method and how to use *return* statements to pass information out of a method. Dividing a program into a set of discrete methods, each designed to perform a specific task or calculation, is a necessary design strategy. Many programs need to solve large and complex problems. Breaking up a program into methods helps you understand these problems and focus on how to solve them one piece at a time. You also need to be able to write methods that selectively perform different actions depending on the circumstances. In this chapter, you'll see how to accomplish this task.

Declaring Boolean Variables

In the world of programming (unlike in the real world), everything is black or white, right or wrong, true or false. For example, if you create an integer variable called *x*, assign the value 99 to *x*, and then ask, "Does *x* contain the value 99?" the answer is definitely true. If you ask, "Is *x* less than 10?" the answer is definitely false. These are examples of *Boolean expressions*. A Boolean expression always evaluates to true or false.

Note The answers to these questions are not definitive for all programming languages. An unassigned variable has an undefined value, and you cannot, for example, say that it is definitely less than 10. Issues such as this one are a common source of errors in C and C++ programs. The Microsoft Visual C# compiler solves this problem by ensuring that you always assign a value to a variable before examining it. If you try to examine the contents of an unassigned variable, your program will not compile.

Microsoft Visual C# provides a data type called *bool*. A *bool* variable can hold one of two values: *true* or *false*. For example, the following three statements declare a *bool* variable called *areYouReady*, assign *true* to the variable, and then write its value to the console:

```
bool areYouReady;
areYouReady = true;
Console.WriteLine(areYouReady); // writes True
```

Using Boolean Operators

A Boolean operator is an operator that performs a calculation whose result is either true or false. C# has several very useful Boolean operators, the simplest of which is the NOT operator, which is represented by the exclamation point (!). The ! operator negates a Boolean value, yielding the opposite of that value. In the preceding example, if the value of the variable *areYouReady* is true, the value of the expression !*areYouReady* is false.

Understanding Equality and Relational Operators

Two Boolean operators that you will frequently use are the equality (==) and inequality (!=) operators. You use these binary operators to find out whether one value is the same as another value of the same type. The following table summarizes how these operators work, using an *int* variable called *age* as an example.

Operator	Meaning	Example	Outcome if age is 42
==	Equal to	age == 100	False
!=	Not equal to	age != 0	True

Closely related to these two operators are the *relational* operators. You use these operators to find out whether a value is less than or greater than another value of the same type. The following table shows how to use these operators.

Operator	Meaning	Example	Outcome if age is 42
<	Less than	age < 21	False
<=	Less than or equal to	age <= 18	False
>	Greater than	age > 16	True
>=	Greater than or equal to	age >= 30	True

Note Don't confuse the *equality* operator == with the *assignment* operator =. The expression x==y compares x with y and has the value *true* if the values are the same. The expression x=y assigns the value of y to x.

Understanding Conditional Logical Operators

C# also provides two other Boolean operators: the logical AND operator, which is represented by the *&&* symbol, and the logical OR operator, which is represented by the *||* symbol. Collectively, these are known as the conditional logical operators. Their purpose is to combine two Boolean expressions or values into a single Boolean result. These binary operators are similar to the equality and relational operators in that the value of the expressions in which they appear is either true or false, but they differ in that the values on which they operate must be either true or false.

The outcome of the *&&* operator is *true* if and only if both of the Boolean expressions it operates on are *true*. For example, the following statement assigns the value *true* to *validPercentage* if and only if the value of *percent* is greater than or equal to 0 and the value of *percent* is less than or equal to 100:

```
bool validPercentage;
validPercentage = (percent >= 0) && (percent <= 100);
```

> **Tip** A common beginner's error is to try to combine the two tests by naming the *percent* variable only once, like this:
>
> ```
> percent >= 0 && <= 100 // this statement will not compile
> ```
>
> Using parentheses helps avoid this type of mistake and also clarifies the purpose of the expression. For example, compare these two expressions:
>
> ```
> validPercentage = percent >= 0 && percent <= 100
> ```
>
> and
>
> ```
> validPercentage = (percent >= 0) && (percent <= 100)
> ```
>
> Both expressions return the same value because the precedence of the *&&* operator is less than that of >= and <=. However, the second expression conveys its purpose in a more readable manner.

The outcome of the *||* operator is *true* if either of the Boolean expressions it operates on is *true*. You use the *||* operator to determine whether any one of a combination of Boolean expressions is *true*. For example, the following statement assigns the value *true* to *invalidPercentage* if the value of *percent* is less than 0 or the value of *percent* is greater than 100:

```
bool invalidPercentage;
invalidPercentage = (percent < 0) || (percent > 100);
```

Short-Circuiting

The *&&* and *||* operators both exhibit a feature called *short-circuiting*. Sometimes it is not necessary to evaluate both operands when ascertaining the result of a conditional logical expression. For example, if the left operand of the *&&* operator evaluates to *false*, the result of the entire expression must be *false* regardless of the value of the right operand. Similarly, if the value of the left operand of the *||* operator evaluates to *true*, the result of the entire expression must be *true*, irrespective of the value of the right operand. In these cases, the *&&* and *||* operators bypass the evaluation of the right operand. Here are some examples:

```
(percent >= 0) && (percent <= 100)
```

In this expression, if the value of *percent* is less than 0, the Boolean expression on the left side of *&&* evaluates to *false*. This value means that the result of the entire expression must be *false*, and the Boolean expression to the right of the *&&* operator is not evaluated.

```
(percent < 0) || (percent > 100)
```

In this expression, if the value of *percent* is less than 0, the Boolean expression on the left side of *||* evaluates to *true*. This value means that the result of the entire expression must be *true* and the Boolean expression to the right of the *||* operator is not evaluated.

If you carefully design expressions that use the conditional logical operators, you can boost the performance of your code by avoiding unnecessary work. Place simple Boolean expressions that can be evaluated easily on the left side of a conditional logical operator and put more complex expressions on the right side. In many cases, you will find that the program does not need to evaluate the more complex expressions.

Summarizing Operator Precedence and Associativity

The following table summarizes the precedence and associativity of all the operators you have learned about so far. Operators in the same category have the same precedence. The operators in categories higher up in the table take precedence over operators in categories lower down.

Category	Operators	Description	Associativity
Primary	( )	Precedence override	Left
	++	Post-increment	
	--	Post-decrement	
Unary	!	Logical NOT	Left
	+	Addition	
	-	Subtraction	
	++	Pre-increment	
	--	Pre-decrement	
Multiplicative	*	Multiply	Left
	/	Divide	
	%	Division remainder (modulus)	
Additive	+	Addition	Left
	-	Subtraction	
Relational	<	Less than	Left
	<=	Less than or equal to	
	>	Greater than	
	>=	Greater than or equal to	
Equality	==	Equal to	Left
	!=	Not equal to	
Conditional AND	&&	Logical AND	Left
Conditional OR	||	Logical OR	Left
Assignment	=		Right

Using *if* Statements to Make Decisions

When you want to choose between executing two different blocks of code depending on the result of a Boolean expression, you can use an *if* statement.

Understanding *if* Statement Syntax

The syntax of an *if* statement is as follows (*if* and *else* are C# keywords):

```
if ( booleanExpression )
    statement-1;
else
    statement-2;
```

If *booleanExpression* evaluates to *true*, *statement-1* runs; otherwise, *statement-2* runs. The *else* keyword and the subsequent *statement-2* are optional. If there is no *else* clause and the *booleanExpression* is *false*, execution continues with whatever code follows the *if* statement.

For example, here's an *if* statement that increments a variable representing the second hand of a stopwatch (minutes are ignored for now). If the value of the *seconds* variable is 59, it is reset to 0; otherwise, it is incremented using the ++ operator:

```
int seconds;
...
if (seconds == 59)
    seconds = 0;
else
    seconds++;
```

Boolean Expressions Only, Please!

The expression in an *if* statement must be enclosed in parentheses. Additionally, the expression must be a Boolean expression. In some other languages (notably C and C++), you can write an integer expression, and the compiler will silently convert the integer value to *true* (nonzero) or *false* (0). C# does not support this behavior, and the compiler reports an error if you write such an expression.

If you accidentally specify the assignment operator, =, instead of the equality test operator, ==, in an *if* statement, the C# compiler recognizes your mistake and refuses to compile your code. For example:

```
int seconds;
...
if (seconds = 59)  // compile-time error
...
if (seconds == 59) // ok
```

Accidental assignments were another common source of bugs in C and C++ programs, which would silently convert the value assigned (59) to a Boolean expression (anything nonzero was considered to be true), with the result that the code following the *if* statement would be performed every time.

Incidentally, you can use a Boolean variable as the expression for an *if* statement, although it must still be enclosed in parentheses, as shown in this example:

```
bool inWord;
...
if (inWord == true) // ok, but not commonly used
...
if (inWord)         // better
```

Using Blocks to Group Statements

Notice that the syntax of the *if* statement shown earlier specifies a single statement after the *if (booleanExpression)* and a single statement after the *else* keyword. Sometimes you'll want to perform more than one statement when a Boolean expression is true. You could group the statements inside a new method and then call the new method, but a simpler solution is to group the statements inside a *block*. A block is simply a sequence of statements grouped between an opening and a closing brace. A block also starts a new scope. You can define variables inside a block, but they will disappear at the end of the block.

In the following example, two statements that reset the *seconds* variable to 0 and increment the *minutes* variable are grouped inside a block, and the whole block executes if the value of *seconds* is equal to 59:

```
int seconds = 0;
int minutes = 0;
...
if (seconds == 59)
{
    seconds = 0;
    minutes++;
}
else
    seconds++;
```

> **Important** If you omit the braces, the C# compiler associates only the first statement (`seconds = 0;`) with the *if* statement. The subsequent statement (`minutes++;`) will not be recognized by the compiler as part of the *if* statement when the program is compiled. Furthermore, when the compiler reaches the *else* keyword, it will not associate it with the previous *if* statement, and it will report a syntax error instead.

Cascading *if* Statements

You can nest *if* statements inside other *if* statements. In this way, you can chain together a sequence of Boolean expressions, which are tested one after the other until one of them evaluates to *true*. In the following example, if the value of *day* is 0, the first test evaluates to *true* and *dayName* is assigned the string *"Sunday"*. If the value of *day* is not 0, the first test fails and control passes to the *else* clause, which runs the second *if* statement and compares the value of *day* with 1. The second *if* statement is reached only if the first test is *false*. Similarly, the third *if* statement is reached only if the first and second tests are *false*.

```
if (day == 0)
    dayName = "Sunday";
else if (day == 1)
    dayName = "Monday";
```

```
else if (day == 2)
    dayName = "Tuesday";
else if (day == 3)
    dayName = "Wednesday";
else if (day == 4)
    dayName = "Thursday";
else if (day == 5)
    dayName = "Friday";
else if (day == 6)
    dayName = "Saturday";
else
    dayName = "unknown";
```

In the following exercise, you'll write a method that uses a cascading *if* statement to compare two dates.

Write *if* statements

1. Start Microsoft Visual Studio 2008 if it is not already running.

2. Open the Selection project, located in the \Microsoft Press\Visual CSharp Step by Step\Chapter 4\Selection folder in your Documents folder.

3. On the *Debug* menu, click *Start Without Debugging*.

   Visual Studio 2008 builds and runs the application. The form contains two *DateTimePicker* controls called *first* and *second*. (These controls display a calendar allowing you to select a date when you click the drop-down arrow.) Both controls are initially set to the current date.

4. Click *Compare*.

   The following text appears in the text box:

```
first == second : False
first != second : True
first <  second : False
first <= second : False
first >  second : True
first >= second : True
```

   The Boolean expression first == second should be *true* because both *first* and *second* are set to the current date. In fact, only the less than operator and the greater than or equal to operator seem to be working correctly.

5. Click *Quit* to return to the Visual Studio 2008 programming environment.

6. Display the code for Window1.xaml.cs in the *Code and Text Editor* window.

7. Locate the *compareClick* method, which looks like this:

```
private int compareClick(object sender, RoutedEventArgs e)
{
    int diff = dateCompare(first.Value, second.Value);
    info.Text = "";
    show("first == second", diff == 0);
    show("first != second", diff != 0);
    show("first < second", diff < 0);
    show("first <= second", diff <= 0);
    show("first > second", diff > 0);
    show("first >= second", diff >= 0);
}
```

This method runs whenever the user clicks the *Compare* button on the form. It retrieves the values of the dates displayed in the *first* and *second DateTimePicker* controls on the form and calls another method called *dateCompare* to compare them. You will examine the *dateCompare* method in the next step.

The *show* method summarizes the results of the comparison in the *info* text box control on the form.

8. Locate the *dateCompare* method, which looks like this:

```
private int dateCompare(DateTime leftHandSide, DateTime rightHandSide)
{
    // TO DO
    return 42;
}
```

This method currently returns the same value whenever it is called, rather than 0, -1, or +1 depending on the values of its parameters. This explains why the application is not working as expected!

The purpose of this method is to examine its arguments and return an integer value based on their relative values; it should return 0 if they have the same value, -1 if the value of the first argument is less than the value of the second argument, and +1 if the value of the first argument is greater than the value of the second argument. (A date is considered greater than another date if it comes after it chronologically.) You need to implement the logic in this method to compare two dates correctly.

9. Remove the // TO DO comment and the *return* statement from the *dateCompare* method.

10. Add the following statements shown in bold type to the body of the *dateCompare* method:

```
private int dateCompare(DateTime leftHandSide, DateTime rightHandSide)
{
    int result;

    if (leftHandSide.Year < rightHandSide.Year)
        result = -1;
    else if (leftHandSide.Year > rightHandSide.Year)
        result = +1;
}
```

If the expression leftHandSide.Year < rightHandSide.Year is *true*, the date in *leftHandSide* must be earlier than the date in *rightHandSide*, so the program sets the *result* variable to -1. Otherwise, if the expression leftHandSide.Year > rightHandSide.Year is *true*, the date in *leftHandSide* must be later than the date in *rightHandSide*, and the program sets the result *variable* to +1.

If the expression leftHandSide.Year < rightHandSide.Year is *false* and the expression leftHandSide.Year > rightHandSide.Year is also *false*, the *Year* property of both dates must be the same, so the program needs to compare the months in each date.

11. Add the following statements shown in bold type to the body of the *dateCompare* method, after the code you entered in the preceding step:

```
private int dateCompare(DateTime leftHandSide, DateTime rightHandSide)
{
    ...

    else if (leftHandSide.Month < rightHandSide.Month)
        result = -1;
    else if (leftHandSide.Month > rightHandSide.Month)
        result = +1;
}
```

These statements follow a similar logic for comparing months to that used to compare years in the preceding step.

If the expression leftHandSide.Month < rightHandSide.Month is *false* and the expression leftHandSide.Month > rightHandSide.Month is also *false*, the *Month* property of both dates must be the same, so the program finally needs to compare the days in each date.

12. Add the following statements to the body of the *dateCompare* method, after the code you entered in the preceding two steps:

```
private int dateCompare(DateTime leftHandSide, DateTime rightHandSide)
{
    ...

    else if (leftHandSide.Day < rightHandSide.Day)
        result = -1;
    else if (leftHandSide.Day > rightHandSide.Day)
        result = +1;
    else
        result = 0;
    return result;
}
```

You should recognize the pattern in this logic by now.

If leftHandSide.Day < rightHandSide.Day and leftHandSide.Day > rightHandSide.Day both are *false*, the value in the *Day* properties in both variables must be the same. The *Month* values and the *Year* values must also be identical, respectively, for the program logic to have reached this far, so the two dates must be the same, and the program sets the value of *result* to 0.

The final statement returns the value stored in the *result* variable.

13. On the *Debug* menu, click *Start Without Debugging*.

The application is rebuilt and restarted. Once again, the two *DateTimePicker* controls, *first* and *second*, are set to the current date.

14. Click *Compare*.

The following text appears in the text box:

```
first == second : True
first != second : False
first <  second : False
first <= second : True
first >  second : False
first >= second : True
```

These are the correct results for identical dates.

15. Click the drop-down arrow for the second *DateTimePicker* control, and then click tomorrow's date.

16. Click *Compare*.

The following text appears in the text box:

```
first == second : False
first != second : True
first <  second : True
first <= second : True
first >  second : False
first >= second : False
```

Again, these are the correct results when the first date is earlier than the second date.

17. Test some other dates, and verify that the results are as you would expect. Click *Quit* when you have finished.

Comparing Dates in Real-World Applications

Now that you have seen how to use a rather long and complicated series of *if* and *else* statements, I should mention that this is not the technique you would use to compare dates in a real-world application. In the Microsoft .NET Framework class library, dates are held using a special type called *DateTime*. If you look at the *dateCompare* method you have written in the preceding exercise, you will see that the two parameters, *leftHandSide* and *rightHandSide*, are *DateTime* values. The logic you have written compares only the date part of these variables—there is also a time element. For two *DateTime* values to be considered equal, they should not only have the same date but also the same time. Comparing dates and times is such a common operation that the *DateTime* type has a built-in method called *Compare* for doing just that. The *Compare* method takes two *DateTime* arguments and compares them, returning a value indicating whether the first argument is less than the second, in which case the result will be negative; whether the first argument is greater than the second, in which case the result will be positive; or whether both arguments represent the same date and time, in which case the result will be 0.

Using *switch* Statements

Sometimes when you write a cascading *if* statement, all the *if* statements look similar because they all evaluate an identical expression. The only difference is that each *if* compares the result of the expression with a different value. For example, consider the following block of code that uses an *if* statement to examine the value in the *day* variable and work out which day of the week it is:

```
if (day == 0)
    dayName = "Sunday";
else if (day == 1)
    dayName = "Monday";
else if (day == 2)
    dayName = "Tuesday";
else if (day == 3)
    ...
else
    dayName = "Unknown";
```

In these situations, often you can rewrite the cascading *if* statement as a *switch* statement to make your program more efficient and more readable.

Understanding *switch* Statement Syntax

The syntax of a *switch* statement is as follows (*switch*, *case*, and *default* are keywords):

```
switch ( controllingExpression )
{
case constantExpression :
    statements
    break;
case constantExpression :
    statements
    break;
...
default :
    statements
    break;
}
```

The *controllingExpression* is evaluated once. Control then jumps to the block of code identi-fied by the *constantExpression* whose value is equal to the result of the *controllingExpression*. (The identifier is called a *case label*.) Execution runs as far as the *break* statement, at which point the *switch* statement finishes and the program continues at the first statement after the closing brace of the *switch* statement. If none of the *constantExpression* values are equal to the value of the *controllingExpression*, the statements below the optional *default* label run.

> **Note** Each *constantExpression* value must be unique, so the *controllingExpression* will match only one of them. If the value of the *controllingExpression* does not match any *constantExpression* value, and there is no *default* label, program execution continues with the first statement after the closing brace of the *switch* statement.

For example, you can rewrite the previous cascading *if* statement as the following *switch* statement:

```
switch (day)
{
case 0 :
    dayName = "Sunday";
    break;
case 1 :
    dayName = "Monday";
    break;
case 2 :
    dayName = "Tuesday";
    break;
...
default :
    dayName = "Unknown";
    break;
}
```

Following the *switch* Statement Rules

The *switch* statement is very useful, but unfortunately, you can't always use it when you may like to. Any *switch* statement you write must adhere to the following rules:

- You can use switch only on primitive data types, such as int or string. With any other types (including float and double), you'll have to use an if statement.

- The case labels must be constant expressions, such as 42 or "42". If you need to calculate your case label values at run time, you must use an if statement.

- The case labels must be unique expressions. In other words, two case labels cannot have the same value.

- You can specify that you want to run the same statements for more than one value by providing a list of case labels and no intervening statements, in which case the code for the final label in the list is executed for all cases in that list. However, if a label has one or more associated statements, execution cannot fall through to subsequent labels, and the compiler generates an error. For example:

```
switch (trumps)
{
case Hearts :
case Diamonds :      // Fall-through allowed - no code between labels
    color = "Red";   // Code executed for Hearts and Diamonds
    break;
case Clubs :
    color = "Black";
case Spades :        // Error - code between labels
    color = "Black";
    break;
}
```

Note The *break* statement is the most common way to stop fall-through, but you can also use a *return* statement or a *throw* statement. The *throw* statement is described in Chapter 6, "Managing Errors and Exceptions."

switch Fall-Through Rules

Because you cannot accidentally fall through from one *case* label to the next if there is any intervening code, you can freely rearrange the sections of a *switch* statement without affecting its meaning (including the *default* label, which by convention is usually placed as the last label but does not have to be).

C and C++ programmers should note that the *break* statement is mandatory for every case in a *switch* statement (even the default case). This requirement is a good thing; it is very common in C or C++ programs to forget the *break* statement, allowing execution to fall through to the next label and leading to bugs that are very difficult to spot.

If you really want to, you can mimic C/C++ fall-through in C# by using a *goto* statement to go to the following *case* or *default* label. Using *goto* in general is not recommended, though, and this book does not show you how to do it!

In the following exercise, you will complete a program that reads the characters of a string and maps each character to its XML representation. For example, the left angle bracket character, <, has a special meaning in XML (it's used to form elements). If you have data that contains this character, it must be translated into the text "<" so that an XML processor knows that it is data and not part of an XML instruction. Similar rules apply to the right angle bracket (>), ampersand (&), single quotation mark ('), and double quotation mark (") characters. You will write a *switch* statement that tests the value of the character and traps the special XML characters as *case* labels.

Write *switch* statements

1. Start Visual Studio 2008 if it is not already running.

2. Open the SwitchStatement project, located in the \Microsoft Press\Visual CSharp Step by Step\Chapter 4\SwitchStatement folder in your Documents folder.

3. On the *Debug* menu, click *Start Without Debugging*.

   Visual Studio 2008 builds and runs the application. The application displays a form containing two text boxes separated by a *Copy* button.

4. Type the following sample text into the upper text box:

```
inRange = (lo <= number) && (hi >= number);
```

5. Click *Copy*.

The statement is copied verbatim into the lower text box, and no translation of the <, &, or > character occurs.

6. Close the form, and return to Visual Studio 2008.

7. Display the code for Window1.xaml.cs in the *Code and Text Editor* window, and locate the *copyOne* method.

The *copyOne* method copies the character specified as its input parameter to the end of the text displayed in the lower text box. At the moment, *copyOne* contains a *switch* statement with a single *default* section. In the following few steps, you will modify this *switch* statement to convert characters that are significant in XML to their XML mapping. For example, the < character will be converted to the string "<".

8. Add the following statements to the *switch* statement after the opening brace for the statement and directly before the *default* label:

```
case '<' :
    target.Text += "&lt;";
    break;
```

If the current character being copied is a >, this code will append the string "<" to the text being output in its place.

9. Add the following statements to the *switch* statement after the *break* statement you have just added and above the *default* label:

```
case '>' :
    target.Text += "&gt;";
    break;
case '&' :
    target.Text += "&amp;";
    break;
```

```
case '\"' :
    target.Text +=  "&#34;";
    break;
case '\'' :
    target.Text += "&#39;";
    break;
```

 Note The single quotation mark (') and double quotation mark (") have a special meaning in C# as well as in XML—they are used to delimit character and string constants. The backslash (\) in the final two *case* labels is an escape character that causes the C# compiler to treat these characters as literals rather than as delimiters.

10. On the *Debug* menu, click *Start Without Debugging*.

11. Type the following text into the upper text box:

```
inRange = (lo <= number) && (hi >= number);
```

12. Click *Copy*.

The statement is copied into the lower text box. This time, each character undergoes the XML mapping implemented in the *switch* statement. The target text box displays the following text:

```
inRange = (lo &lt;= number) &amp;&amp; (hi &gt;= number)
```

13. Experiment with other strings, and verify that all special characters (<, >, &, ", and ') are handled correctly.

14. Close the form.

- If you want to continue to the next chapter

  Keep Visual Studio 2008 open, and turn to Chapter 5.

- If you want to exit Visual Studio 2008 now

  On the *File* menu, click *Exit*. If you see a *Save* dialog box, click *Yes* (if you are using Visual Studio 2008) or *Save* (if you are using Visual C# 2008 Express Edition) and save the project.

Chapter 4 Quick Reference

To	Do this	Example
Determine whether two values are equivalent	Use the == or != operator.	`answer == 42`
Compare the value of two expressions	Use the <, <=, >, or >= operator.	`Age >= 21`
Declare a Boolean variable	Use the *bool* keyword as the type of the variable.	`bool inRange;`
Create a Boolean expression that is true only if two other conditions are true	Use the *&&* operator.	`inRange = (lo <= number)` `        && (number <= hi);`
Create a Boolean expression that is true if either of two other conditions is true	Use the || operator.	`outOfRange = (number < lo)` `           || (hi < number);`
Run a statement if a condition is true	Use an *if* statement.	`If (inRange)` `  process();`

Run more than one statement if a condition is true	Use an *if* statement and a block.	`If (seconds == 59)`
		`{`
		`    seconds = 0;`
		`    minutes++;`
		`}`

Associate different statements with different values of a controlling expression	Use a *switch* statement.	`switch (current)`
		`{`
		`  case 0:`
		`      ...`
		`      break;`
		
		`  case 1:`
		`      ...`
		`      break;`
		`  default :`
		`      ...`
		`      break;`
		`}`

Chapter 5
Using Compound Assignment and Iteration Statements

After completing this chapter, you will be able to:

- Update the value of a variable by using compound assignment operators.

- Write *while*, *for*, and *do* iteration statements.

- Step through a *do* statement and watch as the values of variables change.

In Chapter 4, "Using Decision Statements," you learned how to use the *if* and *switch* constructs to run statements selectively. In this chapter, you'll see how to use a variety of iteration (or *looping*) statements to run one or more statements repeatedly. When you write iteration statements, you usually need to control the number of iterations that you perform. You can achieve this by using a variable, updating its value with each iteration, and stopping the process when the variable reaches a particular value. You'll also learn about the special assignment operators that you should use to update the value of a variable in these circumstances.

Using Compound Assignment Operators

You've already seen how to use arithmetic operators to create new values. For example, the following statement uses the plus operator (+) to display to the console a value that is 42 greater than the variable *answer*:

```
Console.WriteLine(answer + 42);
```

You've also seen how to use assignment statements to change the value of a variable. The following statement uses the assignment operator to change the value of *answer* to 42:

```
answer = 42;
```

If you want to add 42 to the value of a variable, you can combine the assignment operator and the addition operator. For example, the following statement adds 42 to *answer*. After this statement runs, the value of *answer* is 42 more than it was before:

```
answer = answer + 42;
```

Although this statement works, you'll probably never see an experienced programmer write code like this. Adding a value to a variable is so common that C# lets you perform this task in shorthand manner by using the operator +=. To add 42 to *answer*, you can write the following statement:

```
answer += 42;
```

You can use this shortcut to combine any arithmetic operator with the assignment operator, as the following table shows. These operators are collectively known as the *compound assignment operators*.

Don't write this	Write this
variable = variable * number;	variable *= number;
variable = variable / number;	variable /= number;
variable = variable % number;	variable %= number;
variable = variable + number;	variable += number;
variable = variable - number;	variable -= number;

Tip The compound assignment operators share the same precedence and right associativity as the simple assignment operators.

The += operator also functions on strings; it appends one string to the end of another. For example, the following code displays "Hello John" on the console:

```
string name = "John";
string greeting = "Hello ";
greeting += name;
Console.WriteLine(greeting);
```

You cannot use any of the other compound assignment operators on strings.

Note Use the increment (++) and decrement (−−) operators instead of a compound assignment operator when incrementing or decrementing a variable by 1. For example, replace:

```
count += 1;
```

with

```
count++;
```

Writing *while* Statements

You use a *while* statement to run a statement repeatedly while some condition is true. The syntax of a *while* statement is as follows:

```
while ( booleanExpression )
    statement
```

The Boolean expression is evaluated, and if it is true, the statement runs and then the Boolean expression is evaluated again. If the expression is still true, the statement is repeated and then the Boolean expression is evaluated again. This process continues until the Boolean expression evaluates to false, when the *while* statement exits. Execution then continues with the first statement after the *while* statement. A *while* statement shares many syntactic similarities with an *if* statement (in fact, the syntax is identical except for the keyword):

- The expression must be a Boolean expression.

- The Boolean expression must be written inside parentheses.

- If the Boolean expression evaluates to false when first evaluated, the statement does n not run.

- If you want to perform two or more statements under the control of a *while* statement, you must use braces to group those statements in a block.

Here's a *while* statement that writes the values 0 through 9 to the console:

```
int i = 0;
while (i < 10)
{
    Console.WriteLine(i);
    i++;
}
```

All *while* statements should terminate at some point. A common beginner's mistake is forgetting to include a statement to cause the Boolean expression eventually to evaluate to false and terminate the loop, which results in a program that runs forever. In the example, the i++ statement performs this role.

 Note The variable *i* in the *while* loop controls the number of iterations that it performs. This is a very common idiom, and the variable that performs this role is sometimes called the *Sentinel* variable.

In the following exercise, you will write a *while* loop to iterate through the contents of a text file one line at a time and write each line to a text box in a form.

Write a *while* statement

1. Using Microsoft Visual Studio 2008, open the WhileStatement project, located in the \Microsoft Press\Visual CSharp Step by Step\Chapter 5\WhileStatement folder in your Documents folder.

2. On the *Debug* menu, click *Start Without Debugging*.

   Visual Studio 2008 builds and runs the application. The application is a simple text file viewer that you can use to select a file and display its contents.

3. Click *Open File*.

   The *Open* dialog box opens.

4. Move to the \Microsoft Press\Visual CSharp Step by Step\Chapter 5\WhileStatement\ WhileStatement folder in your Documents folder.

5. Select the file Window1.xaml.cs, and then click *Open*.

   The name of the file, Window1.xaml.cs, appears in the small text box on the form, but the contents of the file Window1.xaml.cs do not appear in the large text box. This is because you have not yet implemented the code that reads the contents of the file and displays it. You will add this functionality in the following steps.

6. Close the form and return to Visual Studio 2008.

7. Display the code for the file Window1.xaml.cs in the *Code and Text Editor* window, and locate the *openFileDialogFileOk* method.

   This method runs when the user clicks the *Open* button after selecting a file in the *Open* dialog box. The body of the method is currently implemented as follows:

```
private void openFileDialogFileOk(object sender, System.ComponentModel.
CancelEventArgs e)
{
    string fullPathname = openFileDialog.FileName;
    FileInfo src = new FileInfo(fullPathname);
    filename.Text = src.Name;

    // add while loop here
}
```

The first statement declares a *string* variable called *fullPathname* and initializes it to the *FileName* property of the *openFileDialog* object. This property contains the full name (including the folder) of the source file selected in the *Open* dialog box.

> **Note** The *openFileDialog* object is an instance of the *OpenFileDialog* class. This class provides methods that you can use to display the standard *Windows Open* dialog box, select a file, and retrieve the name and path of the selected file.

The second statement declares a *FileInfo* variable called *src* and initializes it to an object that represents the file selected in the *Open* dialog box. (*FileInfo* is a class provided by the Microsoft .NET Framework that you can use to manipulate files.)

The third statement assigns the *Text* property of the *filename* control to the *Name* property of the *src* variable. The *Name* property of the *src* variable holds the name of the file selected in the *Open* dialog box, but without the name of the folder. This statement displays the name of the file in the text box on the form.

8. Replace the *// add while loop here* comment with the following statement:

```
source.Text = "";
```

The *source* variable refers to the large text box on the form. Setting its *Text* property to the empty string ("") clears any text that is currently displayed in this text box.

9. Type the following statement after the line you just added to the *openFileDialogFileOk* method:

```
TextReader reader = src.OpenText();
```

This statement declares a *TextReader* variable called *reader*. *TextReader* is another class, provided by the .NET Framework, that you can use for reading streams of characters from sources such as files. It is located in the *System.IO* namespace. The *FileInfo* class provides the *OpenText* method for opening a file for reading. This statement opens the file selected by the user in the *Open* dialog box so that the *reader* variable can read the contents of this file.

10. Add the following statement after the previous line you added to the *openFileDialog-FileOk* method:

```
string line = reader.ReadLine();
```

This statement declares a *string* variable called *line* and calls the *reader.ReadLine* method to read the first line from the file into this variable. This method returns either the next line of text or a special value called *null* if there are no more lines to read. (If there are no lines initially, the file must be empty.)

11. Add the following statements to the *openFileDialogFileOk* method after the code you have just entered:

```
while (line != null)
{
    source.Text += line + '\n';
    line = reader.ReadLine();
}
```

This is a *while* loop that iterates through the file one line at a time until there are no more lines available.

The Boolean expression at the start of the *while* loop examines the value in the *line* variable. If it is not null, the body of the loop displays the current line of text by appending it to the end of the *Text* property of the *source* text box, together with a newline character ('\n' —the *ReadLine* method of the *TextReader* object strips out the newline characters as it reads each line, so the code needs to add it back in again). The *while* loop then reads in the next line of text before performing the next iteration. The *while* loop finishes when there is no more text in the file and the *ReadLine* method returns a null value.

12. Add the following statement after the closing brace at the end of the *while* loop:

```
reader.Close();
```

This statement closes the file.

13. On the *Debug* menu, click *Start Without Debugging*.

14. When the form appears, click *Open File*.

15. In the *Open File* dialog box, move to the \Microsoft Press\Visual CSharp Step by Step\ Chapter 5\WhileStatement\WhileStatement folder in your Documents folder. Select the file Window1.xaml.cs, and then click *Open*.

This time the contents of the selected file appear in the text box—you should recognize the code that you have just been editing:

16. Scroll through the text in the text box, and find the *openFileDialogFileOk* method. Verify that this method contains the code you just added.

17. Close the form and return to the Visual Studio 2008 programming environment.

Writing *for* Statements

Most *while* statements have the following general structure:

```
initialization
while (Boolean expression)
{
  statement
  update control variable
}
```

With a *for* statement, you can write a more formal version of this kind of construct by combining the initialization, the Boolean expression, and the update (the loop's "housekeeping"). You'll find the *for* statement useful because it is much harder to forget any one of the three parts. Here is the syntax of a *for* statement:

```
for (initialization; Boolean expression; update control variable)
    statement
```

You can rephrase the *while* loop shown earlier that displays the integers from 0 to 9 as the following *for* loop:

```
for (int i = 0; i < 10; i++)
{
    Console.WriteLine(i);
}
```

The initialization occurs once at the start of the loop. Then, if the Boolean expression evaluates to *true*, the statement runs. The control variable update occurs, and then the Boolean expression is reevaluated. If the condition is still *true*, the statement is executed again, the control variable is updated, the Boolean expression is evaluated again, and so on.

Notice that the initialization occurs only once, that the statement in the body of the loop always executes before the update occurs, and that the update occurs before the Boolean expression reevaluates.

You can omit any of the three parts of a *for* statement. If you omit the Boolean expression, it defaults to *true*. The following *for* statement runs forever:

```
for (int i = 0; ;i++)
{
    Console.WriteLine("somebody stop me!");
}
```

If you omit the initialization and update parts, you have a strangely spelled *while* loop:

```
int i = 0;
for (; i < 10; )
{
    Console.WriteLine(i);
    i++;
}
```

> **Note** The initialization, Boolean expression, and update control variable parts of a *for* statement must always be separated by semicolons, even when they are omitted.

If necessary, you can provide multiple initializations and multiple updates in a *for* loop (you can have only one Boolean expression). To achieve this, separate the various initializations and updates with commas, as shown in the following example:

```
for (int i = 0, j = 10; i <= j; i++, j--)
{
    ...
}
```

As a final example, here is the *while* loop from the preceding exercise recast as a *for* loop.

```
for (string line = reader.ReadLine(); line != null; line = reader.ReadLine())
{
    source.Text += line + '\n';
}
```

> **Tip** It's considered good style to use braces to explicitly delineate the statement block for the body of *if*, *while*, and *for* statements even when the block contains only one statement. By writing the block, you make it easier to add statements to the block at a later date. Without the block, to add another statement, you'd have to remember to add both the extra statement *and* the braces, and it's very easy to forget the braces.

Understanding *for* Statement Scope

You might have noticed that you can declare a variable in the initialization part of a *for* statement. That variable is scoped to the body of the *for* statement and disappears when the *for* statement finishes. This rule has two important consequences. First, you cannot use that variable after the *for* statement has ended because it's no longer in scope. Here's an example:

```
for (int i = 0; i < 10; i++)
{
    ...
}
Console.WriteLine(i); // compile-time error
```

Second, you can write next to each other two or more *for* statements that reuse the same variable name because each variable is in a different scope. Here's an example:

```
for (int i = 0; i < 10; i++)
{
    ...
}

for (int i = 0; i < 20; i += 2) // okay
{
    ...
}
```

Writing *do* Statements

The *while* and *for* statements both test their Boolean expression at the start of the loop. This means that if the expression evaluates to *false* on the very first test, the body of the loop does not run, not even once. The *do* statement is different; its Boolean expression is evaluated after each iteration, so the body always executes at least once.

The syntax of the *do* statement is as follows (don't forget the final semicolon):

```
do
    statement
while (booleanExpression);
```

You must use a *statement* block if the body of the loop comprises more than one statement. Here's a version of the example that writes the values 0 through 9 to the console, this time constructed using a *do* statement:

```
int i = 0;
do
{
    Console.WriteLine(i);
    i++;
}
while (i < 10);
```

The *break* and *continue* Statements

In Chapter 4, you saw the *break* statement being used to jump out of a *switch* statement. You can also use a *break* statement to jump out of the body of an iteration statement. When you break out of a loop, the loop exits immediately and execution continues at the first statement after the loop. Neither the update nor the continuation condition of the loop is rerun.

In contrast, the *continue* statement causes the program to perform the next iteration of the loop immediately (after reevaluating the Boolean expression). Here's another version of the example that writes the values 0 through 9 to the console, this time using *break* and *continue* statements:

```
int i = 0;
while (true)
{
    Console.WriteLine("continue " + i);
    i++;
    if (i < 10)
        continue;
    else
        break;
}
```

This code is absolutely ghastly. Many programming guidelines recommend using *continue* cautiously or not at all because it is often associated with hard-to-understand code. The behavior of *continue* is also quite subtle. For example, if you execute a *continue* statement from inside a *for* statement, the update part runs before performing the next iteration of the loop.

In the following exercise, you will write a *do* statement to convert a positive whole number to its string representation in octal notation.

Examine a *do* statement

1. Using Visual Studio 2008, open the DoStatement project, located in the \Microsoft Press\Visual CSharp Step by Step\Chapter 5\DoStatement folder in your Documents folder.

2. On the *Debug* menu, click *Start Without Debugging*.

   The application displays a form that has two text boxes and a button called *Show Steps*. When you type a positive integer (the program doesn't work with negative integers) in the upper text box and click *Show Steps*, the program takes the number that you have typed in and converts it to a string representing the octal (base 8) value of the same number. The program uses a well-known algorithm that repeatedly divides a number by 8, calculating the remainder at each stage. The lower text box shows the steps used to build this octal presentation.

3. Type *2693* in the upper text box, and then click *Show Steps*.

The lower text box displays the steps used to create the octal representation of 2693 (5205):

4. Close the window to return to the Visual Studio 2008 programming environment.

5. Display the code for Window1.xaml.cs in the *Code and Text Editor* window.

6. Locate the *showStepsClick* method. This method runs when the user clicks the *Show Steps* button on the form.

This method contains the following statements:

```
int amount = int.Parse(number.Text);
steps.Text = "";
string current = "";
do
{
    int nextDigit = amount % 8;
    int digitCode = '0' + nextDigit;
    char digit = Convert.ToChar(digitCode);
    current = digit + current;
    steps.Text += current + "\n";
    amount /= 8;
}
while (amount != 0);
```

The first statement converts the string value in the *Text* property of the *number* text box into an *int* using the *Parse* method of the *int* type:

```
int amount = int.Parse(number.Text);
```

The second statement clears the text displayed in the lower text box (called *steps*) by setting its *Text* property to the empty string:

```
steps.Text = "";
```

The third statement declares a *string* variable called *current* and initializes it to the empty string:

```
string current = "";
```

The real work in this method is performed by the *do* statement, which begins at the fourth statement:

```
do
{
    ...
}
while (amount != 0);
```

The algorithm repeatedly performs integer arithmetic to divide the *amount* variable by 8 and determine the remainder; the remainder after each successive division constitutes the next digit in the string being built. Eventually, when *amount* is reduced to 0, the loop finishes. Notice that the body must run at least once. This behavior is exactly what is required because even the number 0 has one octal digit.

Look more closely at the code, and you will see that the first statement inside the *do* loop is this:

```
int nextDigit = amount % 8;
```

This statement declares an *int* variable called *nextDigit* and initializes it to the remainder after dividing the value in *amount* by 8. This will be a number somewhere between 0 and 7.

The next statement is this:

```
int digitCode = '0' + nextDigit;
```

This statement requires a little explanation! Characters have a unique code according to the character set used by the operating system. In the character sets frequently used by the Microsoft Windows operating system, the code for character '0' has integer value 48. The code for character '1' is 49, the code for character '2' is 50, and so on up to the code for character '9', which has integer value 57. C# allows you to treat a character as an integer and perform arithmetic on it, but when you do so, C# uses the character's code as the value. So the expression '0' + nextDigit will actually result in a value somewhere between 48 and 55 (remember that *nextDigit* will be between 0 and 7), corresponding to the code for the equivalent octal digit.

The third statement inside the *do* loop is

```
char digit = Convert.ToChar(digitCode);
```

This statement declares a *char* variable called *digit* and initializes it to the result of the *Convert.ToChar(digitCode)* method call. The *Convert.ToChar* method takes an integer holding a character code and returns the corresponding character. So, for example, if *digitCode* has the value 54, *Convert.ToChar(digitCode)* will return the character '6'.

To summarize, the first three statements in the *do* loop have determined the character representing the least-significant (rightmost) octal digit corresponding to the number the user typed in. The next task is to prepend this digit to the string being output, like this:

```
current = digit + current;
```

The next statement inside the *do* loop is this:

```
steps.Text += current + "\n";
```

This statement adds to the *Steps* text box the string containing the digits produced so far for the octal representation of the number.

The final statement inside the *do* loop is

```
amount /= 8;
```

This is a compound assignment statement and is equivalent to writing amount = amount / 8;. If the value of *amount* is 2693, the value of *amount* after this statement runs is 336.

Finally, the condition in the *while* clause at the end of the loop is evaluated:

```
while (amount != 0)
```

Because the value of *amount* is not yet 0, the loop performs another iteration.

In the final exercise, you will use the Visual Studio 2008 debugger to step through the previous *do* statement to help you understand how it works.

Step through the *do* statement

1. In the *Code and Text Editor* window displaying the Window1.xaml.cs file, move the cursor to the first statement of the *showStepsClick* method:

```
int amount = int.Parse(number.Text);
```

2. Right-click anywhere in the first statement, and then click *Run To Cursor*.

3. When the form appears, type *2693* in the upper text box, and then click *Show Steps*.

   The program stops, and you are placed in Visual Studio 2008 in debug mode. A yellow arrow in the left margin of the *Code and Text Editor* window indicates the current statement.

4. Display the *Debug* toolbar if it is not visible. (On the *View* menu, point to *Toolbars*, and then click *Debug*.)

5. On the *Debug* toolbar, click the *Windows* drop-down arrow.

 Note The *Windows* icon is the rightmost icon in the *Debug* toolbar.

The following menu appears:

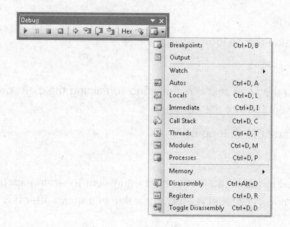

 Note If you are using Microsoft Visual C# 2008 Express Edition, the shortcut menu that appears contains a subset of those shown in the following image.

6. On the drop-down menu, click *Locals*.

The *Locals* window appears (if it wasn't already open). This window displays the name, value, and type of the local variables in the current method, including the *amount* local variable. Notice that the value of *amount* is currently 0:

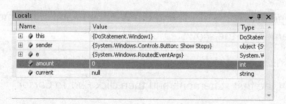

7. On the *Debug* toolbar, click the *Step Into* button.

The debugger runs the statement:

```
int amount = int.Parse(number.Text);
```

The value of *amount* in the *Locals* window changes to 2693, and the yellow arrow moves to the next statement.

8. Click *Step Into* again.

The debugger runs the statement:

```
steps.Text = "";
```

This statement does not affect the *Locals* window because *steps* is a control on the form and not a local variable. The yellow arrow moves to the next statement.

9. Click *Step Into*.

The debugger runs the statement:

```
string current = "";
```

The yellow arrow moves to the opening brace at the start of the *do* loop.

10. Click *Step Into*.

The yellow arrow moves to the first statement inside the *do* loop. The *do* loop contains three local variables of its own: *nextDigit*, *digitCode*, and *digit*. Notice that these local variables appear in the *Locals* window, and that the value of all three variables is 0.

11. Click *Step Into*.

The debugger runs the statement:

```
int nextDigit = amount % 8;
```

The value of *nextDigit* in the *Locals* window changes to 5. This is the remainder after dividing 2693 by 8.

12. Click *Step Into*.

The debugger runs the statement:

```
int digitCode = '0' + nextDigit;
```

The value of *digitCode* in the *Locals* window changes to 53. This is the character code of '5' (48 + 5).

13. Click *Step Into*.

The debugger runs the statement:

```
char digit = Convert.ToChar(digitCode);
```

The value of *digit* changes to '5' in the *Locals* window. The *Locals* window shows *char* values using both the underlying numeric value (in this case, 53) and also the character representation ('5').

Note that in the *Locals* window, the value of the *current* variable is "".

14. Click *Step Into*.

The debugger runs the statement:

```
current = current + digit;
```

The value of *current* changes to "5" in the *Locals* window.

15. Click *Step Into*.

The debugger runs the statement:

```
steps.Text += current + "\n";
```

This statement displays the text "5" in the *steps* text box, followed by a newline character to cause subsequent output to be displayed on the next line in the text box. (The form is currently hidden behind Visual Studio, so you won't be able to see it.)

16. Click *Step Into*.

The debugger runs the statement:

```
amount /= 8;
```

The value of *amount* changes to 336 in the *Locals* window. The yellow arrow moves to the brace at the end of the *do* loop.

17. Click *Step Into*.

The yellow arrow moves to the *while* statement.

18. Click *Step Into*.

The debugger runs the statement:

```
while (amount != 0);
```

The value of *amount* is 336, and the expression 336 != 0 evaluates to *true*, so the *do* loop performs another iteration. The yellow arrow jumps back to the opening brace at the start of the *do* loop.

19. Click *Step Into*.

The yellow arrow moves to the first statement inside the *do* loop again.

20. Repeatedly click *Step Into* to step through the next three iterations of the *do* loop, and watch how the values of the variables change in the *Locals* window.

21. At the end of the fourth iteration of the loop, the value of *amount* is now 0 and the value of *current* is "5205". The yellow arrow is on the continuation condition of the *do* loop:

```
while (amount != 0);
```

The value of *amount* is now 0, so the expression amount != 0 will evaluate to *false*, and the *do* loop will terminate.

22. Click *Step Into*.

The debugger runs the statement:

```
while (amount != 0);
```

As predicted, the *do* loop terminates, and the yellow arrow moves to the closing brace at the end of the *showStepsClick* method.

23. Click the *Continue* button on the *Debug* toolbar.

The form appears, displaying the four steps used to create the octal representation of 2693: "5", "05", "205", and "5205".

24. Close the form to return to the Visual Studio 2008 programming environment.

Congratulations! You have successfully written meaningful *while* and *do* statements and used the Visual Studio 2008 debugger to step through the *do* statement.

- If you want to continue to the next chapter

  Keep Visual Studio 2008 running, and turn to Chapter 6.

- If you want to exit Visual Studio 2008 now

  On the *File* menu, click *Exit*. If you see a *Save* dialog box, click *Yes* (if you are using Visual Studio 2008) or *Save* (if you are using Visual C# 2008 Express Edition) and save the project.

Chapter 5 Quick Reference

To	Do this
Add an amount to a variable	Use the compound addition operator. For example: ```\nvariable += amount;\n```
Subtract an amount from a variable	Use the compound subtraction operator. For example: ```\nvariable -= amount;\n```

Run one or more statements while a condition is true	Use a *while* statement. For example:
	
	```\nint i = 0;\nwhile (i < 10)\n{\n Console.WriteLine(i);\n i++;\n}\n```
	
	Alternatively, use a *for* statement. For example:
	
	```\nfor (int i = 0; i < 10; i++)\n{\n Console.WriteLine(i);\n}\n```

Repeatedly execute statements one or more times	Use a *do* statement. For example: ```\nint i = 0;\ndo\n{\n Console.WriteLine(i);\n i++;\n}\nwhile (i < 10);\n```

Chapter 6
Managing Errors and Exceptions

After completing this chapter, you will be able to:

- Handle exceptions by using the *try*, *catch*, and *finally* statements.

- Control integer overflow by using the *checked* and *unchecked* keywords.

- Raise exceptions from your own methods by using the *throw* keyword.

- Ensure that code always runs, even after an exception has occurred, by using a *finally* block.

You have now seen the core Microsoft Visual C# statements you need to know to read and write methods; declare variables; use operators to create values; write *if* and *switch* statements to run code selectively; and write *while*, *for*, and *do* statements to run code repeatedly. However, the previous chapters haven't considered the possibility (or probability) that things can go wrong. It is very difficult to ensure that a piece of code always works as expected. Failures can occur for a large number of reasons, many of which are beyond your control as a programmer. Any applications that you write must be capable of detecting failures and handling them in a graceful manner. In this final chapter of Part I, "Introducing Microsoft Visual C# and Microsoft Visual Studio 2008," you'll learn how C# throws exceptions to signal that an error has occurred and how to use the *try*, *catch*, and *finally* statements to catch and handle the errors that these exceptions represent. By the end of this chapter, you'll have a solid foundation in C#, on which you will build in Part II, "Understanding the C# Language."

Coping with Errors

It's a fact of life that bad things sometimes happen. Tires get punctured, batteries run down, screwdrivers are never where you left them, and users of your applications behave in an unpredictable manner. Errors can occur at almost any stage when a program runs, so how do you detect them and attempt to recover? Over the years, a number of mechanisms have evolved. A typical approach adopted by older systems such as UNIX involved arranging for the operating system to set a special global variable whenever a method failed. Then, after each call to a method, you checked the global variable to see whether the method succeeded. C# and most other modern object-oriented languages don't handle errors in this way. It's just too painful. They use *exceptions* instead. If you want to write robust C# programs, you need to know about exceptions.

Trying Code and Catching Exceptions

C# makes it easy to separate the error handling code from the code that implements the main flow of the program by using exceptions and exception handlers. To write exception-aware programs, you need to do two things:

1. Write your code inside a *try* block (*try* is a C# keyword). When the code runs, it attempts to execute all the statements inside the *try* block, and if none of the statements generates an exception, they all run, one after the other, to completion. However, if an error condition occurs, execution jumps out of the *try* block and into another piece of code designed to catch and handle the exception—a *catch* handler.

2. Write one or more *catch* handlers (*catch* is another C# keyword) immediately after the *try* block to handle any possible error conditions. A *catch* handler is intended to catch and handle a specific type of exception, and you can have multiple *catch* handlers after a *try* block, each one designed to trap and process a specific exception so that you can provide different handlers for the different errors that could arise in the *try* block. If any one of the statements inside the *try* block causes an error, the runtime generates and throws an exception. The runtime then examines the *catch* handlers after the *try* block and transfers control directly to the first matching handler.

Here's an example of code in a *try* block that attempts to convert strings that a user has typed in some text boxes on a form to integer values, call a method to calculate a value, and write the result to another text box. Converting a string to an integer requires that the string contain a valid representation and not some arbitrary sequence of characters. If the string contains invalid characters, the *int.Parse* method throws a *FormatException*, and execution transfers to the corresponding *catch* handler. When the *catch* handler finishes, the program continues with the first statement after the handler:

```
try
{
    int leftHandSide = int.Parse(lhsOperand.Text);
    int rightHandSide = int.Parse(rhsOperand.Text);
    int answer = doCalculation(leftHandSide, rightHandSide);
    result.Text = answer.ToString();
}
catch (FormatException fEx)
{
    // Handle the exception
    ...
}
```

Handling an Exception

A *catch* handler uses syntax similar to that used by a method parameter to specify the exception to be caught. In the preceding example, when a *FormatException* is thrown, the *fEx* variable is populated with an object containing the details of the exception. The *FormatException* type has a number of properties that you can examine to determine the exact cause of the exception. Many of these properties are common to all exceptions. For example, the *Message* property contains a text description of the error that caused the exception. You can use this information when handling the exception, perhaps recording the details to a log file or displaying a meaningful message to the user and asking the user to try again.

Unhandled Exceptions

What happens if a *try* block throws an exception and there is no corresponding *catch* handler? In the previous example, it is possible that the *lhsOperand* text box contains the string representation of a valid integer, but the integer that it represents is outside the range of valid integers supported by C# (for example, "2147483648"). In this case, the *int.Parse* statement will throw an *OverflowException*, which will not be caught by the *FormatException* *catch* handler. If this occurs, if the *try* block is part of a method, the method immediately exits and execution returns to the calling method. If the calling method uses a *try* block, the runtime attempts to locate a matching *catch* handler after the *try* block in the calling method and execute it. If the calling method does not use a *try* block, or there is no matching *catch* handler, the calling method immediately exits and execution returns to its caller, where the process is repeated. If a matching *catch* handler is eventually found, the handler runs and execution continues with the first statement after the *catch* handler in the catching method.

> **Important** Notice that after catching an exception, execution continues in the method containing the *catch* block that caught the exception. If the exception occurred in a method other than the one containing the *catch* handler, control does *not* return to the method that caused the exception.

If, after cascading back through the list of calling methods, the runtime is unable to find a matching *catch* handler, the program terminates with an unhandled exception. If you are running the application in Microsoft Visual Studio 2008 in debug mode (you selected *Start Debugging* on the *Debug* menu to run the application), the following information dialog box appears and the application drops into the debugger, allowing you to determine the cause of the exception:

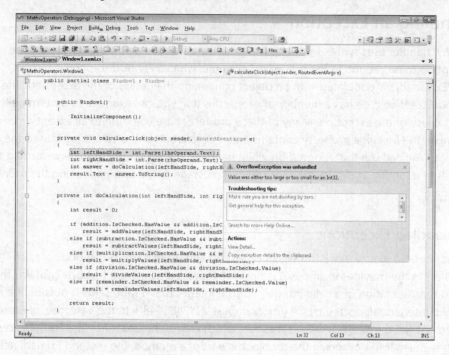

Using Multiple *catch* Handlers

The previous discussion highlights how different errors throw different kinds of exceptions to represent different kinds of failures. To cope with these situations, you can supply multiple *catch* handlers, one after the other, like this:

```
try
{
    int leftHandSide = int.Parse(lhsOperand.Text);
    int rightHandSide = int.Parse(rhsOperand.Text);
    int answer = doCalculation(leftHandSide, rightHandSide);
    result.Text = answer.ToString();
}
catch (FormatException fEx)
{
    //...
}
catch (OverflowException oEx)
{
    //...
}
```

Catching Multiple Exceptions

The exception-catching mechanism provided by C# and the Microsoft .NET Framework is quite comprehensive. Many different exceptions are defined in the .NET Framework, and any programs you write will be able to throw most of them! It is highly unlikely that you will want to write *catch* handlers for every possible exception that your code can throw. So how do you ensure that your programs catch and handle all possible exceptions?

The answer to this question lies in the way the different exceptions are related to one another. Exceptions are organized into families called inheritance hierarchies. (You will learn about inheritance in Chapter 12, "Working with Inheritance.") *FormatException* and *OverflowException* both belong to a family called *SystemException*, as do a number of other exceptions. Rather than catching each of these exceptions individually, you can create a handler that catches *SystemException*. *SystemException* is a member of a family simply called *Exception*, which is the great-granddaddy of all exceptions. If you catch *Exception*, the handler traps every possible exception that can occur.

> **Note** The *Exception* family includes a wide variety of exceptions, many of which are intended for use by various parts of the .NET Framework. Some of these are somewhat esoteric, but it is still useful to understand how to catch them.

The next example shows how to catch all possible system exceptions:

```
try
{
    int leftHandSide = int.Parse(lhsOperand.Text);
    int rightHandSide = int.Parse(rhsOperand.Text);
    int answer = doCalculation(leftHandSide, rightHandSide);
    result.Text = answer.ToString();
}
catch (Exception ex) // this is a general catch handler
{
    //...
}
```

> **Tip** If you want to catch *Exception*, you can actually omit its name from the *catch* handler because it is the default exception:
>
> ```
> catch
> {
>     // ...
> }
> ```
>
> However, this is not always recommended. The exception object passed in to the *catch* handler can contain useful information concerning the exception, which is not accessible when using this version of the *catch* construct.

There is one final question you should be asking at this point: What happens if the same exception matches multiple *catch* handlers at the end of a *try* block? If you catch *FormatException* and *Exception* in two different handlers, which one will run (or will both execute)?

When an exception occurs, the first handler found by the runtime that matches the exception is used, and the others are ignored. What this means is that if you place a handler for *Exception* before a handler for *FormatException*, the *FormatException* handler will never run. Therefore, you should place more-specific *catch* handlers above a general *catch* handler after a *try* block. If none of the specific *catch* handlers matches the exception, the general *catch* handler will.

In the following exercise, you will write a *try* block and catch an exception.

Write a *try/catch* statement

1. Start Visual Studio 2008 if it is not already running.

2. Open the MathsOperators solution located in the \Microsoft Press\Visual CSharp Step By Step\Chapter 6\MathsOperators folder in your Documents folder.

   This is a variation on the program that you first saw in Chapter 2, "Working with Variables, Operators, and Expressions." It was used to demonstrate the different arithmetic operators.

3. On the *Debug* menu, click *Start Without Debugging*.

   The form appears. You are now going to enter some text that is deliberately not valid in the left operand text box. This operation will demonstrate the lack of robustness in the current version of the program.

4. Type **John** in the left operand text box, and then click *Calculate*.

   A dialog box reports an unhandled exception; the text you entered in the left operand text box caused the application to fail.

> **Note** The *Debug* button does not appear if you are using Microsoft Visual C# 2008 Express Edition.

You might see a different version of this dialog box (shown later) depending on how you have configured problem reporting in Control Panel. If you see this dialog box, simply click the *Close the program* link whenever the instructions in the following steps refer to the *Close Program* button, and click the *Debug the program* link whenever the instructions refer to the *Debug* button. (If you are using Windows XP rather than Windows Vista, you will see a different dialog box with Debug, Send Error Report, and Don't Send buttons. Click the Don't Send button to close the program.)

5. If you are using Visual Studio 2008, click *Debug*. In the *Visual Studio Just-In-Time Debugger* dialog box, in the *Possible Debuggers* list box, select *MathsOperators – Microsoft Visual Studio: Visual Studio 2008*, and then click *Yes*:

6. If you are using Visual C# 2008 Express Edition, click *Close Program*. On the *Debug* menu, click *Start Debugging*. Type **John** in the left operand text box, and then click *Calculate*.

7. Whether you are using Visual Studio 2008 or Visual C# 2008 Express Edition, the Visual Studio 2008 debugger starts and highlights the line of code that caused the exception and displays some additional information about the exception:

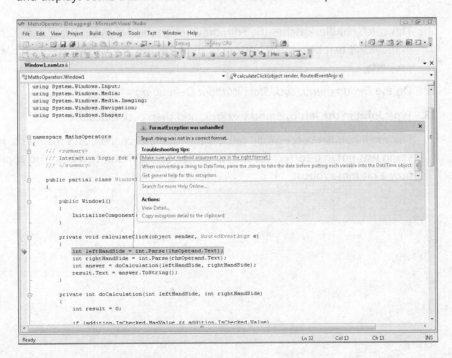

You can see that the exception was thrown by the call to *int.Parse* inside the *calculateClick* method. The problem is that this method is unable to parse the text "John" into a valid number.

> **Note** You can view the code that caused an exception only if you actually have the source code available on your computer.

8. On the *Debug* menu, click *Stop Debugging*.

9. Display the code for the file Window1.xaml.cs in the *Code and Text Editor* window, and locate the *calculateClick* method.

10. Add a *try* block (including braces) around the four statements inside this method, as shown in bold type here:

```
try
{
    int leftHandSide = int.Parse(lhsOperand.Text);
    int rightHandSide = int.Parse(rhsOperand.Text);
    int answer = doCalculation(leftHandSide, rightHandSide);
    result.Text = answer.ToString();
}
```

11. Add a *catch* block immediately after the closing brace for this new *try* block, as follows:

```
catch (FormatException fEx)
{
    result.Text = fEx.Message;
}
```

This *catch* handler catches the *FormatException* thrown by *int.Parse* and then displays in the *result* text box at the bottom of the form the text in the exception's *Message* property.

12. On the *Debug* menu, click *Start Without Debugging*.

13. Type **John** in the left operand text box, and then click *Calculate*.

The *catch* handler successfully catches the *FormatException*, and the message "Input string was not in a correct format" is written to the *Result* text box. The application is now a bit more robust.

14. Replace John with the number **10**, type **Sharp** in the right operand text box, and then click *Calculate*.

    Notice that because the *try* block surrounds the statements that parse both text boxes, the same exception handler handles user input errors in both text boxes.

15. Click *Quit* to return to the Visual Studio 2008 programming environment.

Using Checked and Unchecked Integer Arithmetic

In Chapter 2, you learned how to use binary arithmetic operators such as + and * on primitive data types such as *int* and *double*. You also saw that the primitive data types have a fixed size. For example, a C# *int* is 32 bits. Because *int* has a fixed size, you know exactly the range of value that it can hold: it is –2147483648 to 2147483647.

> **Tip** If you want to refer to the minimum or maximum value of *int* in code, you can use the *int. MinValue* or *int.MaxValue* property.

The fixed size of the *int* type creates a problem. For example, what happens if you add 1 to an *int* whose value is currently 2147483647? The answer is that it depends on how the application is compiled. By default, the C# compiler generates code that allows the calculation to overflow silently. In other words, you get the wrong answer. (In fact, the calculation wraps around to the largest negative integer value, and the result generated is –2147483648.) The reason for this behavior is performance: integer arithmetic is a common operation in almost every program, and adding the overhead of overflow checking to each integer expression could lead to very poor performance. In many cases, the risk is acceptable because you know (or hope!) that your *int* values won't reach their limits. If you don't like this approach, you can turn on overflow checking.

Tip You can activate and disable overflow checking in Visual Studio 2008 by setting the project properties. On the *Project* menu, click *YourProject Properties* (where *YourProject* is the name of your project). In the project properties dialog box, click the *Build* tab. Click the *Advanced* button in the lower-right corner of the page. In the *Advanced Build Settings* dialog box, select or clear the *Check for arithmetic overflow/underflow* check box.

Regardless of how you compile an application, you can use the *checked* and *unchecked* keywords to turn on and off integer arithmetic overflow checking selectively in parts of an application that you think need it. These keywords override the compiler option specified for the project.

Writing Checked Statements

A checked statement is a block preceded by the *checked* keyword. All integer arithmetic in a checked statement always throws an *OverflowException* if an integer calculation in the block overflows, as shown in this example:

```
int number = int.MaxValue;
checked
{
    int willThrow = number++;
    Console.WriteLine("this won't be reached");
}
```

Important Only integer arithmetic directly inside the *checked* block is subject to overflow checking. For example, if one of the checked statements is a method call, checking does not apply to code that runs in the method that is called.

You can also use the *unchecked* keyword to create an *unchecked* block statement. All integer arithmetic in an *unchecked* block is not checked and never throws an *OverflowException*. For example:

```
int number = int.MaxValue;
unchecked
{
    int wontThrow = number++;
    Console.WriteLine("this will be reached");
}
```

Writing Checked Expressions

You can also use the *checked* and *unchecked* keywords to control overflow checking on integer expressions by preceding just the individual parenthesized expression with the *checked* or *unchecked* keyword, as shown in this example:

```
int wontThrow = unchecked(int.MaxValue + 1);
int willThrow = checked(int.MaxValue + 1);
```

The compound operators (such as += and -=) and the increment (++) and decrement (--) operators are arithmetic operators and can be controlled by using the *checked* and *unchecked* keywords. Remember, x += y; is the same as x = x + y;.

> **Important** You cannot use the *checked* and *unchecked* keywords to control floating-point (non-integer) arithmetic. The *checked* and *unchecked* keywords apply only to integer arithmetic using data types such as *int* and *long*. Floating-point arithmetic never throws *OverflowException*—not even when you divide by 0.0. (The .NET Framework has a representation for infinity.)

In the following exercise, you will see how to perform checked arithmetic when using Visual Studio 2008.

Use checked expressions

1. Return to Visual Studio 2008.

2. On the *Debug* menu, click *Start Without Debugging*.

   You will now attempt to multiply two large values.

3. Type **9876543** in the left operand text box, type **9876543** in the right operand text box, select the *Multiplication* option, and then click *Calculate*.

   The value −1195595903 appears in the *Result* text box on the form. This is a negative value, which cannot possibly be correct. This value is the result of a multiplication operation that silently overflowed the 32-bit limit of the *int* type.

4. Click *Quit*, and return to the Visual Studio 2008 programming environment.

5. In the *Code and Text Editor* window displaying Window1.xaml.cs, locate the *multiplyValues* method. It looks like this:

```
private int multiplyValues(int leftHandSide, int rightHandSide)
{
    expression.Text = leftHandSide.ToString() + " * " + rightHandSide.ToString();
    return leftHandSide * rightHandSide;
}
```

The *return* statement contains the multiplication operation that is silently overflowing.

6. Edit the *return* statement so that the return value is checked, like this:

```
return checked(leftHandSide * rightHandSide);
```

The multiplication is now checked and will throw an *OverflowException* rather than silently returning the wrong answer.

7. Locate the *calculateClick* method.

8. Add the following *catch* handler immediately after the existing *FormatException catch* handler in the *calculateClick* method:

```
catch (OverflowException oEx)
{
    result.Text = oEx.Message;
}
```

> **Tip** The logic of this *catch* handler is the same as that for the *FormatException catch* handler. However, it is still worth keeping these handlers separate rather than simply writing a generic *Exception catch* handler because you might decide to handle these exceptions differently in the future.

9. On the *Debug* menu, click *Start Without Debugging* to build and run the application.

10. Type **9876543** in the left operand text box, type **9876543** in the right operand text box, select the *Multiplication* option, and then click *Calculate*.

The second *catch* handler successfully catches the *OverflowException* and displays the message "Arithmetic operation resulted in an overflow" in the *Result* text box.

11. Click *Quit* to return to the Visual Studio 2008 programming environment.

Throwing Exceptions

Suppose you are implementing a method called *monthName* that accepts a single *int* argument and returns the name of the corresponding month. For example, *monthName(1)* returns "January", *monthName(2)* returns "February", and so on. The question is: What should the method return when the integer argument is less than 1 or greater than 12? The best answer is that the method shouldn't return anything at all; it should throw an exception. The .NET Framework class libraries contain lots of exception classes specifically designed for situations such as this. Most of the time, you will find that one of these classes describes your exceptional condition. (If not, you can easily create your own exception class, but you need to know a bit more about the C# language before you can do that.) In this case, the existing .NET Framework *ArgumentOutOfRangeException* class is just right. You can throw an exception by using the *throw* statement, as shown in the following example:

```
public static string monthName(int month)
{
    switch (month)
    {
        case 1 :
            return "January";
        case 2 :
            return "February";
        ...
        case 12 :
            return "December";
        default :
            throw new ArgumentOutOfRangeException("Bad month");
    }
}
```

The *throw* statement needs an exception object to throw. This object contains the details of the exception, including any error messages. This example uses an expression that creates a new *ArgumentOutOfRangeException* object. The object is initialized with a string that will populate its *Message* property by using a constructor. Constructors are covered in detail in Chapter 7, "Creating and Managing Classes and Objects."

In the following exercises, you will add to the MathsOperators project code for throwing an exception.

Throw your own exception

1. Return to Visual Studio 2008.

2. On the *Debug* menu, click *Start Without Debugging*.

3. Type **24** in the left operand text box, type **36** in the right operand text box, and then click *Calculate*.

    The value *0* appears in the *Result* text box. The fact that you have not selected an operator option is not immediately obvious. It would be useful to write a diagnostic message in the *Result* text box.

4. Click *Quit* to return to the Visual Studio 2008 programming environment.

5. In the *Code and Text Editor* window displaying Window1.xaml.cs, locate and examine the *doCalculation* method. It looks like this:

```
private int doCalculation(int leftHandSide, int rightHandSide) {
    int result = 0;

    if (addition.IsChecked.HasValue && addition.IsChecked.Value)
        result = addValues(leftHandSide, rightHandSide);
```

```
        else if (subtraction.IsChecked.HasValue && subtraction.IsChecked.Value)
            result = subtractValues(leftHandSide, rightHandSide);
        else if (multiplication.IsChecked.HasValue &&  multiplication.IsChecked.Value)
            result = multiplyValues(leftHandSide, rightHandSide);
        else if (division.IsChecked.HasValue && division.IsChecked.Value)
            result = divideValues(leftHandSide, rightHandSide);
        else if (remainder.IsChecked.HasValue && remainder.IsChecked.Value)
            result = remainderValues(leftHandSide, rightHandSide);

        return result;
    }
```

The *addition*, *subtraction*, *multiplication*, *division*, and *remainder* fields are the radio buttons that appear on the form. Each radio button has a property called *IsChecked* that indicates whether the user has selected it. The *IsChecked* property is an example of a *nullable* value, which means it can either contain a specific value or be in an undefined state. (You learn more about nullable values in Chapter 8, "Understanding Values and References.") The *IsChecked.HasValue* property indicates whether the radio button is in a defined state, and if it is, the *IsChecked.Value* property indicates what this state is. The *IsChecked.Value* property is a Boolean that has the value *true* if the radio button is selected or *false* otherwise. The cascading *if* statement examines each radio button in turn to find which one is selected. (The radio buttons are mutually exclusive, so the user can select only one radio button at most.) If none of the buttons is selected, none of the *if* statements will be true and the *result* variable will remain at its initial value (0). This variable holds the value that is returned by the method.

You could try to solve the problem by adding one more *else* statement to the *if-else* cascade to write a message to the *result* text box on the form. However, this solution is not a good idea because it is not really the purpose of this method to output messages. It is better to separate the detection and signaling of an error from the catching and handling of that error.

6. Add another *else* statement to the list of *if-else* statements (immediately before the *return* statement), and throw an *InvalidOperationException* exactly as follows:

```
else
    throw new InvalidOperationException("no operator selected");
```

7. On the *Debug* menu, click *Start Without Debugging* to build and run the application.

8. Type **24** in the left operand text box, type **36** in the right operand text box, and then click *Calculate*.

   An exception dialog box appears. The application has thrown an exception, but your code does not catch it yet.

9. Click *Close program*.

   The application terminates, and you return to Visual Studio 2008.

Now that you have written a *throw* statement and verified that it throws an exception, you will write a *catch* handler to handle this exception.

Catch your own exception

1. In the *Code and Text Editor* window displaying Window1.xaml.cs, locate the *calculateClick* method.

2. Add the following *catch* handler immediately below the existing two *catch* handlers in the *calculateClick* method:

```
catch (InvalidOperationException ioEx)
{
    result.Text = ioEx.Message;
}
```

This code catches the *InvalidOperationException* that is thrown when no operator radio button is selected.

3. On the *Debug* menu, click *Start Without Debugging*.

4. Type **24** in the left operand text box, type **36** in the right operand text box, and then click *Calculate*.

The message "no operator selected" appears in the *Result* text box.

5. Click *Quit*.

The application is now a lot more robust than it was. However, several exceptions could still arise that would not be caught and that might cause the application to fail. For example, if you attempt to divide by 0, an unhandled *DivideByZeroException* will be thrown. (Integer division by 0 does throw an exception, unlike floating-point division by 0.) One way to solve this is to write an ever larger number of *catch* handlers inside the *calculateClick* method. However, a better solution is to add a general *catch* handler that catches *Exception* at the end of the list of *catch* handlers. This will trap all unhandled exceptions.

Tip The decision of whether to catch all unhandled exceptions explicitly in a method depends on the nature of the application you are building. In some cases, it makes sense to catch exceptions as close as possible to the point at which they occur. In other situations, it is more useful to let an exception propagate back to the method that invoked the routine that threw the exception.

Catch unhandled exceptions

1. In the *Code and Text Editor* window displaying Window1.xaml.cs, locate the *calculateClick* method.

2. Add the following *catch* handler to the end of the list of existing *catch* handlers:

```
catch (Exception ex)
{
    result.Text = ex.Message;
}
```

This *catch* handler will catch all hitherto unhandled exceptions, whatever their specific type.

3. On the *Debug* menu, click *Start Without Debugging*.

You will now attempt to perform some calculations known to cause exceptions and confirm that they are all handled correctly.

4. Type **24** in the left operand text box, type **36** in the right operand text box, and then click *Calculate*.

Confirm that the diagnostic message "no operator selected" still appears in the *Result* text box. This message was generated by the *InvalidOperationException* handler.

5. Type **John** in the left operand text box, and then click *Calculate*.

Confirm that the diagnostic message "Input string was not in a correct format" appears in the *Result* text box. This message was generated by the *FormatException* handler.

6. Type **24** in the left operand text box, type **0** in the right operand text box, select the *Divide* radio button, and then click *Calculate*.

Confirm that the diagnostic message "Attempted to divide by zero" appears in the *Result* text box. This message was generated by the general *Exception* handler.

7. Click *Quit*.

Using a *finally* Block

It is important to remember that when an exception is thrown, it changes the flow of execution through the program. This means you can't guarantee that a statement will always run when the previous statement finishes because the previous statement might throw an exception. Look at the following example. It's very easy to assume that the call to *reader.Close* will always occur. After all, it's right there in the code:

```
TextReader reader = src.OpenText();
string line;
while ((line = reader.ReadLine()) != null)
{
    source.Text += line + "\n";
}
reader.Close();
```

Sometimes it's not an issue if one particular statement does not run, but on many occasions it can be a big problem. If the statement releases a resource that was acquired in a previous statement, failing to execute this statement results in the resource being retained. This example is just such a case: If the call to *src.OpenText* succeeds, it acquires a resource (a file handle) and you must ensure that you call *reader.Close* to release the resource. If you don't, sooner or later you'll run out of file handles and be unable to open more files. (If you find file handles too trivial, think of database connections instead.)

The way to ensure that a statement is always run, whether or not an exception has been thrown, is to write that statement inside a *finally* block. A *finally* block occurs immediately after a *try* block or immediately after the last *catch* handler after a *try* block. As long as the program enters the *try* block associated with a *finally* block, the *finally* block will always be run, even if an exception occurs. If an exception is thrown and caught locally, the exception handler executes first, followed by the *finally* block. If the exception is not caught locally (the runtime has to search through the list of calling methods to find a handler), the *finally* block runs first. In any case, the *finally* block always executes.

The solution to the *reader.Close* problem is as follows:

```
TextReader reader = null;
try
{
    reader = src.OpenText();
    string line;
    while ((line = reader.ReadLine()) != null)
    {
        source.Text += line + "\n";
    }
}
finally
{
    if (reader != null)
    {
        reader.Close();
    }
}
```

Even if an exception is thrown, the *finally* block ensures that the *reader.Close* statement always executes. You'll see another way to solve this problem in Chapter 14, "Using Garbage Collection and Resource Management."

- If you want to continue to the next chapter

  Keep Visual Studio 2008 running, and turn to Chapter 7.

- If you want to exit Visual Studio 2008 now

  On the *File* menu, click *Exit*. If you see a *Save* dialog box, click *Yes* (if you are using Visual Studio 2008) or *Save* (if you are using Visual C# 2008 Express Edition) and save the project.

Chapter 6 Quick Reference

To	Do this
Throw an exception	Use a *throw* statement. For example: ```throw new FormatException(source);```

Ensure that integer arithmetic is always checked for overflow	Use the *checked* keyword. For example:
	
	```int number = Int32.MaxValue;```
	```checked```
	```{```
	``` number++;```
	```}```

Catch a specific exception	Write a *catch* handler that catches the specific exception class. For example:
	
	```try```
	```{```
	``` ...```
	```}```
	```catch (FormatException fEx)```
	```{```
	``` ...```
	```}```

Catch all exceptions in a single *catch* handler	Write a *catch* handler that catches *Exception*. For example:
	
	```try```
	```{```
	``` ...```
	```}```
	```catch (Exception ex)```
	```{```
	``` ...```
	```}```

Ensure that some code will always be run, even if an exception is thrown	Write the code inside a *finally* block. For example:
	
	```try```
	```{```
	``` ...```
	```}```
	```finally```
	```{```
	``` // always run```
	```}```

Part II
Understanding the C# Language

Chapter 7
Creating and Managing Classes and Objects

After completing this chapter, you will be able to:

- Define a class containing a related set of methods and data items.

- Control the accessibility of members by using the *public* and *private* keywords.

- Create objects by using the *new* keyword to invoke a constructor.

- Write and call your own constructors.

- Create methods and data that can be shared by all instances of the same class by using the *static* keyword.

In Part I, "Introducing Microsoft Visual C# and Microsoft Visual Studio 2008," you learned how to declare variables, use operators to create values, call methods, and write many of the statements you need when implementing a method. You now know enough to progress to the next stage—combining methods and data into your own classes.

The Microsoft .NET Framework contains thousands of classes, and you have used a number of them already, including *Console* and *Exception*. Classes provide a convenient mechanism for modeling the entities manipulated by applications. An *entity* can represent a specific item, such as a customer, or something more abstract, such as a transaction. Part of the design process of any system is concerned with determining the entities that are important and then performing an analysis to see what information they need to hold and what functions they should perform. You store the information that a class holds as fields and use methods to implement the operations that a class can perform.

The chapters in Part II, "Understanding the C# Language," provide you with all you need to know to create your own classes.

Understanding Classification

Class is the root word of the term *classification*. When you design a class, you systematically arrange information into a meaningful entity. This arranging is an act of classification and is something that everyone does—not just programmers. For example, all cars share common behaviors (they can be steered, stopped, accelerated, and so on) and common attributes (they have a steering wheel, an engine, and so on). People use the word *car* to mean objects that share these common behaviors and attributes. As long as everyone agrees on what a word means, this system works well; you can express complex but precise ideas in a concise form. Without classification, it's hard to imagine how people could think or communicate at all.

Given that classification is so deeply ingrained in the way we think and communicate, it makes sense to try to write programs by classifying the different concepts inherent in a problem and its solution and then modeling these classes in a programming language. This is exactly what you can do with modern object-oriented programming languages, such as Microsoft Visual C#.

The Purpose of Encapsulation

Encapsulation is an important principle when defining classes. The idea is that a program that uses a class should not have to worry how that class actually works internally; the program simply creates an instance of a class and calls the methods of that class. As long as those methods do what they say they will do, the program does not care how they are implemented. For example, when you call the *Console.WriteLine* method, you don't want to be bothered with all the intricate details of how the *Console* class physically arranges for data to be written to the screen. A class might need to maintain all sorts of internal state information to perform its various methods. This additional state information and activity is hidden from the program that is using the class. Therefore, encapsulation is sometimes referred to as information hiding. Encapsulation actually has two purposes:

- To combine methods and data inside a class; in other words, to support classification

- To control the accessibility of the methods and data; in other words, to control the use of the class

Defining and Using a Class

In C#, you use the *class* keyword to define a new class. The data and methods of the class occur in the body of the class between a pair of braces. Here is a C# class called *Circle* that contains one method (to calculate the circle's area) and one piece of data (the circle's radius):

```
class Circle
{
    double Area()
    {
        return Math.PI * radius * radius;
    }

    int radius;
}
```

> **Note** The *Math* class contains methods for performing mathematical calculations and fields containing mathematical constants. The *Math.PI* field contains the value 3.14159265358979323846, which is an approximation of the value of pi.

The body of a class contains ordinary methods (such as *Area*) and fields (such as *radius*)—remember that variables in a class are called fields. You've already seen how to declare variables in Chapter 2, "Working with Variables, Operators, and Expressions," and how to write methods in Chapter 3, "Writing Methods and Applying Scope"; in fact, there's almost no new syntax here.

Using the *Circle* class is similar to using other types that you have already met; you create a variable specifying *Circle* as its type, and then you initialize the variable with some valid data. Here is an example:

```
Circle c;          // Create a Circle variable
c = new Circle();  // Initialize it
```

Note the use of the *new* keyword. Previously, when you initialized a variable such as an *int* or a *float,* you simply assigned it a value:

```
int i;
i = 42;
```

You cannot do the same with variables of class types. One reason is that C# just doesn't provide the syntax for assigning literal class values to variables. (What is the *Circle* equivalent of 42?) Another reason concerns the way in which memory for variables of class types is allocated and managed by the runtime—this is discussed further in Chapter 8, "Understanding Values and References." For now, just accept that the *new* keyword creates a new instance of a class (more commonly called an object).

You can, however, directly assign an instance of a class to another variable of the same type, like this:

```
Circle c;
c = new Circle();
Circle d;
d = c;
```

However, this is not as straightforward as it first appears, for reasons that I cover in Chapter 8.

> **Important** Don't get confused between the terms *class* and *object*. A class is the definition of a type. An object is an instance of that type, created when the program runs.

Controlling Accessibility

Surprisingly, the *Circle* class is currently of no practical use. When you encapsulate your methods and data inside a class, the class forms a boundary to the outside world. Fields (such as *radius*) and methods (such as *Area*) defined in the class can be seen by other methods inside the class but not by the outside world—they are private to the class. So, although you can create a *Circle* object in a program, you cannot access its *radius* field or call its *Area* method, which is why the class is not of much use—yet! However, you can modify the definition of a field or method with the *public* and *private* keywords to control whether it is accessible from the outside:

- A method or field is said to be private if it is accessible only from the inside of the class. To declare that a method or field is private, you write the keyword private before its declaration. This is actually the default, but it is good practice to state explicitly that fields and methods are private to avoid any confusion.

- A method or field is said to be public if it is accessible from both the inside and the outside of the class. To declare that a method or field is public, you write the keyword public before its declaration.

Here is the *Circle* class again. This time *Area* is declared as a public method and *radius* is declared as a private field:

```
class Circle
{
    public double Area()
    {
        return Math.PI * radius * radius;
    }

    private int radius;
}
```

 Note C++ programmers should note that there is no colon after the *public* and *private* keywords. You must repeat the keyword for every field and method declaration.

Note that *radius* is declared as a private field; it is not accessible from outside the class. However, *radius* is accessible from inside the *Circle* class. This is why the *Area* method can access the *radius* field; *Area* is inside the *Circle* class, so the body of *Area* has access to *radius*. This means that the class is still of limited value because there is no way of initializing the *radius* field. To fix this, you can use a constructor.

 Tip The fields in a class are automatically initialized to *0*, *false*, or *null* depending on their type. However, it is still good practice to provide an explicit means of initializing fields.

Naming and Accessibility

The following recommendations relate to the naming conventions for fields and methods based on the accessibility of class members:

- Identifiers that are *public* should start with a capital letter. For example, *Area* starts with "A" (not "a") because it's *public*. This system is known as the *PascalCase* naming scheme (it was first used in the Pascal language).

- Identifiers that are not *public* (which include local variables) should start with a lowercase letter. For example, *radius* starts with "r" (not "R") because it's *private*. This system is known as the *camelCase* naming scheme.

There's only one exception to this rule: class names should start with a capital letter, and constructors must match the name of their class exactly; therefore, a *private* constructor must start with a capital letter.

Important Don't declare two *public* class members whose names differ only in case. If you do, your class will not be usable from other languages that are not case sensitive, such as Microsoft Visual Basic.

Working with Constructors

When you use the *new* keyword to create an object, the runtime has to construct that object by using the definition of the class. The runtime has to grab a piece of memory from the operating system, fill it with the fields defined by the class, and then invoke a constructor to perform any initialization required.

A *constructor* is a special method that runs automatically when you create an instance of a class. It has the same name as the class, and it can take parameters, but it cannot return a value (not even *void*). Every class must have a constructor. If you don't write one, the compiler automatically generates a default constructor for you. (However, the compiler-generated default constructor doesn't actually do anything.) You can write your own default constructor quite easily—just add a public method with the same name as the class that does not return a value. The following example shows the *Circle* class with a default constructor that initializes the *radius* field to 0:

```
class Circle
{
    public Circle()  // default constructor
    {
        radius = 0;
    }
}
```

```
        public double Area()
        {
            return Math.PI * radius * radius;
        }

        private int radius;
}
```

> **Note** In C# parlance, the *default* constructor is a constructor that does not take any parameters. It does not matter whether the compiler generates it or you write it; it is still the default constructor. You can also write nondefault constructors (constructors that *do* take parameters), as you will see in the upcoming section titled "Overloading Constructors.".

Note that the constructor is marked as *public*. If this keyword is omitted, the constructor will be private (just like any other methods and fields). If the constructor is private, it cannot be used outside the class, which prevents you from being able to create *Circle* objects from methods that are not part of the *Circle* class. You might therefore think that private constructors are not that valuable. However, they do have their uses, but they are beyond the scope of the current discussion.

You can now use the *Circle* class and exercise its *Area* method. Notice how you use dot notation to invoke the *Area* method on a *Circle* object:

```
Circle c;
c = new Circle();
double areaOfCircle = c.Area();
```

Overloading Constructors

You're almost finished, but not quite. You can now declare a *Circle* variable, point it to a newly created *Circle* object, and then call its *Area* method. However, there is still one last problem. The area of all *Circle* objects will always be 0 because the default constructor sets the radius to 0 and it stays at 0; the *radius* field is private, and there is no way of changing its value after it has been initialized. One way to solve this problem is to realize that a constructor is just a special kind of method and that it—like all methods—can be overloaded. Just as there are several versions of the *Console.WriteLine* method, each of which takes different parameters, so too you can write different versions of a constructor. You can add a constructor to the *Circle* class, with the radius as its parameter, like this:

```
class Circle
{

    public Circle()  // default constructor
    {
        radius = 0;
    }
```

```
public Circle(int initialRadius) // overloaded constructor
{
    radius = initialRadius;
}

public double Area()
{
    return Math.PI * radius * radius;
}

private int radius;
}
```

> **Note** The order of the constructors in a class is immaterial; you can define constructors in whatever order you feel most comfortable with.

You can then use this constructor when creating a new *Circle* object, like this:

```
Circle c;
c = new Circle(45);
```

When you build the application, the compiler works out which constructor it should call based on the parameters that you specify to the *new* operator. In this example, you passed an *int*, so the compiler generates code that invokes the constructor that takes an *int* parameter.

You should be aware of a quirk of the C# language: if you write your own constructor for a class, the compiler does not generate a default constructor. Therefore, if you've written your own constructor that accepts one or more parameters and you also want a default constructor, you'll have to write the default constructor yourself.

Partial Classes

A class can contain a number of methods, fields, and constructors, as well as other items discussed in later chapters. A highly functional class can become quite large. With C#, you can split the source code for a class into separate files so that you can organize the definition of a large class into smaller, easier to manage pieces. This feature is used by Microsoft Visual Studio 2008 for Windows Presentation Foundation (WPF) applications, where the source code that the developer can edit is maintained in a separate file from the code that is generated by Visual Studio whenever the layout of a form changes.

When you split a class across multiple files, you define the parts of the class by using the *partial* keyword in each file. For example, if the *Circle* class is split between two files called circ1.cs (containing the constructors) and circ2.cs (containing the methods and fields), the contents of circ1.cs look like this:

```
partial class Circle
{
    public Circle()  // default constructor
    {
        radius = 0;
    }

    public Circle(int initialRadius) // overloaded constructor
    {
        radius = initialRadius;
    }
}
```

The contents of circ2.cs look like this:

```
partial class Circle
{
    public double Area()
    {
        return Math.PI * radius * radius;
    }

    private int radius;
}
```

When you compile a class that has been split into separate files, you must provide all the files to the compiler.

In the following exercise, you will declare a class that models a point in two-dimensional space. The class will contain two private fields for holding the *x*- and *y*-coordinates of a point and will provide constructors for initializing these fields. You will create instances of the class by using the *new* keyword and calling the constructors.

Write constructors and create objects

1. Start Visual Studio 2008 if it is not already running.

2. Open the Classes project located in the \Microsoft Press\Visual CSharp Step by Step\ Chapter 7\Classes folder in your Documents folder.

3. In *Solution Explorer*, double-click the file Program.cs to display it in the *Code and Text Editor* window.

4. Locate the *Main* method in the *Program* class.

   The *Main* method calls the *Entrance* method, wrapped in a *try* block and followed by a *catch* handler. With this *try/catch* block, you can write the code that would typically go inside *Main* in the *Entrance* method instead, safe in the knowledge that it will catch and handle any exceptions.

5. Display the file Point.cs in the *Code and Text Editor* window.

   This file defines a class called *Point*, which you will use to represent the location of a point defined by a pair of *x*- and *y*-coordinates. The *Point* class is currently empty.

6. Return to the Program.cs file, and locate the *Entrance* method of the *Program* class. Edit the body of the *Entrance* method, replacing the // to do comment with the following statement:

```
Point origin = new Point();
```

7. On the *Build* menu, click *Build Solution*.

   The code builds without error because the compiler automatically generates the code for a default constructor for the *Point* class. However, you cannot see the C# code for this constructor because the compiler does not generate any source language statements.

8. Return to the *Point* class in the file Point.cs. Replace the // to do comment with a *public* constructor that accepts two *int* arguments called *x* and *y* and that calls the *Console.WriteLine* method to display the values of these arguments to the console, as shown in bold type in the following code example. The *Point* class should look like this:

```
class Point
{
    public Point(int x, int y)
    {
        Console.WriteLine("x:{0}, y:{1}", x, y);
    }
}
```

> **Note** Remember that the *Console.WriteLine* method uses *{0}* and *{1}* as placeholders. In the statement shown, *{0}* will be replaced with the value of *x*, and *{1}* will be replaced with the value of *y* when the program runs.

9. On the *Build* menu, click *Build Solution*.

   The compiler now reports an error:

```
'Classes.Point' does not contain a constructor that takes '0 ' arguments
```

The call to the default constructor in *Entrance* no longer works because there is no longer a default constructor. You have written your own constructor for the *Point* class, so the compiler no longer generates the default constructor. You will now fix this by writing your own default constructor.

10. Edit the *Point* class, and add a *public* default constructor that calls *Console.WriteLine* to write the string *"default constructor called"* to the console, as shown in bold type in the following code example. The *Point* class should now look like this:

```
class Point
{
    public Point()
    {
        Console.WriteLine("default constructor called");
    }

    public Point(int x, int y)
    {
        Console.WriteLine("x:{0}, y:{1}", x, y);
    }
}
```

11. On the *Build* menu, click *Build Solution*.

The program should now build successfully.

12. In the Program.cs file, edit the body of the *Entrance* method. Declare a variable called *bottomRight* of type Point, and initialize it to a new *Point* object by using the constructor with two arguments, as shown in bold type in the following code. Supply the values 1024 and 1280, representing the coordinates at the lower-right corner of the screen based on the resolution 1024 × 1280. The *Entrance* method should now look like this:

```
static void Entrance()
{
    Point origin = new Point();
    Point bottomRight = new Point(1024, 1280);
}
```

13. On the *Debug* menu, click *Start Without Debugging*.

The program builds and runs, displaying the following messages to the console:

```
default constructor called
x:1024, y:1280
```

14. Press the Enter key to end the program and return to Visual Studio 2008.

You will now add two *int* fields to the *Point* class to represent the *x*- and *y*-coordinates of a point, and you will modify the constructors to initialize these fields.

15. Edit the *Point* class in the Point.cs file, and add two *private* instance fields called *x* and *y* of type *int*, as shown in bold type in the following code. The *Point* class should now look like this:

```
class Point
{
    public Point()
    {
        Console.WriteLine("default constructor called");
    }

    public Point(int x, int y)
    {
        Console.WriteLine("x:{0}, y:{1}", x, y);
    }

    private int x, y;
}
```

You will now edit the second *Point* constructor to initialize the *x* and *y* fields to the values of the *x* and *y* parameters. There is a potential trap when you do this. If you are not careful, the constructor will look like this:

```
public Point(int x, int y) // Don't type this!
{
    x = x;
    y = y;
}
```

Although this code will compile, these statements appear to be ambiguous. How does the compiler know in the statement *x = x;* that the first *x* is the field and the second *x* is the parameter? It doesn't! A method parameter with the same name as a field hides the field for all statements in the method. All this constructor actually does is assign the parameters to themselves; it does not modify the fields at all. This is clearly not what you want.

The solution is to use the *this* keyword to qualify which variables are parameters and which are fields. Prefixing a variable with *this* means "the field in this object."

16. Modify the *Point* constructor that takes two parameters, and replace the *Console. WriteLine* statement with the following code shown in bold type:

```
public Point(int x, int y)
{
    this.x = x;
    this.y = y;
}
```

17. Edit the default *Point* constructor to initialize the *x* and *y* fields to –1, as follows in bold type. Note that although there are no parameters to cause confusion, it is still good practice to qualify the field references with *this*:

```
public Point()
{
    this.x = -1;
    this.y = -1;
}
```

18. On the *Build* menu, click *Build Solution*. Confirm that the code compiles without errors or warnings. (You can run it, but it does not produce any output yet.)

Methods that belong to a class and that operate on the data belonging to a particular instance of a class are called *instance methods*. (There are other types of methods that you will meet later in this chapter.) In the following exercise, you will write an instance method for the *Point* class, called *DistanceTo*, that calculates the distance between two points.

Write and call instance methods

1. In the Classes project in Visual Studio 2008, add the following public instance method called *DistanceTo* to the *Point* class between the constructors and the private variables. The method accepts a single *Point* argument called *other* and returns a *double*.

   The *DistanceTo* method should look like this:

```
class Point
{
    ...

    public double DistanceTo(Point other)
    {
    }
    ...
}
```

In the following steps, you will add code to the body of the *DistanceTo* instance method to calculate and return the distance between the *Point* object being used to make the call and the *Point* object passed as a parameter. To do this, you must calculate the difference between the *x*-coordinates and the *y*-coordinates.

2. In the *DistanceTo* method, declare a local *int* variable called *xDiff*, and initialize it to the difference between *this.x* and *other.x*, as shown here in bold type:

```
public double DistanceTo(Point other)
{
    int xDiff = this.x - other.x;
}
```

3. Declare another local *int* variable called *yDiff*, and initialize it to the difference between *this.y* and *other.y*, as shown here in bold type:

```
public double DistanceTo(Point other)
{
    int xDiff = this.x - other.x;
    int yDiff = this.y - other.y;
}
```

To calculate the distance, you can use the Pythagorean theorem. Work out the square root of the sum of the square of *xDiff* and the square of *yDiff*. The *System.Math* class provides the *Sqrt* method that you can use to calculate square roots.

4. Add the *return* statement shown in bold type in the following code to the end of the *DistanceTo* method to perform the calculation:

```
public double DistanceTo(Point other)
{
    int xDiff = this.x - other.x;
    int yDiff = this.y - other.y;
    return Math.Sqrt((xDiff * xDiff) + (yDiff * yDiff));
}
```

You will now test the *DistanceTo* method.

5. Return to the *Entrance* method in the *Program* class. After the statements that declare and initialize the *origin* and *bottomRight Point* variables, declare a variable called *distance* of type *double*. Initialize this *double* variable to the result obtained when you call the *DistanceTo* method on the *origin* object, passing the *bottomRight* object to it as an argument.

The *Entrance* method should now look like this:

```
static void Entrance()
{
    Point origin = new Point();
    Point bottomRight = new Point(1024, 1280);
    double distance = origin.DistanceTo(bottomRight);
}
```

 Note IntelliSense should display the *DistanceTo* method when you type the period character after *origin*.

6. Add to the *Entrance* method another statement that writes the value of the *distance* variable to the console by using the *Console.WriteLine* method.

The completed *Entrance* method should look like this:

```
static void Entrance()
{
    Point origin = new Point();
    Point bottomRight = new Point(1024, 1280);
```

```
            double distance = origin.DistanceTo(bottomRight);
            Console.WriteLine("Distance is: {0}", distance);
    }
```

7. On the *Debug* menu, click *Start Without Debugging*.

8. Confirm that the value 1640.60537607311 is written to the console window.

9. Press Enter to close the application and return to Visual Studio 2008.

Understanding *static* Methods and Data

In the preceding exercise, you used the *Sqrt* method of the *Math* class; similarly, when looking at the *Circle* class, you read the *PI* field of the *Math* class. If you think about it, the way
in which you called the *Sqrt* method or read the *PI* field was slightly odd. You invoked the
method on the class itself, not on an object of type *Math*. It is like trying to write *Point.
DistanceTo* rather than *origin.DistanceTo* in the code you added in the preceding exercise. So
what's happening, and how does this work?

You will often find that not all methods naturally belong to an instance of a class; they are
utility methods inasmuch as they provide a useful function that is independent of any specific
class instance. The *Sqrt* method is just such an example. If *Sqrt* were an instance method of
Math, you'd have to create a *Math* object to call *Sqrt* on:

```
Math m = new Math();
double d = m.Sqrt(42.24);
```

This would be cumbersome. The *Math* object would play no part in the calculation of the
square root. All the input data that *Sqrt* needs is provided in the parameter list, and the result
is passed back to the caller by using the method's return value. Objects are not really needed
here, so forcing *Sqrt* into an instance straitjacket is just not a good idea. As well as the *Sqrt*
method and the *PI* field, the *Math* class contains many other mathematical utility methods,
such as *Sin*, *Cos*, *Tan*, and *Log*.

In C#, all methods must be declared inside a class. However, if you declare a method or a
field as *static*, you can call the method or access the field by using the name of the class. No
instance is required. This is how the *Sqrt* method of the real *Math* class is declared:

```
class Math
{
    public static double Sqrt(double d) { ... }
    ...
}
```

When you define a *static* method, it does not have access to any instance fields defined for
the class; it can use only fields that are marked as *static*. Furthermore, it can directly invoke

only other methods in the class that are marked as *static*; nonstatic (instance) methods require you first to create an object on which to call them.

Creating a Shared Field

As mentioned in the preceding section, you can also use the *static* keyword when defining a field. With this feature, you can create a single field that is shared among all objects created from a single class. (Nonstatic fields are local to each instance of an object.) In the following example, the *static* field *NumCircles* in the *Circle* class is incremented by the *Circle* constructor every time a new *Circle* object is created:

```
class Circle
{
    public Circle()  // default constructor
    {
        radius = 0;
        NumCircles++;
    }

    public Circle(int initialRadius) // overloaded constructor
    {
        radius = initialRadius;
        NumCircles++;
    }

    ...
    private int radius;
    public static int NumCircles = 0;
}
```

All *Circle* objects share the same *NumCircles* field, so the statement *NumCircles++;* increments the same data every time a new instance is created. You access the *NumCircles* field by specifying the *Circle* class rather than a *Circle* object. For example:

```
Console.WriteLine("Number of Circle objects: {0}", Circle.NumCircles);
```

> **Tip** *static* methods are also called *class* methods. However, *static* fields aren't usually called *class* fields; they're just called *static* fields (or sometimes *static* variables).

Creating a *static* Field by Using the *const* Keyword

By prefixing the field with the *const* keyword, you can declare that a field is static but that its value can never change. *const* is short for "constant." A *const* field does not use the *static* keyword in its declaration but is nevertheless static. However, for reasons that are beyond the scope of this book, you can declare a field as *const* only when the field is an enumeration, a numeric type such as *int* or *double*, or a string. (You learn about enumerations in Chapter 9,

"Creating Value Types with Enumerations and Structs.") For example, here's how the *Math* class declares *PI* as a *const* field:

```
class Math
{
    ...
    public const double PI = 3.14159265358979323846;
}
```

static Classes

Another feature of the C# language is the ability to declare a class as *static*. A *static* class can contain only *static* members. (All objects that you create using the class share a single copy of these members.) The purpose of a *static* class is purely to act as a holder of utility methods and fields. A *static* class cannot contain any instance data or methods, and it does not make sense to try to create an object from a *static* class by using the *new* operator. In fact, you can't actually create an instance of an object using a *static* class by using *new* even if you want to. (The compiler will report an error if you try.) If you need to perform any initialization, a *static* class can have a default constructor as long as it is also declared as *static*. Any other types of constructor are illegal and will be reported as such by the compiler.

If you were defining your own version of the *Math* class, one containing only *static* members, it could look like this:

```
public static class Math
{
    public static double Sin(double x) {...}
    public static double Cos(double x) {...}
    public static double Sqrt(double x) {...}
    ...
}
```

Note, however, that the real *Math* class is not defined this way because it actually does have some instance methods.

In the final exercise in this chapter, you will add a *private static* field to the *Point* class and initialize the field to 0. You will increment this count in both constructors. Finally, you will write a *public static* method to return the value of this *private static* field. With this field, you can find out how many *Point* objects have been created.

Write *static* members, and call *static* methods

1. Using Visual Studio 2008, display the *Point* class in the *Code and Text Editor* window.

2. Add a *private static* field called *objectCount* of type *int* to the end of the *Point* class. Initialize it to 0 as you declare it, like this:

```
class Point
{
    ...;
    private static int objectCount = 0;
}
```

 Note You can write the keywords *private* and *static* in any order. The preferred order is *private* first, *static* second.

3. Add a statement to both *Point* constructors to increment the *objectCount* field, as shown in bold type in the following code example.

Each time an object is created, its constructor is called. As long as you increment the *objectCount* in each constructor (including the default constructor), *objectCount* will hold the number of objects created so far. This strategy works only because *object-Count* is a shared *static* field. If *objectCount* were an instance field, each object would have its own personal *objectCount* field that would be set to 1.

The *Point* class should now look like this:

```
class Point
{
    public Point()
    {
        this.x = -1;
        this.y = -1;
        objectCount++;
    }

    public Point(int x, int y)
    {
        this.x = x;
        this.y = y;
        objectCount++;
    }

    private int x, y;
    private static int objectCount = 0;
}
```

Notice that you cannot prefix *static* fields and methods with the *this* keyword because they do not belong to the current instance of the class. (They do not actually belong to any instance.)

The question now is this: How can users of the *Point* class find out how many *Point* objects have been created? At the moment, the *objectCount* field is *private* and not available outside the class. A poor solution would be to make the *objectCount* field publicly accessible. This strategy would break the encapsulation of the class; you would then have no guarantee that its value was correct because anyone could change the value in the field. A much better idea is to provide a *public static* method that returns the value of the *objectCount* field. This is what you will do now.

4. Add a *public static* method to the *Point* class called *ObjectCount* that returns an *int* but does not take any parameters. In this method, return the value of the *objectCount* field, as follows in bold type:

```
class Point
{
    ...
    public static int ObjectCount()
    {
        return objectCount;
    }
    ...
}
```

5. Display the *Program* class in the *Code and Text Editor* window, and locate the *Entrance* method.

6. Add a statement to the *Entrance* method to write the value returned from the *ObjectCount* method of the *Point* class to the screen, as shown in bold type in the following code example. The *Entrance* method should look like this:

```
static void Entrance()
{
    Point origin = new Point();
    Point bottomRight = new Point(600, 800);
    double distance = origin.distanceTo(bottomRight);
    Console.WriteLine("Distance is: {0}", distance);
    Console.WriteLine("No of Point objects: {0}", Point.ObjectCount());
}
```

The *ObjectCount* method is called by referencing *Point*, the name of the class, and not the name of a *Point* variable (such as *origin* or *bottomRight*). Because two *Point* objects have been created by the time *ObjectCount* is called, the method should return the value 2.

7. On the *Debug* menu, click *Start Without Debugging*.

Confirm that the value 2 is written to the console window (after the message displaying the value of the *distance* variable).

8. Press Enter to finish the program and return to Visual Studio 2008.

Congratulations. You have successfully created a class and used constructors to initialize the fields in a class. You have created instance and *static* methods, and you have called both of these types of methods. You have also implemented instance and *static* fields. You have seen how to make fields and methods accessible by using the *public* keyword and how to hide them using the *private* keyword.

Anonymous Classes

An *anonymous class* is a class that does not have a name. This sounds rather strange but is actually quite handy in some situations that you will see later in this book, especially when using query expressions. (You learn about query expressions in Chapter 20, "Querying In-Memory Data by Using Query Expressions.") For the time being, just accept the fact that they are useful.

You create an anonymous class simply by using the *new* keyword and a pair of braces defining the fields and values that you want the class to contain, like this:

```
myAnonymousObject = new { Name = "John", Age = 42 };
```

This class contains two public fields called *Name* (initialized to the string "John") and *Age* (initialized to the integer 42). The compiler infers the types of the fields from the types of the data you specify to initialize them.

When you define an anonymous class, the compiler generates its own name for the class, but it won't tell you what it is. Anonymous classes therefore raise a potentially interesting conundrum: If you don't know the name of the class, how can you create an object of the appropriate type and assign an instance of the class to it? In the code example shown earlier, what should the type of the variable *myAnonymousObject* be? The answer is that you don't know—that is the point of anonymous classes! However, this is not a problem if you declare *myAnonymousObject* as an implicitly typed variable by using the *var* keyword, like this:

```
var myAnonymousObject = new { Name = "John", Age = 42 };
```

Remember that the *var* keyword causes the compiler to create a variable of the same type as the expression used to initialize it. In this case, the type of the expression is whatever name the compiler happens to generate for the anonymous class.

You can access the fields in the object by using the familiar dot notation, like this:

```
Console.WriteLine("Name: {0} Age: {1}", myAnonymousObject.Name, myAnonymousObject.
Age};
```

You can even create other instances of the same anonymous class but with different values:

```
var anotherAnonymousObject = new { Name = "Diana", Age = 43 };
```

The C# compiler uses the names, types, number, and order of the fields to determine whether two instances of an anonymous class have the same type. In this case, variables *myAnonymousObject* and *anotherAnonymousObject* have the same number of fields, with the same name and type, in the same order, so both variables are instances of the same anonymous class. This means that you can perform assignment statements such as this:

```
anotherAnonymousObject = myAnonymousObject;
```

> **Note** Be warned that this assignment statement might not accomplish what you expect. You'll learn more about assigning object variables in Chapter 8, "Understanding Values and References."

There are quite a lot of restrictions on the contents of an anonymous class. Anonymous classes can contain only public fields, the fields must all be initialized, they cannot be static, and you cannot specify any methods.

- If you want to continue to the next chapter

  Keep Visual Studio 2008 running, and turn to Chapter 8.

- If you want to exit Visual Studio 2008 now

  On the *File* menu, click *Exit*. If you see a *Save* dialog box, click *Yes* (if you are using Visual Studio 2008) or *Save* (if you are using Microsoft Visual C# 2008 Express Edition) and save the project.

Chapter 7 Quick Reference

To	Do this

Declare a class	Write the keyword *class*, followed by the name of the class, followed by an opening and closing brace. The methods and fields of the class are declared between the opening and closing braces. For example:
	
	`class Point`
	`{`
	`    ...`
	`}`

Declare a constructor	Write a method whose name is the same as the name of the class and that has no return type (not even *void*). For example:

```
class Point
{
    public Point(int x, int y)
    {
        ...
    }
}
```

Call a constructor	Use the *new* keyword, and specify the constructor with an appropriate set of parameters. For example:

```
Point origin = new Point(0, 0);
```

Declare a *static* method	Write the keyword *static* before the declaration of the method. For example:

```
class Point
{
    public static int ObjectCount()
    {
        ...
    }
}
```

Call a *static* method	Write the name of the class, followed by a period, followed by the name of the method. For example:

```
int pointsCreatedSoFar = Point.ObjectCount();
```

Declare a *static* field	Write the keyword *static* before the declaration of the field. For example:

```
class Point
{
    ...
    private static int objectCount;
}
```

Declare a *const* field	Write the keyword *const* before the declaration of the field, and omit the *static* keyword. For example:

```
class Math
{
    ...
    public const double PI = ...;
}
```

Access a *static* field	Write the name of the class, followed by a period, followed by the name of the *static* field. For example:

```
double area = Math.PI * radius * radius;
```

Chapter 8
Understanding Values and References

After completing this chapter, you will be able to:

- Explain the differences between a value type and a reference type.

- Modify the way in which arguments are passed as method parameters by using the *ref* and *out* keywords.

- Box a value by initializing or assigning a variable of type *object*.

- Unbox a value by casting the object reference that refers to the boxed value.

In Chapter 7, "Creating and Managing Classes and Objects," you learned how to declare your own classes and how to create objects by using the *new* keyword. You also saw how to initialize an object by using a constructor. In this chapter, you will learn about how the characteristics of the primitive types, such as *int*, *double*, and *char*, differ from the characteristics of class types.

Copying Value Type Variables and Classes

Collectively, the types such as *int*, *float*, *double*, and *char* are called *value types*. When you declare a variable as a value type, the compiler generates code that allocates a block of memory big enough to hold a corresponding value. For example, declaring an *int* variable causes the compiler to allocate 4 bytes of memory (32 bits). A statement that assigns a value (such as 42) to the *int* causes the value to be copied into this block of memory.

Class types, such as *Circle* (described in Chapter 7), are handled differently. When you declare a *Circle* variable, the compiler *does not* generate code that allocates a block of memory big enough to hold a *Circle*; all it does is allot a small piece of memory that can potentially hold the address of (or a reference to) another block of memory containing a *Circle*. (An address specifies the location of an item in memory.) The memory for the actual *Circle* object is allocated only when the *new* keyword is used to create the object. A class is an example of a *reference type*. Reference types hold references to blocks of memory.

> **Note** Most of the primitive types of the C# language are value types except for *string*, which is a reference type. The description of reference types such as classes in this chapter applies to the *string* type as well. In fact, the *string* keyword in C# is just an alias for the *System.String* class.

You need to fully understand the difference between value types and reference types. Consider the situation in which you declare a variable named *i* as an *int* and assign it the value 42. If you declare another variable called *copyi* as an *int* and then assign *i* to *copyi*, *copyi* will hold the same value as *i* (42). However, even though *copyi* and *i* happen to hold the same value, there are still two blocks of memory containing the value 42: one block for *i* and the other block for *copyi*. If you modify the value of *i*, the value of *copyi* does not change. Let's see this in code:

```
int i = 42;      // declare and initialize i
int copyi = i; // copyi contains a copy of the data in i
i++;             // incrementing i has no effect on copyi
```

The effect of declaring a variable *c* as a *Circle* (the name of a class) is very different. When you declare *c* as a *Circle*, *c* can refer to a *Circle* object. If you declare *refc* as another *Circle*, it can also refer to a *Circle* object. If you assign *c* to *refc*, *refc* will refer to the same *Circle* object that *c* does; there is only one *Circle* object, and *refc* and *c* both refer to it. What has happened here is that the compiler has allocated two blocks of memory, one for *c* and one for *refc*, but the address contained in each block points to the same location in memory that stores the actual *Circle* object. Let's see this in code:

```
Circle c = new Circle(42);
Circle refc = c;
```

The following graphic illustrates both examples. The at sign (@) in the *Circle* objects represents a reference to an address in memory:

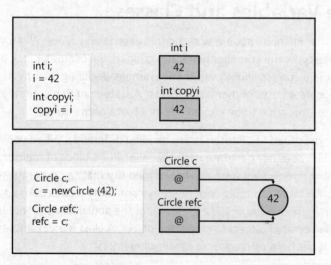

The difference explained here is very important. In particular, it means that the behavior of method parameters depends on whether they are value types or reference types. You'll explore this difference in the following exercise.

Note If you actually want to copy the contents of the *c* variable into *refc* rather than just copying the reference, you must make *refc* refer to a new instance of the *Circle* class and then copy the data field by field from *c* into *refc*, like this:

```
Circle refc = new Circle();
refc.radius = c.radius;       // Don't try this
```

However, if any members of the *Circle* class are private (like the *radius* field), you will not be able to copy this data. Instead, you should make the data in the private fields accessible by exposing them as properties. You will learn how to do this in Chapter 15, "Implementing Properties to Access Fields."

Use value parameters and reference parameters

1. Start Microsoft Visual Studio 2008 if it is not already running.

2. Open the Parameters project located in the \Microsoft Press\Visual CSharp Step by Step\Chapter 8\Parameters folder in your Documents folder.

   The project contains three C# code files named Pass.cs, Program.cs, and WrappedInt.cs.

3. Display the Pass.cs file in the *Code and Text Editor* window. Add a *public static* method called *Value* to the *Pass* class, replacing the `// to do` comment, as shown in bold type in the following code example. This method should accept a single *int* parameter (a value type) called *param* and have the return type *void*. The body of the *Value* method should simply assign 42 to *param*.

```
namespace Parameters
{
    class Pass
    {
        public static void Value(int param)
        {
            param = 42;
        }
    }
}
```

4. Display the Program.cs file in the *Code and Text Editor* window, and then locate the *Entrance* method of the *Program* class.

   The *Entrance* method is called by the *Main* method when the program starts running. As explained in Chapter 7, the method call is wrapped in a *try* block and followed by a *catch* handler.

5. Add four statements to the *Entrance* method to perform the following tasks:

   - Declare a local *int* variable called *i*, and initialize it to 0.
   - Write the value of *i* to the console by using *Console.WriteLine*.
   - Call *Pass.Value*, passing *i* as an argument.
   - Write the value of *i* to the console again.

With the calls to *Console.WriteLine* before and after the call to *Pass.Value*, you can see whether the call to *Pass.Value* actually modifies the value of *i*. The completed *Entrance* method should look exactly like this:

```
static void Entrance()
{
    int i = 0;
    Console.WriteLine(i);
    Pass.Value(i);
    Console.WriteLine(i);
}
```

6.   On the *Debug* menu, click *Start Without Debugging* to build and run the program.

7.   Confirm that the value 0 is written to the console window twice.

The assignment statement inside the *Pass.Value* method that updates the parameter uses a copy of the argument passed in, and the original argument *i* is completely unaffected.

8.   Press the Enter key to close the application.

You will now see what happens when you pass an *int* parameter that is wrapped inside a class.

9.   Display the WrappedInt.cs file in the *Code and Text Editor* window. Add a *public* instance field called *Number* of type *int* to the *WrappedInt* class, as shown in bold type here.

```
namespace Parameters
{
    class WrappedInt
    {
        public int Number;
    }
}
```

10.   Display the Pass.cs file in the *Code and Text Editor* window. Add a *public static* method called *Reference* to the *Pass* class. This method should accept a single *WrappedInt* parameter called *param* and have the return type *void*. The body of the *Reference* method should assign 42 to *param.Number*, like this:

```
public static void Reference(WrappedInt param)
{
    param.Number = 42;
}
```

11.   Display the Program.cs file in the *Code and Text Editor* window. Add four more statements to the *Entrance* method to perform the following tasks:

   ■ Declare a local *WrappedInt* variable called *wi*, and initialize it to a new *WrappedInt* object by calling the default constructor.

   ■ Write the value of *wi.Number* to the console.

- Call the *Pass.Reference* method, passing *wi* as an argument.
- Write the value of *wi.Number* to the console again.

As before, with the calls to *Console.WriteLine*, you can see whether the call to *Pass. Reference* modifies the value of *wi.Number*. The *Entrance* method should now look exactly like this (the new statements are shown in bold type):

```
static void Entrance()
{
    int i = 0;
    Console.WriteLine(i);
    Pass.Value(i);
    Console.WriteLine(i);

    WrappedInt wi = new WrappedInt();
    Console.WriteLine(wi.Number);
    Pass.Reference(wi);
    Console.WriteLine(wi.Number);
}
```

12. On the *Debug* menu, click *Start Without Debugging* to build and run the application.

As before, the first two values written to the console window are 0 and 0, before and after the call to *Pass.Value*. However, the next two values correspond to the value of *wi.Number* before and after *Pass.Reference*, and you should see that the values 0 and 42 are written to the console window.

13. Press the Enter key to close the application and return to Visual Studio 2008.

In the previous exercise, the value of *wi.Number* is initialized to 0 by the compiler-generated default constructor. The *wi* variable contains a reference to the newly created *WrappedInt* object (which contains an *int*). The *wi* variable is then copied as an argument to the *Pass. Reference* method. Because *WrappedInt* is a class (a reference type), *wi* and *param* both refer to the same *WrappedInt* object. Any changes made to the contents of the object through the *param* variable in the *Pass.Reference* method are visible by using the *wi* variable when the method completes. The following diagram illustrates what happens when a *WrappedInt* object is passed as an argument to the *Pass.Reference* method:

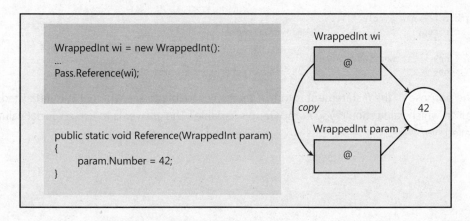

Understanding Null Values and Nullable Types

When you declare a variable, it is always a good idea to initialize it. With value types, it is common to see code such as this:

```
int i = 0;
double d = 0.0;
```

Remember that to initialize a reference variable such as a class, you can create a new instance of the class and assign the reference variable to the new object, like this:

```
Circle c = new Circle(42);
```

This is all very well, but what if you don't actually want to create a new object—perhaps the purpose of the variable is simply to store a reference to an existing object. In the following code example, the *Circle* variable *copy* is initialized, but later it is assigned a reference to another instance of the *Circle* class:

```
Circle c = new Circle(42);
Circle copy = new Circle(99);    // Some random value, for initializing copy
...
copy = c;                        // copy and c refer to the same object
```

After assigning *c* to *copy*, what happens to the original *Circle* object with a radius of 99 that you used to initialize *copy*? Nothing refers to it anymore. In this situation, the runtime can reclaim the memory by performing an operation known as *garbage collection*. You will learn about garbage collection in Chapter 14, "Using Garbage Collection and Resource Management." The important thing to understand for now is that garbage collection is a potentially expensive operation.

You could argue that if a variable is going to be assigned a reference to another object at some point in a program, there is no point initializing it. But this is poor programming practice and can lead to problems in your code. For example, you will inevitably meet the situation where you want to refer a variable to an object only if that variable does not already contain a reference:

```
Circle c = new Circle(42);
Circle copy;                     // Uninitialized !!!
...
if (copy == // what goes here?)
    copy = c;                    // copy and c refer to the same object
```

The purpose of the *if* statement is to test the *copy* variable to see whether it is initialized, but to which value should you compare this variable? The answer is to use a special value called *null*.

In C#, you can assign the *null* value to any reference variable. The *null* value simply means that the variable does not refer to an object in memory. You can use it like this:

```
Circle c = new Circle(42);
Circle copy = null;          // Initialized
...
if (copy == null)
    copy = c;                // copy and c refer to the same object
```

Using Nullable Types

The *null* value is useful for initializing reference types, but *null* is itself a reference, and you cannot assign it to a value type. The following statement is therefore illegal in C#:

```
int i = null; // illegal
```

However, C# defines a modifier that you can use to declare that a variable is a *nullable* value type. A nullable value type behaves in a similar manner to the original value type, but you can assign the *null* value to it. You use the question mark (?)to indicate that a value type is nullable, like this:

```
int? i = null; // legal
```

You can ascertain whether a nullable variable contains *null* by testing it in the same way as a reference type:

```
if (i == null)
    ...
```

You can assign an expression of the appropriate value type directly to a nullable variable. The following examples are all legal:

```
int? i = null;
int j = 99;
i = 100;      // Copy a value type constant to a nullable type
i = j;        // Copy a value type variable to a nullable type
```

You should note that the converse is not true. You cannot assign a nullable value to an ordinary value type variable, so given the definitions of variables *i* and *j* from the preceding example, the following statement is not allowed:

```
j = i;    // Illegal
```

This also means that you cannot use a nullable variable as a parameter to a method that expects an ordinary value type. If you recall, the *Pass.Value* method from the preceding exercise expects an ordinary *int* parameter, so the following method call will not compile:

```
int? i = 99;
Pass.Value(i);   // Compiler error
```

Understanding the Properties of Nullable Types

Nullable types expose a pair of properties that you can use and that you have already met in Chapter 6, "Managing Errors and Exceptions." The *HasValue* property indicates whether a nullable type contains a value or is *null*, and you can retrieve the value of a non-null nullable type by reading the *Value* property, like this:

```
int? i = null;
...
if (!i.HasValue)
    i = 99;
else
    Console.WriteLine(i.Value);
```

Recall from Chapter 4, "Using Decision Statements," that the NOT operator (!) negates a Boolean value. This code fragment tests the nullable variable *i*, and if it does not have a value (it is *null*), it assigns it the value 99; otherwise, it displays the value of the variable. In this example, using the *HasValue* property does not provide any benefit over testing for a *null* value directly. Additionally, reading the *Value* property is a long-winded way of reading the contents of the variable. However, these apparent shortcomings are caused by the fact that *int?* is a very simple nullable type. You can create more complex value types and use them to declare nullable variables where the advantages of using the *HasValue* and *Value* properties become more apparent. You will see some examples in Chapter 9, "Creating Value Types with Enumerations and Structures."

> **Note** The *Value* property of a nullable type is read-only. You can use this property to read the value of a variable but not to modify it. To update a nullable variable, use an ordinary assignment statement.

Using *ref* and *out* Parameters

Ordinarily, when you pass an argument to a method, the corresponding parameter is initialized with a copy of the argument. This is true regardless of whether the parameter is a value type (such as an *int*), a nullable type (such as *int?*), or a reference type (such as a *WrappedInt*). This arrangement means it's impossible for any change to the parameter to affect the value of the argument passed in. For example, in the following code, the value output to the console is 42 and not 43. The *DoWork* method increments a *copy* of the argument (*arg*) and *not* the original argument:

```
static void DoWork(int param)
{
    param++;
}
```

```
static void Main()
{
    int arg = 42;
    DoWork(arg);
    Console.WriteLine(arg); // writes 42, not 43
}
```

In the preceding exercise, you saw that if the parameter to a method is a reference type, any changes made by using that parameter change the data referenced by the argument passed in. The key point is that, although the data that was referenced changed, the parameter itself did not—it still references the same object. In other words, although it is possible to modify the object that the argument refers to through the parameter, it's not possible to modify the argument itself (for example, to set it to refer to a completely different object). Most of the time, this guarantee is very useful and can help to reduce the number of bugs in a program. Occasionally, however, you might want to write a method that actually needs to modify an argument. C# provides the *ref* and *out* keywords so that you can do this.

Creating *ref* Parameters

If you prefix a parameter with the *ref* keyword, the parameter becomes an alias for (or a reference to) the actual argument rather than a copy of the argument. When using a *ref* parameter, anything you do to the parameter you also do to the original argument because the parameter and the argument both reference the same object. When you pass an argument to a *ref* parameter, you must also prefix the argument with the *ref* keyword. This syntax provides a useful visual indication that the argument might change. Here's the preceding example again, this time modified to use the *ref* keyword:

```
static void DoWork(ref int param) // using ref
{
    param++;
}

static void Main()
{
    int arg = 42;
    DoWork(ref arg);        // using ref
    Console.WriteLine(arg);  // writes 43
}
```

This time, you pass to the *DoWork* method a reference to the original argument rather than a copy of the original argument, so any changes the method makes by using this reference also change the original argument. That's why the value 43 is displayed on the console.

The rule that you must assign a value to a variable before you can use the variable still applies to *ref* arguments. For example, in the following example, *arg* is not initialized, so this code will not compile. This failure is because *param++* inside *DoWork* is really *arg++*, and *arg++* is allowed only if *arg* has a defined value:

```
static void DoWork(ref int param)
{
    param++;
}

static void Main()
{
    int arg;                    // not initialized
    DoWork(ref arg);
    Console.WriteLine(arg);
}
```

Creating *out* Parameters

The compiler checks whether a *ref* parameter has been assigned a value before calling the method. However, there may be times when you want the method to initialize the parameter. With the *out* keyword, you can do this.

The *out* keyword is very similar to the *ref* keyword. You can prefix a parameter with the *out* keyword so that the parameter becomes an alias for the argument. As when using *ref*, anything you do to the parameter, you also do to the original argument. When you pass an argument to an *out* parameter, you must also prefix the argument with the *out* keyword.

The keyword *out* is short for *output*. When you pass an *out* parameter to a method, the method *must* assign a value to it. The following example does not compile because *DoWork* does not assign a value to *param*:

```
static void DoWork(out int param)
{
    // Do nothing
}
```

However, the following example does compile because *DoWork* assigns a value to *param*.

```
static void DoWork(out int param)
{
    param = 42;
}
```

Because an *out* parameter must be assigned a value by the method, you're allowed to call the method without initializing its argument. For example, the following code calls *DoWork* to initialize the variable *arg*, which is then displayed on the console:

```
static void DoWork(out int param)
{
    param = 42;
}
```

```
static void Main()
{
    int arg;                    // not initialized
    DoWork(out arg);
    Console.WriteLine(arg); // writes 42
}
```

You will examine *ref* parameters in the next exercise.

Use *ref* parameters

1. Return to the Parameters project in Visual Studio 2008.

2. Display the Pass.cs file in the *Code and Text Editor* window.

3. Edit the *Value* method to accept its parameter as a *ref* parameter.

   The *Value* method should look like this:

   ```
   class Pass
   {
       public static void Value(ref int param)
       {
           param = 42;
       }
       ...
   }
   ```

4. Display the Program.cs file in the *Code and Text Editor* window.

5. Edit the third statement of the *Entrance* method so that the *Pass.Value* method call passes its argument as a *ref* parameter.

   The *Entrance* method should now look like this:

   ```
   class Application
   {
       static void Entrance()
       {
           int i = 0;
           Console.WriteLine(i);
           Pass.Value(ref i);
           Console.WriteLine(i);
           ...
       }
   }
   ```

6. On the *Debug* menu, click *Start Without Debugging* to build and run the program.

   This time, the first two values written to the console window are 0 and 42. This result shows that the call to the *Pass.Value* method has successfully modified the argument *i*.

7. Press the Enter key to close the application and return to Visual Studio 2008.

> **Note** You can use the *ref* and *out* modifiers on reference type parameters as well as on value type parameters. The effect is exactly the same. The parameter becomes an alias for the argument. If you reassigned the parameter to a newly constructed object, you would also actually be reassigning the argument to the newly constructed object.

How Computer Memory Is Organized

Computers use memory to hold programs being executed and the data that these programs use. To understand the differences between value and reference types, it is helpful to understand how data is organized in memory.

Operating systems and runtimes frequently divide the memory used for holding data in two separate chunks, each of which is managed in a distinct manner. These two chunks of memory are traditionally called *the stack* and *the heap*. The stack and the heap serve very different purposes:

- When you call a method, the memory required for its parameters and its local variables is always acquired from the stack. When the method finishes (because it either returns or throws an exception), the memory acquired for the parameters and local variables is automatically released back to the stack and is available for reuse when another method is called.

- When you create an object (an instance of a class) by using the *new* keyword, the memory required to build the object is always acquired from the heap. You have seen that the same object can be referenced from several places by using reference variables. When the last reference to an object disappears, the memory used by the object becomes available for reuse (although it might not be reclaimed immediately). Chapter 14 includes a more detailed discussion of how heap memory is reclaimed.

> **Note** All value types are created on the stack. All reference types (objects) are created on the heap (although the reference itself is on the stack). Nullable types are actually reference types, and they are created on the heap.

The names *stack* and *heap* come from the way in which the runtime manages the memory:

- Stack memory is organized like a stack of boxes piled on top of one another. When a method is called, each parameter is put in a box that is placed on top of the stack. Each local variable is likewise assigned a box, and these are placed on top of the boxes already on the stack. When a method finishes, all its boxes are removed from the stack.

- Heap memory is like a large pile of boxes strewn around a room rather than stacked neatly on top of each other. Each box has a label indicating whether it is in use. When a new object is created, the runtime searches for an empty box and allocates it to the object. The reference to the object is stored in a local variable on the stack. The runtime keeps track of the number of references to each box (remember that two variables can refer to the same object). When the last reference disappears, the runtime marks the box as not in use, and at some point in the future it will empty the box and make it available for reuse.

Using the Stack and the Heap

Now let's examine what happens when the following *Method* is called:

```
void Method(int param)
{
    Circle c;
    c = new Circle(param);
    ...
}
```

Suppose the value passed into *param* is the value 42. When the method is called, a block of memory (just enough for an *int*) is allocated from the stack and initialized with the value 42. As execution moves inside the method, another block of memory big enough to hold a reference (a memory address) is also allocated from the stack but left uninitialized (this is for the *Circle* variable, *c*). Next, another piece of memory big enough for a *Circle* object is allocated from the heap. This is what the *new* keyword does. The *Circle* constructor runs to convert this raw heap memory to a *Circle* object. A reference to this *Circle* object is stored in the variable *c*. The following graphic illustrates the situation:

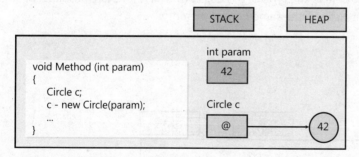

At this point, you should note two things:

- Although the object is stored on the heap, the reference to the object (the variable *c*) is stored on the stack.

- Heap memory is not infinite. If heap memory is exhausted, the *new* operator will throw an *OutOfMemoryException* and the object will not be created.

> **Note** The *Circle* constructor could also throw an exception. If it does, the memory allocated to the *Circle* object will be reclaimed and the value returned by the constructor will be *null*.

When the method ends, the parameters and local variables go out of scope. The memory acquired for *c* and for *param* is automatically released back to the stack. The runtime notes that the *Circle* object is no longer referenced and at some point in the future will arrange for its memory to be reclaimed by the heap (see Chapter 14).

The *System.Object* Class

One of the most important reference types in the Microsoft .NET Framework is the *Object* class in the *System* namespace. To fully appreciate the significance of the *System.Object* class requires that you understand inheritance, which is described in Chapter 12, "Working with Inheritance." For the time being, simply accept that all classes are specialized types of *System.Object* and that you can use *System.Object* to create a variable that can refer to any reference type. *System.Object* is such an important class that C# provides the *object* keyword as an alias for *System.Object*. In your code, you can use *object* or you can write *System.Object*; they mean exactly the same thing.

> **Tip** Use the *object* keyword in preference to *System.Object*. It's more direct, and it's consistent with other keywords that are synonyms for classes (such as *string* for *System.String* and some others that you'll discover in Chapter 9).

In the following example, the variables *c* and *o* both refer to the same *Circle* object. The fact that the type of *c* is *Circle* and the type of *o* is *object* (the alias for *System.Object*) in effect provides two different views of the same item in memory:

```
Circle c;
c = new Circle(42);
object o;
o = c;
```

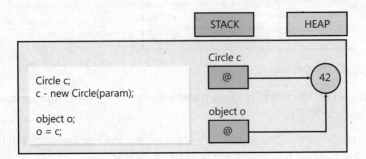

Boxing

As you have just seen, variables of type *object* can refer to any object of any reference type. However, variables of type *object* can also refer to a value type. For example, the following two statements initialize the variable *i* (of type *int*, a value type) to 42 and then initialize the variable *o* (of type *object*, a reference type) to *i*:

```
int i = 42;
object o = i;
```

The second statement requires a little explanation to appreciate what is actually happening. Remember that *i* is a value type and that it exists in the stack. If the reference inside *o* referred directly to *i*, the reference would refer to the stack. However, all references must refer to objects on the heap; creating references to items on the stack could seriously compromise the robustness of the runtime and create a potential security flaw, so it is not allowed. Therefore, the runtime allocates a piece of memory from the heap, copies the value of integer *i* to this piece of memory, and then refers the object *o* to this copy. This automatic copying of an item from the stack to the heap is called *boxing*. The following graphic shows the result:

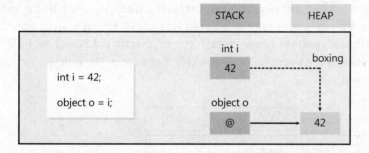

 Important If you modify the original value of a variable, the value on the heap will not change. Likewise, if you modify the value on the heap, the original value of the variable will not change.

Unboxing

Because a variable of type *object* can refer to a boxed copy of a value, it's only reasonable to allow you to get at that boxed value through the variable. You might expect to be able to access the boxed *int* value that a variable *o* refers to by using a simple assignment statement such as this:

```
int i = o;
```

However, if you try this syntax, you'll get a compile-time error. If you think about it, it's pretty sensible that you can't use the int i = o; syntax. After all, o could be referencing absolutely anything and not just an *int*. Consider what would happen in the following code if this statement were allowed:

```
Circle c = new Circle();
int i = 42;
object o;

o = c;  // o refers to a circle
i = o;  // what is stored in i?
```

To obtain the value of the boxed copy, you must use what is known as a *cast*, an operation that checks whether it is safe to convert one type to another and then does the conversion. You prefix the *object* variable with the name of the type in parentheses, as in this example:

```
int i = 42;
object o = i;  // boxes
i = (int)o;    // compiles okay
```

The effect of this cast is subtle. The compiler notices that you've specified the type *int* in the cast. Next, the compiler generates code to check what o actually refers to at run time. It could be absolutely anything. Just because your cast says o refers to an *int*, that doesn't mean it actually does. If o really does refer to a boxed *int* and everything matches, the cast succeeds and the compiler-generated code extracts the value from the boxed *int*. (In this example, the boxed value is then stored in *i*.) This is called *unboxing*. The following diagram shows what is happening:

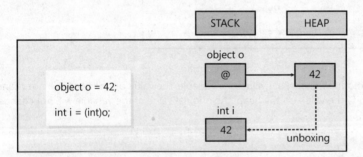

However, if o does not refer to a boxed *int*, there is a type mismatch, causing the cast to fail. The compiler-generated code throws an *InvalidCastException* at run time. Here's an example of an unboxing cast that fails:

```
Circle c = new Circle(42);
object o = c;       // doesn't box because Circle is a reference variable
int i = (int)o;     // compiles okay but throws an exception at run time
```

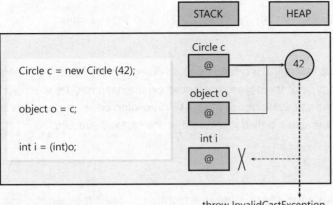

throw InvalidCastException

You will use boxing and unboxing in later exercises. Keep in mind that boxing and unboxing are expensive operations because of the amount of checking required and the need to allocate additional heap memory. Boxing has its uses, but injudicious use can severely impair the performance of a program. You will see an alternative to boxing in Chapter 18, "Introducing Generics."

Casting Data Safely

By using a cast, you can specify that, *in your opinion*, the data referenced by an object has a specific type and that it is safe to reference the object by using that type. The key phrase here is "in your opinion." The C# compiler will trust you when it builds your application, but the runtime is more suspicious and will actually check that this is the case when your application runs. If the type of object in memory does not match the cast, the runtime will throw an *InvalidCastException*, as described in the preceding section. You should be prepared to catch this exception and handle it appropriately if it occurs.

However, catching an exception and attempting to recover in the event that the type of an object is not what you expected it to be is a rather cumbersome approach. C# provides two more very useful operators that can help you perform casting in a much more elegant manner, the *is* and *as* operators.

The *is* Operator

You can use the *is* operator to verify that the type of an object is what you expect it to be, like this:

```
WrappedInt wi = new WrappedInt();
...
object o = wi;
if (o is WrappedInt)
```

```
{
    WrappedInt temp = (WrappedInt)o;  // This is safe; o is a WrappedInt
    ...
}
```

The *is* operator takes two operands: a reference to an object on the left and the name of a type on the right. If the type of the object referenced on the heap has the specified type, *is* evaluates to *true*; otherwise, it evaluates to *false*. The preceding code attempts to cast the reference to the *object* variable o only if it knows that the cast will succeed.

The *as* Operator

The *as* operator fulfills a similar role to *is* but in a slightly truncated manner. You use the *as* operator like this:

```
WrappedInt wi = new WrappedInt();
...
object o = wi;
WrappedInt temp = o as WrappedInt;
if (temp != null)
    ...   // Cast was successful
```

Like the *is* operator, the *as* operator takes an object and a type as its operands. The runtime attempts to cast the object to the specified type. If the cast is successful, the result is returned, and, in this example, it is assigned to the *WrappedInt* variable *temp*. If the cast is unsuccessful, the *as* operator evaluates to the *null* value and assigns that to *temp* instead.

There is a little more to the *is* and *as* operators than described here, and you will meet them again in Chapter 12, "Working with Inheritance."

Pointers and Unsafe Code

This section is purely for your information and is aimed at developers who are familiar with C or C++. If you are new to programming, feel free to skip this section!

If you have already written programs in languages such as C or C++, much of the discussion in this chapter concerning object references might be familiar. Although neither C nor C++ has explicit reference types, both languages have a construct that provides similar functionality—pointers.

A *pointer* is a variable that holds the address of, or a reference to, an item in memory (on the heap or on the stack). A special syntax is used to identify a variable as a pointer. For example, the following statement declares the variable *pi* as a pointer to an integer:

```
int *pi;
```

Although the variable *pi* is declared as a pointer, it does not actually point anywhere until you initialize it. For example, to use *pi* to point to the integer variable *i*, you can use the following statements, and the address operator (&), which returns the address of a variable:

```
int *pi;
int i = 99;
...
pi = &i;
```

You can access and modify the value held in the variable *i* through the pointer variable *pi* like this:

```
*pi = 100;
```

This code updates the value of the variable *i* to 100 because *pi* points to the same memory location as the variable *i*.

One of the main problems that developers learning C and C++ have is understanding the syntax used by pointers. The * operator has at least two meanings (in addition to being the arithmetic multiplication operator), and there is often great confusion about when to use & rather than *. The other issue with pointers is that it is very easy to point somewhere invalid, or to forget to point somewhere at all, and then try to reference the data pointed to. The result will be either garbage or a program that fails with an error because the operating system detects an attempt to access an illegal address in memory. There is also a whole range of security flaws in many existing systems resulting from the mismanagement of pointers; some environments (not Microsoft Windows) fail to enforce checks that a pointer does not refer to memory that belongs to another process, opening up the possibility that confidential data could be compromised.

Reference variables were added to C# to avoid all these problems. If you really want to, you can continue to use pointers in C#, but you must mark the code as *unsafe*. The *unsafe* keyword can be used to mark a block of code, or an entire method, as shown here:

```
public static void Main(string [] args)
{
    int x = 99, y = 100;
    unsafe
    {
        swap (&x, &y);
    }
    Console.WriteLine("x is now {0}, y is now {1}", x, y);
}
```

```
public static unsafe void swap(int *a, int *b)
{
    int temp;
    temp = *a;
    *a = *b;
    *b = temp;
}
```

When you compile programs containing unsafe code, you must specify the */unsafe* option.

Unsafe code also has a bearing on how memory is managed; objects created in unsafe code are said to be unmanaged. We discuss this issue in more detail in Chapter 14.

In this chapter, you have learned some important differences between value types that hold their value directly on the stack and reference types that refer indirectly to their objects on the heap. You have also learned how to use the *ref* and *out* keywords on method parameters to gain access to the arguments. You have seen how assigning a value (such as the *int* 42) to a variable of the *System.Object* class creates a boxed copy of the value on the heap and then causes the *System.Object* variable to refer to this boxed copy. You have also seen how assigning a variable of a value type (such as an *int*) to a variable of the *System.Object* class copies (or unboxes) the value in the *System.Object* class to the memory used by the *int*.

- If you want to continue to the next chapter

  Keep Visual Studio 2008 running, and turn to Chapter 9.

- If you want to exit Visual Studio 2008 now

  On the *File* menu, click *Exit*. If you see a *Save* dialog box, click *Yes* (if you are using Visual Studio 2008) or *Save* (if you are using Visual C# 2008 Express Edition) and save the project.

Chapter 8 Quick Reference

To	Do this

Copy a value type variable	Simply make the copy. Because the variable is a value type, you will have two copies of the same value. For example:
	
	`int i = 42;`
	`int copyi = i;`

Copy a reference type variable	Simply make the copy. Because the variable is a reference type, you will have two references to the same object. For example:
	
	`Circle c = new Circle(42);`
	`Circle refc = c;`

Declare a variable that can hold a value type or the *null* value	Declare the variable using the ? modifier with the type. For example:
	
	``
	int? i = null;
	``

Pass an argument to a *ref* parameter	Prefix the argument with the *ref* keyword. This makes the parameter an alias for the actual argument rather than a copy of the argument. For example:
	
	``
	static void Main()
	{
	    int arg = 42;
	    DoWork(ref arg);
	    Console.WriteLine(arg);
	}
	``

Pass an argument to an *out* parameter	Prefix the argument with the *out* keyword. This makes the parameter an alias for the actual argument rather than a copy of the argument. For example:
	
	``
	static void Main()
	{
	    int arg = 42;
	    DoWork(out arg);
	    Console.WriteLine(arg);
	}
	``

Box a value	Initialize or assign a variable of type object to the value. For example: ``` Object o = 42; ```

Unbox a value	Cast the object reference that refers to the boxed value to the type of the value variable. For example:
	
	``
	int i = (int)o;
	``

Cast an object safely	Use the *is* operator to test whether the cast is valid. For example:
	
	``
	WrappedInt wi = new WrappedInt();
	...
	object o = wi;
	if (o is WrappedInt)
	{
	    WrappedInt temp = (WrappedInt)o;
	    ...
	}
	``
	
	Alternatively, use the *as* operator to perform the cast, and test whether the result is *null*. For example:
	
	``
	WrappedInt wi = new WrappedInt();
	...
	object o = wi;
	WrappedInt temp = o as WrappedInt;
	if (temp != null)
	    ...
	``

Chapter 9
Creating Value Types with Enumerations and Structures

After completing this chapter, you will be able to:

- Declare an enumeration type.

- Create and use an enumeration type.

- Declare a structure type.

- Create and use a structure type.

In Chapter 8, "Understanding Values and References," you learned about the two fundamental kinds of types that exist in Microsoft Visual C#: *value types* and *reference types*. A value type variable holds its value directly on the stack, whereas a reference type variable holds a reference to an object on the heap. In Chapter 7, "Creating and Managing Classes and Objects," you learned how to create your own reference types by defining classes. In this chapter, you'll learn how to create your own value types.

C# supports two kinds of value types: *enumerations* and *structures*. We'll look at each of them in turn.

Working with Enumerations

Suppose you want to represent the seasons of the year in a program. You could use the integers 0, 1, 2, and 3 to represent spring, summer, fall, and winter, respectively. This system would work, but it's not very intuitive. If you used the integer value 0 in code, it wouldn't be obvious that a particular 0 represented spring. It also wouldn't be a very robust solution. For example, if you declare an *int* variable named *season*, there is nothing to stop you from assigning it any legal integer value apart from 0, 1, 2, or 3. C# offers a better solution. You can create an enumeration (sometimes called an *enum* type), whose values are limited to a set of symbolic names.

Declaring an Enumeration

You define an enumeration by using the *enum* keyword, followed by a set of symbols identifying the legal values that the type can have, enclosed between braces. Here's how to declare an enumeration named *Season* whose literal values are limited to the symbolic names *Spring*, *Summer*, *Fall*, and *Winter*:

```
enum Season { Spring, Summer, Fall, Winter }
```

Using an Enumeration

Once you have declared an enumeration, you can use it in exactly the same way as any other type. If the name of your enumeration is *Season*, you can create variables of type *Season*, fields of type *Season*, and parameters of type *Season*, as shown in this example:

```
enum Season { Spring, Summer, Fall, Winter }

class Example
{
    public void Method(Season parameter)
    {
        Season localVariable;
        ...
    }

    private Season currentSeason;
}
```

Before you can use the value of an enumeration variable, it must be assigned a value. You can assign to an enumeration variable only a value that is defined by the enumeration. For example:

```
Season colorful = Season.Fall;
Console.WriteLine(colorful);  // writes out 'Fall'
```

 Note Like all value types, you can create a nullable version of an enumeration variable by using the ? modifier. You can then assign the *null* value, as well the values defined by the enumeration, to the variable:

```
Season? colorful = null;
```

Notice that you have to write *Season.Fall* rather than just *Fall*. All enumeration literal names are scoped by their enumeration. This is useful because it allows different enumerations to coincidentally contain literals with the same name. Also, notice that when you display an enumeration variable by using *Console.WriteLine*, the compiler generates code that writes out the name of the literal whose value matches the value of the variable.

If needed, you can explicitly convert an enumeration variable to a string that represents its current value by using the built-in *ToString* method that all enumerations automatically contain. For example:

```
string name = colorful.ToString();
Console.WriteLine(name);      // also writes out 'Fall'
```

Many of the standard operators that you can use on integer variables can also be used on enumeration variables (except the *bitwise* and *shift* operators, which are covered in Chapter 16, "Using Indexers"). For example, you can compare two enumeration variables of the same type for equality by using the equality operator (==), and you can even perform arithmetic on an enumeration variable (although the result might not always be meaningful!).

Choosing Enumeration Literal Values

Internally, an enumeration associates an integer value with each element. By default, the numbering starts at 0 for the first element and goes up in steps of 1. It's possible to retrieve the underlying integer value of an enumeration variable. To do this, you must cast it to its underlying type. Remember from the discussion of unboxing in Chapter 8 that casting a type converts the data from one type to another as long as the conversion is valid and meaningful. For example, the following code example will write out the value *2* and not the word *Fall* (*Spring* is *0*, *Summer 1*, *Fall 2*, and *Winter 3*):

```
enum Season { Spring, Summer, Fall, Winter }
...
Season colorful = Season.Fall;
Console.WriteLine((int)colorful); // writes out '2'
```

If you prefer, you can associate a specific integer constant (such as 1) with an enumeration literal (such as *Spring*), as in the following example:

```
enum Season { Spring = 1, Summer, Fall, Winter }
```

> **Important** The integer value with which you initialize an enumeration literal must be a compile-time constant value (such as 1).

If you don't explicitly give an enumeration literal a constant integer value, the compiler gives it a value that is 1 greater than the value of the previous enumeration literal except for the very first enumeration literal, to which the compiler gives the default value *0*. In the preceding example, the underlying values of *Spring*, *Summer*, *Fall*, and *Winter* are 1, 2, 3, and 4.

You are allowed to give more than one enumeration literal the same underlying value. For example, in the United Kingdom, *Fall* is referred to as *Autumn*. You can cater to both cultures as follows:

```
enum Season { Spring, Summer, Fall, Autumn = Fall, Winter }
```

Choosing an Enumeration's Underlying Type

When you declare an enumeration, the enumeration literals are given values of type *int*. You can also choose to base your enumeration on a different underlying integer type. For example, to declare that *Season*'s underlying type is a *short* rather than an *int*, you can write this:

```
enum Season : short { Spring, Summer, Fall, Winter }
```

The main reason for doing this is to save memory; an *int* occupies more memory than a *short*, and if you do not need the entire range of values available to an *int*, using a smaller data type can make sense.

You can base an enumeration on any of the eight integer types: *byte*, *sbyte*, *short*, *ushort*, *int*, *uint*, *long*, or *ulong*. The values of all the enumeration literals must fit inside the range of the chosen base type. For example, if you base an enumeration on the *byte* data type, you can have a maximum of 256 literals (starting at 0).

Now that you know how to declare an enumeration, the next step is to use it. In the following exercise, you will work with a Console application to declare and use an enumeration that represents the months of the year.

Create and use an enumeration

1. Start Microsoft Visual Studio 2008 if it is not already running.

2. Open the *StructsAndEnums* project, located in the \Microsoft Press\Visual CSharp Step by Step\Chapter 9\StructsAndEnums folder in your Documents folder.

3. In the *Code and Text Editor* window, display the Month.cs file.

   The source file contains an empty namespace named *StructsAndEnums*.

4. Add an enumeration named *Month* inside the *StructsAndEnums* namespace for modeling the months of the year, as shown in bold here. The 12 enumeration literals for *Month* are *January* through *December*.

```
namespace StructsAndEnums
{
    enum Month
    {
        January, February, March, April,
        May, June, July, August,
        September, October, November, December
    }
}
```

5. Display the Program.cs file in the *Code and Text Editor* window.

   As in the exercises in previous chapters, the *Main* method calls the *Entrance* method and traps any exceptions that occur.

6. In the *Code and Text Editor* window, add a statement to the *Entrance* method to declare a variable named *first* of type *Month* and initialize it to *Month.January*. Add another statement to write the value of the first variable to the console.

   The *Entrance* method should look like this:

   ```
   static void Entrance()
   {
       Month first = Month.January;
       Console.WriteLine(first);
   }
   ```

> **Note** When you type the period following *Month*, IntelliSense will automatically display all the values in the *Month* enumeration.

7. On the *Debug* menu, click *Start Without Debugging*.

   Visual Studio 2008 builds and runs the program. Confirm that the word *January* is written to the console.

8. Press Enter to close the program and return to the Visual Studio 2008 programming environment.

9. Add two more statements to the *Entrance* method to increment the *first* variable and display its new value to the console, as shown in bold here:

   ```
   static void Entrance()
   {
       Month first = Month.January;
       Console.WriteLine(first);
       first++;
       Console.WriteLine(first);
   }
   ```

10. On the *Debug* menu, click *Start Without Debugging*.

    Visual Studio 2008 builds and runs the program. Confirm that the words *January* and *February* are written to the console.

    Notice that performing a mathematical operation (such as the increment operation) on an enumeration variable changes the internal integer value of the variable. When the variable is written to the console, the corresponding enumeration value is displayed.

11. Press Enter to close the program and return to the Visual Studio 2008 programming environment.

12. Modify the first statement in the *Entrance* method to initialize the *first* variable to *Month.December*, as shown in bold here:

```
static void Entrance()
{
    Month first = Month.December;
    Console.WriteLine(first);
    first++;
    Console.WriteLine(first);
}
```

13. On the *Debug* menu, click *Start Without Debugging*.

   Visual Studio 2008 builds and runs the program. This time the word *December* is written to the console, followed by the number *12*. Although you can perform arithmetic on an enumeration, if the results of the operation are outside the range of values defined for the enumerator, all the runtime can do is treat the variable as the corresponding integer value.

14. Press Enter to close the program and return to the Visual Studio 2008 programming environment.

Working with Structures

You saw in Chapter 8 that classes define reference types that are always created on the heap. In some cases, the class can contain so little data that the overhead of managing the heap becomes disproportionate. In these cases, it is better to define the type as a structure. A structure is a value type. Because structures are stored on the stack, as long as the structure is reasonably small, the memory management overhead is often reduced.

A structure can have its own fields, methods, and constructors just like a class, but not like an enumeration.

Common Structure Types

You might not have realized it, but you have already used structures in previous exercises in this book. In C#, the primitive numeric types *int*, *long*, and *float* are aliases for the structures *System.Int32*, *System.Int64*, and *System.Single*, respectively. These structures have fields and methods, and you can actually call methods on variables and literals of these types. For example, all of these structures provide a *ToString* method that can convert a numeric value to its string representation. The following statements are all legal statements in C#:

```
int i = 99;
Console.WriteLine(i.ToString());
Console.WriteLine(55.ToString());
float f = 98.765F;
```

```
Console.WriteLine(f.ToString());
Console.WriteLine(98.765F.ToString());
```

You don't see this use of the *ToString* method very often, because the *Console.WriteLine* method calls it automatically when it is needed. Use of the static methods exposed by these structures is much more common. For example, in earlier chapters you used the static *int.Parse* method to convert a string to its corresponding integer value. What you are actually doing is invoking the *Parse* method of the *Int32* structure:

```
string s = "42";
int i = int.Parse(s);   // exactly the same as Int32.Parse
```

These structures also include some useful static fields. For example, *Int32.MaxValue* is the maximum value that an *int* can hold, and *Int32.MinValue* is the smallest value you can store in an *int*.

The following table shows the primitive types in C# and their equivalent types in the Microsoft .NET Framework. Notice that the *string* and *object* types are classes (reference types) rather than structures.

Keyword	Type equivalent	Class or structure
bool	*System.Boolean*	Structure
byte	*System.Byte*	Structure
decimal	*System.Decimal*	Structure
double	*System.Double*	Structure
float	*System.Single*	Structure
int	*System.Int32*	Structure
long	*System.Int64*	Structure
object	*System.Object*	Class
sbyte	*System.SByte*	Structure
short	*System.Int16*	Structure
string	*System.String*	Class
uint	*System.UInt32*	Structure
ulong	*System.UInt64*	Structure
ushort	*System.UInt16*	Structure

Declaring a Structure

To declare your own structure value type, you use the *struct* keyword followed by the name of the type, followed by the body of the structure between opening and closing braces. For example, here is a structure named *Time* that contains three *public int* fields named *hours*, *minutes*, and *seconds*:

```
struct Time
{
    public int hours, minutes, seconds;
}
```

As with classes, making the fields of a structure *public* is not advisable in most cases; there is no way to ensure that *public* fields contain valid values. For example, anyone could set the value of *minutes* or *seconds* to a value greater than 60. A better idea is to make the fields *private* and provide your structure with constructors and methods to initialize and manipulate these fields, as shown in this example:

```
struct Time
{
    public Time(int hh, int mm, int ss)
    {
        hours = hh % 24;
        minutes = mm % 60;
        seconds = ss % 60;
    }

    public int Hours()
    {
        return hours;
    }
    ...
    private int hours, minutes, seconds;
}
```

> **Note** By default, you cannot use many of the common operators on your own structure types. For example, you cannot use operators such as the equality operator (==) and the inequality operator (!=) on your own structure type variables. However, you can explicitly declare and implement operators for your own structure types. The syntax for doing this is covered in Chapter 21, "Operator Overloading."

Use structures to implement simple concepts whose main feature is their value. For example, an *int* is a value type because its main feature is its value. If you have two *int* variables that contain the same value (such as 42), one is as good as the other. When you copy a value type variable, you get two copies of the value. In contrast, when you copy a reference type variable, you get two references to the same object. In summary, use structures for small data values where it's just as or nearly as efficient to copy the value as it would be to copy an address. Use classes for more complex data so that you have the option of copying only the address of the actual value when you want to improve the efficiency of your code.

Understanding Structure and Class Differences

A structure and a class are syntactically very similar, but there are a few important differences. Let's look at some of these differences:

- You can't declare a default constructor (a constructor with no parameters) for a structure. The following example would compile if Time were a class, but because Time is a structure, it does not:

```
struct Time
{
    public Time() { ... } // compile-time error
    ...
}
```

The reason you can't declare your own default constructor for a structure is that the compiler *always* generates one. In a class, the compiler generates the default constructor only if you don't write a constructor yourself. The compiler-generated default constructor for a structure always sets the fields to *0, false,* or *null*—just as for a class. Therefore, you should ensure that a structure value created by the default constructor behaves logically and makes sense with these default values. If you don't want to use these default values, you can initialize fields to different values by providing a nondefault constructor. However, if you don't initialize a field in your nondefault structure constructor, the compiler won't initialize it for you. This means that you must explicitly initialize all the fields in all your nondefault structure constructors or you'll get a compile-time error. For example, although the following example would compile and silently initialize *seconds* to *0* if *Time* were a class, because *Time* is a structure, it fails to compile:

```
struct Time
{
    public Time(int hh, int mm)
    {
        hours = hh;
        minutes = mm;
    }   // compile-time error: seconds not initialized
    ...
    private int hours, minutes, seconds;
}
```

■ In a class, you can initialize instance fields at their point of declaration. In a structure, you cannot. The following example would compile if *Time* were a class, but because *Time* is a structure, it causes a compile-time error:

```
struct Time
{
    ...
    private int hours = 0; // compile-time error
    private int minutes;
    private int seconds;
}
```

The following table summarizes the main differences between a structure and a class.

Question	Structure	Class
Is this a value type or a reference type?	A structure is a value type.	A class is a reference type.
Do instances live on the stack or the heap?	Structure instances are called *values* and live on the stack.	Class instances are called *objects* and live on the heap.
Can you declare a default constructor?	No	Yes
If you declare your own constructor, will the compiler still generate the default constructor?	Yes	No
If you don't initialize a field in your own constructor, will the compiler automatically initialize it for you?	No	Yes
Are you allowed to initialize instance fields at their point of declaration?	No	Yes

There are other differences between classes and structures concerning inheritance. These differences are covered in Chapter 12, "Working with Inheritance." Now that you know how to declare structures, the next step is to use them to create values.

Declaring Structure Variables

After you have defined a structure type, you can use it in exactly the same way as any other type. For example, if you have defined the *Time* structure, you can create variables, fields, and parameters of type *Time*, as shown in this example:

```
struct Time
{
    ...
    private int hours, minutes, seconds;
}
```

```
class Example
{
    public void Method(Time parameter)
    {
        Time localVariable;
        ...
    }

    private Time currentTime;
}
```

 Note You can create a nullable version of a structure variable by using the ? modifier. You can then assign the *null* value to the variable:

```
Time? currentTime = null;
```

Understanding Structure Initialization

Earlier in this chapter, you saw how the fields in a structure are initialized by using a constructor. However, because structures are value types, you can create structure variables without calling a constructor, as shown in the following example:

```
Time now;
```

In this example, the variable is created but its fields are left in their uninitialized state. Any attempt to access the values in these fields will result in a compiler error. The following graphic depicts the state of the fields in the *now* variable:

	STACK
Time now;	now.hours ?
	now.minutes ?
	now.seconds ?

If you call a constructor, the various rules of structure constructors described earlier guarantee that all the fields in the structure will be initialized:

```
Time now = new Time();
```

This time, the default constructor initializes the fields in the structure, as shown in the following graphic:

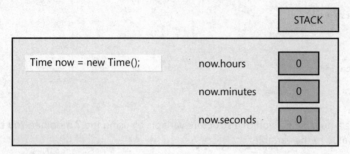

Note that in both cases, the *Time* variable is created on the stack.

If you've written your own structure constructor, you can also use that to initialize a structure variable. As explained earlier in this chapter, a structure constructor must always explicitly initialize all its fields. For example:

```
struct Time
{
    public Time(int hh, int mm)
    {
        hours = hh;
        minutes = mm;
        seconds = 0;
    }
    ...
    private int hours, minutes, seconds;
}
```

The following example initializes *now* by calling a user-defined constructor:

```
Time now = new Time(12, 30);
```

The following graphic shows the effect of this example:

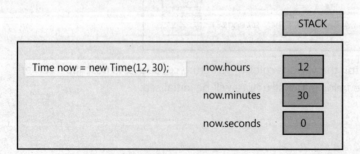

Copying Structure Variables

You're allowed to initialize or assign one structure variable to another structure variable, but only if the structure variable on the right side is completely initialized (that is, if all its fields are initialized). The following example compiles because *now* is fully initialized. (The graphic shows the results of performing the assignment.)

```
Time now = new Time(12, 30);
Time copy = now;
```

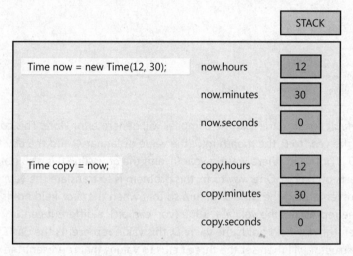

The following example fails to compile because *now* is not initialized:

```
Time now;
Time copy = now; // compile-time error: now has not been assigned
```

When you copy a structure variable, each field on the left side is set directly from the corresponding field on the right side. This copying is done as a fast, single operation that copies the contents of the entire structure and that never throws an exception. Compare this behavior with the equivalent action if *Time* were a class, in which case both variables (*now* and *copy*) would end up referencing the *same* object on the heap.

 Note C++ programmers should note that this copy behavior cannot be customized.

It's time to put this knowledge into practice. In the following exercise, you will create and use a structure to represent a date.

Create and use a structure type

1. In the *StructsAndEnums* project, display the Date.cs file in the *Code and Text Editor* window.

2. Add a structure named *Date* inside the *StructsAndEnums* namespace.

   This structure should contain three private fields: one named *year* of type *int*, one named *month* of type *Month* (using the enumeration you created in the preceding exercise), and one named *day* of type *int*. The *Date* structure should look exactly as follows:

   ```
   struct Date
   {
       private int year;
       private Month month;
       private int day;
   }
   ```

   Consider the default constructor that the compiler will generate for *Date*. This constructor will set the *year* to *0*, the *month* to *0* (the value of January), and the *day* to *0*. The *year* value 0 is not valid (there was no year 0), and the *day* value 0 is also not valid (each month starts on day 1). One way to fix this problem is to translate the *year* and *day* values by implementing the *Date* structure so that when the *year* field holds the value *Y*, this value represents the year *Y* + 1900 (you can pick a different century if you prefer), and when the *day* field holds the value *D*, this value represents the day *D* + 1. The default constructor will then set the three fields to values that represent the date 1 January 1900.

3. Add a *public* constructor to the *Date* structure. This constructor should take three parameters: an *int* named *ccyy* for the *year*, a *Month* named *mm* for the *month*, and an *int* named *dd* for the *day*. Use these three parameters to initialize the corresponding fields. A *year* field of *Y* represents the year *Y* + 1900, so you need to initialize the *year* field to the value *ccyy* – 1900. A *day* field of *D* represents the day *D* + 1, so you need to initialize the *day* field to the value *dd* – 1.

   The *Date* structure should now look like this (the constructor is shown in bold):

   ```
   struct Date
   {
       public Date(int ccyy, Month mm, int dd)
       {
           this.year = ccyy - 1900;
           this.month = mm;
           this.day = dd - 1;
       }

       private int year;
       private Month month;
       private int day;
   }
   ```

4. Add a *public* method named *ToString* to the *Date* structure after the constructor. This method takes no arguments and returns a string representation of the date. Remember, the value of the *year* field represents *year* + 1900, and the value of the *day* field represents *day* + 1.

> **Note** The *ToString* method is a little different from the methods you have seen so far. Every type, including structures and classes that you define, automatically has a *ToString* method whether or not you want it. Its default behavior is to convert the data in a variable to a string representation of that data. Sometimes, the default behavior is meaningful; other times, it is less so. For example, the default behavior of the *ToString* method generated for the *Date* class simply generates the string *"StructsAndEnums.Date"*. To quote Zaphod Beeblebrox in *The Restaurant at the End of the Universe* (Douglas Adams, Del Rey, 2005), this is "shrewd, but dull." You need to define a new version of this method that overrides the default behavior by using the *override* keyword. Overriding methods are discussed in more detail in Chapter 12.

The *ToString* method should look like this:

```
public override string ToString()
{
    return this.month + " " + (this.day + 1) + " " + (this.year + 1900);
}
```

> **Note** The + signs inside the parentheses are the arithmetic addition operator. The others are the string concatenation operator. Without the parentheses, all occurrences of the + sign would be treated as the string concatenation operator because the expression being evaluated is a string. It can be a little confusing when the same symbol in a single expression denotes different operators!

5. Display the Program.cs file in the *Code and Text Editor* window.

6. Add a statement to the end of the *Entrance* method to declare a local variable named *defaultDate*, and initialize it to a *Date* value constructed by using the default *Date* constructor. Add another statement to *Entrance* to write the *defaultDate* variable to the console by calling *Console.WriteLine*.

> **Note** The *Console.WriteLine* method automatically calls the *ToString* method of its argument to format the argument as a string.

The *Entrance* method should now look like this:

```
static void Entrance()
{
    ...
    Date defaultDate = new Date();
    Console.WriteLine(defaultDate);
}
```

7. On the *Debug* menu, click *Start Without Debugging* to build and run the program. Verify that the date *January 1 1900* is written to the console. (The original output of the *Entrance* method will be displayed first.)

8. Press the Enter key to return to the Visual Studio 2008 programming environment.

9. In the *Code and Text Editor* window, return to the *Entrance* method, and add two more statements. The first statement should declare a local variable named *halloween* and initialize it to October 31 2008. The second statement should write the value of *halloween* to the console.

The *Entrance* method should now look like this:

```
static void Entrance()
{
    ...
    Date halloween = new Date(2008, Month.October, 31);
    Console.WriteLine(halloween);
}
```

 Note When you type the *new* keyword, IntelliSense will automatically detect that there are two constructors available for the *Date* type.

10. On the *Debug* menu, click *Start Without Debugging*. Confirm that the date *October 31 2008* is written to the console below the previous information.

11. Press Enter to close the program.

You have successfully used the *enum* and *struct* keywords to declare your own value types and then used these types in code.

■ If you want to continue to the next chapter:

Keep Visual Studio 2008 running, and turn to Chapter 10.

■ If you want to exit Visual Studio 2008 now:

On the *File* menu, click *Exit*. If you see a *Save* dialog box, click *Yes* (if you are using Visual Studio 2008) or *Save* (if you are using Visual C# 2008 Express Edition) and save the project.

Chapter 9 Quick Reference

To	Do this
Declare an enumeration	Write the keyword *enum*, followed by the name of the type, followed by a pair of braces containing a comma-separated list of the enumeration literal names. For example: ```enum Season { Spring, Summer, Fall, Winter }```
Declare an enumeration variable	Write the name of the enumeration on the left followed by the name of the variable, followed by a semicolon. For example: ```Season currentSeason;```

Assign an enumeration variable to a value	Write the name of the enumeration literal in combination with the name of the enumeration to which it belongs. For example:
	
	```currentSeason = Spring; // error```
	```currentSeason = Season.Spring; // correct```

Declare a structure type	Write the keyword *struct*, followed by the name of the structure type, followed by the body of the structure (the constructors, methods, and fields). For example:
	
	```struct Time```
	```{```
	``` public Time(int hh, int mm, int ss)```
	``` { ... }```
	``` ...```
	``` private int hours, minutes, seconds;```
	```}```

Declare a structure variable	Write the name of the structure type, followed by the name of the variable, followed by a semicolon. For example: ```Time now;```
Initialize a structure variable to a value	Initialize the variable to a structure value created by calling the structure constructor. For example: ```Time lunch = new Time(12, 30, 0);```

Chapter 10
Using Arrays and Collections

After completing this chapter, you will be able to:

- Declare, initialize, copy, and use array variables.
- Declare, initialize, copy, and use variables of various collection types.

You have already seen how to create and use variables of many different types. However, all the examples of variables you have seen so far have one thing in common—they hold information about a single item (an *int*, a *float*, a *Circle*, a *Time*, and so on). What happens if you need to manipulate sets of items? One solution would be to create a variable for each item in the set, but this leads to a number of further questions: How many variables do you need? How should you name them? If you need to perform the same operation on each item in the set (such as increment each variable in a set of integers), how would you avoid very repetitive code? This solution assumes that you know, when you write the program, how many items you will need, but how often is this the case? For example, if you are writing an application that reads and processes records from a database, how many records are in the database, and how likely is this number to change?

Arrays and collections provide mechanisms that solve the problems posed by these questions.

What Is an Array?

An *array* is an unordered sequence of elements. All the elements in an array have the same type (unlike the fields in a structure or class, which can have different types). The elements of an array live in a contiguous block of memory and are accessed by using an integer index (unlike fields in a structure or class, which are accessed by name).

Declaring Array Variables

You declare an array variable by specifying the name of the element type, followed by a pair of square brackets, followed by the variable name. The square brackets signify that the variable is an array. For example, to declare an array of *int* variables named *pins*, you would write:

```
int[] pins; // Personal Identification Numbers
```

Microsoft Visual Basic programmers should note that you use square brackets and not parentheses. C and C++ programmers should note that the size of the array is not part of the declaration. Java programmers should note that you must place the square brackets *before* the variable name.

Note You are not restricted to primitive types as array elements. You can also create arrays of structures, enumerations, and classes. For example, you can create an array of *Time* structures like this:

```
Time[] times;
```

Tip It is often useful to give array variables plural names, such as *places* (where each element is a *Place*), *people* (where each element is a *Person*), or *times* (where each element is a *Time*).

Creating an Array Instance

Arrays are reference types, regardless of the type of their elements. This means that an array variable *refers* to the contiguous block of memory holding the array elements on the heap (just as a class variable refers to an object on the heap) and does not hold its array elements directly on the stack (as a structure does). (To review values and references and the differences between the stack and the heap, see Chapter 8, "Understanding Values and References.") Remember that when you declare a class variable, memory is not allocated for the object until you create the instance by using *new*. Arrays follow the same rules—when you declare an array variable, you do not declare its size. You specify the size of an array only when you actually create an array instance.

To create an array instance, you use the *new* keyword followed by the name of the element type, followed by the size of the array you're creating between square brackets. Creating an array also initializes its elements by using the now familiar default values (*0*, *null*, or *false*, depending on whether the type is numeric, a reference, or a Boolean, respectively). For example, to create and initialize a new array of four integers for the *pins* variable declared earlier, you write this:

```
pins = new int[4];
```

The following graphic illustrates the effects of this statement:

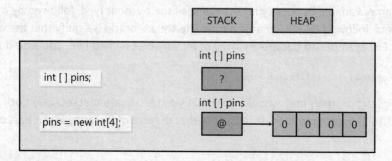

The size of an array instance does not have to be a constant; it can be calculated at run time, as shown in this example:

```
int size = int.Parse(Console.ReadLine());
int[] pins = new int[size];
```

You're allowed to create an array whose size is 0. This might sound bizarre, but it's useful in situations where the size of the array is determined dynamically and could be 0. An array of size 0 is not a *null* array.

It's also possible to create multidimensional arrays. For example, to create a two-dimensional array, you create an array that requires two integer indexes. Detailed discussion of multidimensional arrays is beyond the scope of this book, but here's an example:

```
int[,] table = new int[4,6];
```

Initializing Array Variables

When you create an array instance, all the elements of the array instance are initialized to a default value depending on their type. You can modify this behavior and initialize the elements of an array to specific values if you prefer. You achieve this by providing a comma-separated list of values between a pair of braces. For example, to initialize *pins* to an array of four *int* variables whose values are *9*, *3*, *7*, and *2*, you would write this:

```
int[] pins = new int[4]{ 9, 3, 7, 2 };
```

The values between the braces do not have to be constants. They can be values calculated at run time, as shown in this example:

```
Random r = new Random();
int[] pins = new int[4]{ r.Next() % 10, r.Next() % 10,
                         r.Next() % 10, r.Next() % 10 };
```

> **Note** The *System.Random* class is a pseudorandom number generator. The *Next* method returns a nonnegative random integer in the range *0* to *Int32.MaxValue* by default. The *Next* method is overloaded, and other versions enable you to specify the minimum value and maximum value of the range. The default constructor for the *Random* class seeds the random number generator with a time-dependent seed value, which reduces the possibility of the class duplicating a sequence of random numbers. An overloaded version of the constructor enables you to provide your own seed value. That way you can generate a repeatable sequence of random numbers for testing purposes.

The number of values between the braces must exactly match the size of the array instance being created:

```
int[] pins = new int[3]{ 9, 3, 7, 2 }; // compile-time error
int[] pins = new int[4]{ 9, 3, 7 };    // compile-time error
int[] pins = new int[4]{ 9, 3, 7, 2 }; // okay
```

When you're initializing an array variable, you can actually omit the *new* expression and the size of the array. The compiler calculates the size from the number of initializers and generates code to create the array. For example:

```
int[] pins = { 9, 3, 7, 2 };
```

If you create an array of structures, you can initialize each structure in the array by calling the structure constructor, as shown in this example:

```
Time[] schedule = { new Time(12,30), new Time(5,30) };
```

Creating an Implicitly Typed Array

The element type when you declare an array must match the type of elements that you attempt to store in the array. For example, if you declare *pins* to be an array of *int*, as shown in the preceding examples, you cannot store a *double*, *string*, *struct*, or anything that is not an *int* in this array. If you specify a list of initializers when declaring an array, you can let the C# compiler infer the actual type of the elements in the array for you, like this:

```
var names = new[]{"John", "Diana", "James", "Francesca"};
```

In this example, the C# compiler determines that the *names* variable is an array of strings. It is worth pointing out a couple of syntactic quirks in this declaration. First, you omit the square brackets from the type; the *names* variable in this example is declared simply as *var*, and not *var[]*. Second, you must specify the *new* operator and square brackets before the initializer list.

If you use this syntax, you must ensure that all the initializers have the same type. This next example will cause the compile-time error "No best type found for implicitly typed array":

```
var bad = new[]{"John", "Diana", 99, 100};
```

However, in some cases, the compiler will convert elements to a different type if doing so makes sense. In the following code, the *numbers* array is an array of *double* because the constants *3.5* and *99.999* are both *double*, and the C# compiler can convert the integer values *1* and *2* to *double* values:

```
var numbers = new[]{1, 2, 3.5, 99.999};
```

Generally, it is best to avoid mixing types and hoping that the compiler will convert them for you.

Implicitly typed arrays are most useful when you are working with anonymous types, described in Chapter 7, "Creating and Managing Classes and Objects." The following code creates an array of anonymous objects, each containing two fields specifying the name and age of the members of my family (yes, I am younger than my wife):

```
var names = new[] { new { Name = "John", Age = 42 },
                    new { Name = "Diana", Age = 43 },
                    new { Name = "James", Age = 15 },
                    new { Name = "Francesca", Age = 13 } };
```

The fields in the anonymous types must be the same for each element of the array.

Accessing an Individual Array Element

To access an individual array element, you must provide an index indicating which element you require. For example, you can read the contents of element 2 of the *pins* array into an *int* variable by using the following code:

```
int myPin;
myPin = pins[2];
```

Similarly, you can change the contents of an array by assigning a value to an indexed element:

```
myPin = 1645;
pins[2] = myPin;
```

Array indexes are zero-based. The initial element of an array lives at index 0 and not index 1. An index value of *1* accesses the second element.

All array element access is bounds-checked. If you specify an index that is less than 0 or greater than or equal to the length of the array, the compiler throws an *IndexOutOfRangeException*, as in this example:

```
try
{
    int[] pins = { 9, 3, 7, 2 };
    Console.WriteLine(pins[4]); // error, the 4th and last element is at index 3
}
catch (IndexOutOfRangeException ex)
{
    ...
}
```

Iterating Through an Array

Arrays have a number of useful built-in properties and methods. (All arrays inherit methods and properties from the *System.Array* class in the Microsoft .NET Framework.) You can examine the *Length* property to discover how many elements an array contains and iterate through all the elements of an array by using a *for* statement. The following sample code writes the array element values of the *pins* array to the console:

```
int[] pins = { 9, 3, 7, 2 };
for (int index = 0; index < pins.Length; index++)
{
    int pin = pins[index];
    Console.WriteLine(pin);
}
```

> **Note** *Length* is a property and not a method, which is why there are no parentheses when you call it. You will learn about properties in Chapter 15, "Implementing Properties to Access Fields."

It is common for new programmers to forget that arrays start at element 0 and that the last element is numbered *Length* – 1. C# provides the *foreach* statement to enable you to iterate through the elements of an array without worrying about these issues. For example, here's the preceding *for* statement rewritten as an equivalent *foreach* statement:

```
int[] pins = { 9, 3, 7, 2 };
foreach (int pin in pins)
{
    Console.WriteLine(pin);
}
```

The *foreach* statement declares an iteration variable (in this example, *int pin*) that automatically acquires the value of each element in the array. The type of this variable must match the type of the elements in the array. The *foreach* statement is the preferred way to iterate through an array; it expresses the intention of the code directly, and all of the *for* loop scaffolding drops away. However, in a few cases, you'll find that you have to revert to a *for* statement:

- A foreach statement always iterates through the whole array. If you want to iterate through only a known portion of an array (for example, the first half) or to bypass certain elements (for example, every third element), it's easier to use a for statement.

- A foreach statement always iterates from index 0 through index *Length* – 1. If you want to iterate backward, it's easier to use a for statement.

- If the body of the loop needs to know the index of the element rather than just the value of the element, you'll have to use a for statement.

- If you need to modify the elements of the array, you'll have to use a for statement. This is because the iteration variable of the foreach statement is a read-only copy of each element of the array.

You can declare the iteration variable as a *var* and let the C# compiler work out the type of the variable from the type of the elements in the array. This is especially useful if you don't actually know the type of the elements in the array, such as when the array contains anonymous objects. The following example demonstrates how you can iterate through the array of family members shown earlier:

```
var names = new[] { new { Name = "John", Age = 42 },
                    new { Name = "Diana", Age = 43 },
                    new { Name = "James", Age = 15 },
                    new { Name = "Francesca", Age = 13 } };
foreach (var familyMember in names)
{
    Console.WriteLine("Name: {0}, Age: {1}", familyMember.Name, familyMember.Age);
}
```

Copying Arrays

Arrays are reference types. (Remember that an array is an instance of the *System.Array* class.) An array variable contains a reference to an array instance. This means that when you copy an array variable, you end up with two references to the same array instance—for example:

```
int[] pins = { 9, 3, 7, 2 };
int[] alias = pins; //  alias and pins refer to the same array instance
```

In this example, if you modify the value at *pins[1]*, the change will also be visible by reading *alias[1]*.

If you want to make a copy of the array instance (the data on the heap) that an array variable refers to, you have to do two things. First you need to create a new array instance of the same type and the same length as the array you are copying, as in this example:

```
int[] pins = { 9, 3, 7, 2 };
int[] copy = new int[4];
```

This works, but if you later modify the code to change the length of the original array, you must remember to also change the size of the copy. It's better to determine the length of an array by using its *Length* property, as shown in this example:

```
int[] pins = { 9, 3, 7, 2 };
int[] copy = new int[pins.Length];
```

The values inside *copy* are now all initialized to their default value, *0*.

The second thing you need to do is set the values inside the new array to the same values as the original array. You could do this by using a *for* statement, as shown in this example:

```
int[] pins = { 9, 3, 7, 2 };
int[] copy = new int[pins.Length];
for (int i = 0; i < copy.Length; i++)
{
    copy[i] = pins[i];
}
```

Copying an array is actually a common requirement of many applications—so much so that the *System.Array* class provides some useful methods that you can employ to copy an array rather than writing your own code. For example, the *CopyTo* method copies the contents of one array into another array given a specified starting index:

```
int[] pins = { 9, 3, 7, 2 };
int[] copy = new int[pins.Length];
pins.CopyTo(copy, 0);
```

Another way to copy the values is to use the *System.Array* static method named *Copy*. As with *CopyTo*, you must initialize the target array before calling *Copy*:

```
int[] pins = { 9, 3, 7, 2 };
int[] copy = new int[pins.Length];
Array.Copy(pins, copy, copy.Length);
```

Yet another alternative is to use the *System.Array* instance method named *Clone*. You can call this method to create an entire array and copy it in one action:

```
int[] pins = { 9, 3, 7, 2 };
int[] copy = (int[])pins.Clone();
```

 Note The *Clone* method actually returns an *object*, which is why you must cast it to an array of the appropriate type when you use it. Furthermore, all four ways of copying shown earlier create a *shallow* copy of an array—if the elements in the array being copied contain references, the *for* loop as coded and the three preceding methods simply copy the references rather than the objects being referred to. After copying, both arrays refer to the same set of objects. If you need to create a deep copy of such an array, you must use appropriate code in a *for* loop.

What Are Collection Classes?

Arrays are useful, but they have their limitations. Fortunately, arrays are only one way to collect elements of the same type. The Microsoft .NET Framework provides several classes that also collect elements together in other specialized ways. These are the collection classes, and they live in the *System.Collections* namespace and sub-namespaces.

The basic collection classes accept, hold, and return their elements as objects—that is, the element type of a collection class is an *object*. To understand the implications of this, it is helpful to contrast an array of *int* variables (*int* is a value type) with an array of objects (*object* is a reference type). Because *int* is a value type, an array of *int* variables holds its *int* values directly, as shown in the following graphic:

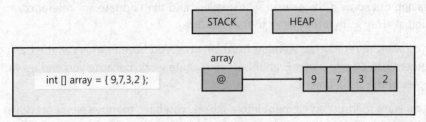

Now consider the effect when the array is an array of objects. You can still add integer values to this array. (In fact, you can add values of any type to it.) When you add an integer value, it is automatically boxed, and the array element (an object reference) refers to the boxed copy of the integer value. (For a refresher on boxing, refer to Chapter 8.) This is illustrated in the following graphic:

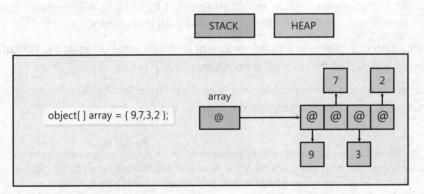

The element type of all the collection classes shown in this chapter is an *object*. This means that when you insert a value into a collection, it is always boxed, and when you remove a value from a collection, you must unbox it by using a cast. The following sections provide a very quick overview of four of the most useful collection classes. Refer to the Microsoft .NET Framework Class Library documentation for more details on each class.

Note There are collection classes that don't always use *object* as their element type and that can hold value types as well as references, but you need to know a bit more about C# before we can talk about them. You will meet these collection classes in Chapter 18, "Introducing Generics."

The *ArrayList* Collection Class

ArrayList is a useful class for shuffling elements around in an array. There are certain occasions when an ordinary array can be too restrictive:

- If you want to resize an array, you have to create a new array, copy the elements (leaving out some if the new array is smaller), and then update any references to the original array so that they refer to the new array.

- If you want to remove an element from an array, you have to move all the trailing elements up by one place. Even this doesn't quite work, because you end up with two copies of the last element.

- If you want to insert an element into an array, you have to move elements down by one place to make a free slot. However, you lose the last element of the array!

Here's how you can overcome these restrictions using the *ArrayList* class:

- You can remove an element from an ArrayList by using its Remove method. The ArrayList automatically reorders its elements.

- You can add an element to the end of an ArrayList by using its Add method. You supply the element to be added. The ArrayList resizes itself if necessary.

- You can insert an element into the middle of an ArrayList by using its Insert method. Again, the ArrayList resizes itself if necessary.

- You can reference an existing element in an ArrayList object by using ordinary array notation, with square brackets and the index of the element.

> **Note** As with arrays, if you use *foreach* to iterate through an *ArrayList*, you cannot use the iteration variable to modify the contents of the *ArrayList*. Additionally, you cannot call the *Remove*, *Add*, or *Insert* method in a *foreach* loop that iterates through an *ArrayList*.

Here's an example that shows how you can create, manipulate, and iterate through the contents of an *ArrayList*:

```
using System;
using System.Collections;
...
ArrayList numbers = new ArrayList();
...
// fill the ArrayList
foreach (int number in new int[12]{10, 9, 8, 7, 7, 6, 5, 10, 4, 3, 2, 1})
{
    numbers.Add(number);
}
...
// insert an element in the penultimate position in the list, and move the last item up
```

```
// (the first parameter is the position;
// the second parameter is the value being inserted)
numbers.Insert(numbers.Count-1, 99);

// remove first element whose value is 7 (the 4th element, index 3)
numbers.Remove(7);
// remove the element that's now the 7th element, index 6 (10)
numbers.RemoveAt(6);
...
// iterate remaining 10 elements using a for statement
for (int i = 0; i < numbers.Count; i++)
{
    int number = (int)numbers[i]; // notice the cast, which unboxes the value
    Console.WriteLine(number);
}
...
// iterate remaining 10 using a foreach statement
foreach (int number in numbers)  // no cast needed
{
    Console.WriteLine(number);
}
```

The output of this code is shown here:

```
10
9
8
7
6
5
4
3
2
99
1
10
9
8
7
6
5
4
3
2
99
1
```

Note The way you determine the number of elements for an *ArrayList* is different from querying the number of items in an array. When using an *ArrayList*, you examine the *Count* property, and when using an array, you examine the *Length* property.

The *Queue* Collection Class

The *Queue* class implements a first-in, first-out (FIFO) mechanism. An element is inserted into the queue at the back (the enqueue operation) and is removed from the queue at the front (the dequeue operation).

Here's an example of a queue and its operations:

```
using System;
using System.Collections;
...
Queue numbers = new Queue();
...
// fill the queue
foreach (int number in new int[4]{9, 3, 7, 2})
{
    numbers.Enqueue(number);
    Console.WriteLine(number + " has joined the queue");
}
...
// iterate through the queue
foreach (int number in numbers)
{
    Console.WriteLine(number);
}
...
// empty the queue
while (numbers.Count > 0)
{
    int number = (int)numbers.Dequeue(); // cast required to unbox the value
    Console.WriteLine(number + " has left the queue");
}
```

The output from this code is:

```
9 has joined the queue
3 has joined the queue
7 has joined the queue
2 has joined the queue
9
3
7
2
9 has left the queue
3 has left the queue
7 has left the queue
2 has left the queue
```

The *Stack* Collection Class

The *Stack* class implements a last-in, first-out (LIFO) mechanism. An element joins the stack at the top (the push operation) and leaves the stack at the top (the pop operation). To visualize this, think of a stack of dishes: new dishes are added to the top and dishes are removed from the top, making the last dish to be placed on the stack the first one to be removed. (The dish at the bottom is rarely used and will inevitably require washing before you can put any food on it as it will be covered in grime!) Here's an example:

```
using System;
using System.Collections;
...
Stack numbers = new Stack();
...
// fill the stack
foreach (int number in new int[4]{9, 3, 7, 2})
{
    numbers.Push(number);
    Console.WriteLine(number + " has been pushed on the stack");
}
...
// iterate through the stack
foreach (int number in numbers)
{
    Console.WriteLine(number);
}
...
// empty the stack
while (numbers.Count > 0)
{
    int number = (int)numbers.Pop();
    Console.WriteLine(number +  " has been popped off the stack");
}
```

The output from this program is:

```
9 has been pushed on the stack
3 has been pushed on the stack
7 has been pushed on the stack
2 has been pushed on the stack
2
7
3
9
2 has been popped off the stack
7 has been popped off the stack
3 has been popped off the stack
9 has been popped off the stack
```

The *Hashtable* Collection Class

The array and *ArrayList* types provide a way to map an integer index to an element. You provide an integer index inside square brackets (for example, [4]), and you get back the element at index 4 (which is actually the fifth element). However, sometimes you might want to provide a mapping where the type you map from is not an *int* but rather some other type, such as *string*, *double*, or *Time*. In other languages, this is often called an *associative array*. The *Hashtable* class provides this functionality by internally maintaining two *object* arrays, one for the *keys* you're mapping from and one for the *values* you're mapping to. When you insert a key/value pair into a *Hashtable*, it automatically tracks which key belongs to which value and enables you to retrieve the value that is associated with a specified key quickly and easily. There are some important consequences of the design of the *Hashtable* class:

- A Hashtable cannot contain duplicate keys. If you call the Add method to add a key that is already present in the keys array, you'll get an exception. You can, however, use the square brackets notation to add a key/value pair (as shown in the following example), without danger of an exception, even if the key has already been added. You can test whether a Hashtable already contains a particular key by using the ContainsKey method.

- Internally, a Hashtable is a sparse data structure that operates best when it has plenty of memory to work in. The size of a Hashtable in memory can grow quite quickly as you insert more elements.

- When you use a foreach statement to iterate through a Hashtable, you get back a DictionaryEntry. The DictionaryEntry class provides access to the key and value elements in both arrays through the Key property and the Value properties.

Here is an example that associates the ages of members of my family with their names and then prints the information:

```
using System;
using System.Collections;
...
Hashtable ages = new Hashtable();
...
// fill the Hashtable
ages["John"] = 42;
ages["Diana"] = 43;
ages["James"] = 15;
ages["Francesca"] = 13;
...
// iterate using a foreach statement
// the iterator generates a DictionaryEntry object containing a key/value pair
foreach (DictionaryEntry element in ages)
{
    string name = (string)element.Key;
    int age = (int)element.Value;
    Console.WriteLine("Name: {0}, Age: {1}", name, age);
}
```

The output from this program is:

```
Name: James, Age: 15
Name: John, Age: 42
Name: Francesca, Age: 13
Name: Diana, Age: 43
```

The *SortedList* Collection Class

The *SortedList* class is very similar to the *Hashtable* class in that it enables you to associate keys with values. The main difference is that the keys array is always sorted. (It is called a *SortedList*, after all.)

When you insert a key/value pair into a *SortedList*, the key is inserted into the keys array at the correct index to keep the keys array sorted. The value is then inserted into the values array at the same index. The *SortedList* class automatically ensures that keys and values are kept synchronized, even when you add and remove elements. This means that you can insert key/value pairs into a *SortedList* in any sequence; they are always sorted based on the value of the keys.

Like the *Hashtable* class, a *SortedList* cannot contain duplicate keys. When you use a *foreach* statement to iterate through a *SortedList*, you get back a *DictionaryEntry*. However, the *DictionaryEntry* objects will be returned sorted by the *Key* property.

Here is the same example that associates the ages of members of my family with their names and then prints the information, but this version has been adjusted to use a *SortedList* rather than a *Hashtable*:

```
using System;
using System.Collections;
...
SortedList ages = new SortedList();
...
// fill the SortedList
ages["John"] = 42;
ages["Diana"] = 43;
ages["James"] = 15;
ages["Francesca"] = 13;
...
// iterate using a foreach statement
// the iterator generates a DictionaryEntry object containing a key/value pair
foreach (DictionaryEntry element in ages)
{
    string name = (string)element.Key;
    int age = (int)element.Value;
    Console.WriteLine("Name: {0}, Age: {1}", name, age);
}
```

The output from this program is sorted alphabetically by the names of my family members:

```
Name: Diana, Age: 43
Name: Francesca, Age: 13
Name: James, Age: 15
Name: John, Age: 42
```

Using Collection Initializers

The examples in the preceding subsections have shown you how to add individual elements to a collection by using the method most appropriate to that collection (*Add* for an *ArrayList*, *Enqueue* for a *Queue*, *Push* for a *Stack*, and so on). You can also initialize *some* collection types when you declare them, using a syntax very similar to that supported by arrays. For example, the following statement creates and initializes the *numbers ArrayList* object shown earlier, demonstrating an alternative technique to repeatedly calling the *Add* method:

```
ArrayList numbers = new ArrayList(){10, 9, 8, 7, 7, 6, 5, 10, 4, 3, 2, 1};
```

Internally, the C# compiler actually converts this initialization to a series of calls to the *Add* method. Consequently, you can use this syntax only for collections that actually support the *Add* method. (The *Stack* and *Queue* classes do not.)

For more complex collections such as *Hashtable* that take key/value pairs, you can specify each key/value pair as an anonymous type in the initializer list, like this:

```
Hashtable ages = new Hashtable(){{"John", 42}, {"Diana", 43}, {"James", 15}, {"Francesca", 13}};
```

The first item in each pair is the key, and the second is the value.

Comparing Arrays and Collections

Here's a summary of the important differences between arrays and collections:

- An array declares the type of the elements that it holds, whereas a collection doesn't. This is because the collections store their elements as objects.

- An array instance has a fixed size and cannot grow or shrink. A collection can dynamically resize itself as required.

- An array can have more than one dimension. A collection is linear.

Note The items in a collection can be other collections, enabling you to mimic a multidimensional array, although a collection containing other collections can be somewhat confusing to use.

Using Collection Classes to Play Cards

The next exercise presents a Microsoft Windows Presentation Foundation (WPF) application that simulates dealing a pack of cards to four players. Cards will either be in the pack or be in one of four hands dealt to the players. The pack and hands of cards are implemented as *ArrayList* objects. You might think that these should be implemented as an array—after all, there are always 52 cards in a pack and 13 cards in a hand. This is true, but it overlooks the fact that when you deal the cards to players' hands, the cards are no longer in the pack. If you use an array to implement a pack, you'll have to record how many slots in the array actually hold a *PlayingCard* and how many have been dealt to players. Similarly, when you return cards from a player's hand to the pack, you'll have to record which slots in the hand no longer contain a *PlayingCard*.

You will study the code and then write two methods: one to shuffle a pack of cards and one to return the cards in a hand to the pack.

Deal the cards

1. Start Microsoft Visual Studio 2008 if it is not already running.

2. Open the *Cards* project, located in the \Microsoft Press\Visual CSharp Step by Step\ Chapter 10\Cards folder in your Documents folder.

3. On the *Debug* menu, click *Start Without Debugging*.

   Visual Studio 2008 builds and runs the program. The form displays the cards in the hands of the four players (North, South, West, and East). There are also two buttons: one to deal the cards and one to return the cards to the pack.

4. On the form, click *Deal*.

   The 52 cards in the pack are dealt to the four hands, 13 cards per hand, as shown here:

As you can see, the cards have not yet been shuffled. You will implement the *Shuffle* method in the next exercise.

5. Click *Return to Pack*.

Nothing happens because the method to return the cards to the pack has also not yet been written.

6. Click *Deal* again.

This time the cards in each of the hands disappear, because before the cards are dealt, each hand is reset. Because there are no cards left in the pack (the method to return cards to the pack has not been written yet either), there is nothing to deal.

7. Close the form to return to the Visual Studio 2008 programming environment.

Now that you know which parts are missing from this application, you will add them.

Shuffle the pack

1. Display the Pack.cs file in the *Code and Text Editor* window.

2. Scroll through the code, and examine it.

The *Pack* class represents a pack of cards. It contains a private *ArrayList* field named *cards*. Notice also that the *Pack* class has a constructor that creates and adds the 52 playing cards to the *ArrayList* by using the *Accept* method defined by this class. The methods in this class constitute the typical operations that you would perform on a pack of cards (*Shuffle*, *Deal*).

3. Display the PlayingCard.cs file in the *Code and Text Editor* window, and examine its contents.

Playing cards are represented by the *PlayingCard* class. A playing card exposes two fields of note: *suit* (which is an enumerated type and is one of *Clubs*, *Diamonds*, *Hearts*, or *Spades*) and *pips* (which indicates the numeric value of the card).

4. Return to the Pack.cs file and locate the *Shuffle* method in the *Pack* class.

The method is not currently implemented. There are a number of ways you can simulate shuffling a pack of cards. Perhaps the simplest technique is to choose each card in sequence and swap it with another card selected at random. The .NET Framework contains a class named *Random* that you can use to generate random integer numbers.

5. Declare a local variable of type *Random* named *random*, and initialize it to a newly created *Random* object by using the default *Random* constructor, as shown here in bold. The *Shuffle* method should look like this:

```
public void Shuffle()
{
    Random random = new Random();
}
```

6. Add a *for* statement with an empty body that iterates an *int i* from 0 up to the number of elements inside the cards *ArrayList*, as shown here in bold:

```
public void Shuffle()
{
    Random random = new Random();
    for (int i = 0; i < cards.Count; i++)
    {
    }
}
```

The next step is to choose a random index between 0 and *cards.Count – 1*. You will then swap the card at index *i* with the card at this random index. You can generate a positive random integer by calling the *Random.Next* instance method. You can specify an upper limit for the random number generated by *Random.Next* as a parameter.

Notice that you have to use a *for* statement here. A *foreach* statement would not work because you need to modify each element in the *ArrayList* and a *foreach* loop limits you to read-only access.

7. Inside the *for* statement, declare a local variable named *cardToSwap*, and initialize it to a random number between 0 and *cards.Count – 1* (inclusive), as shown here in bold:

```
public void Shuffle()
{
    Random random = new Random();
    for (int i = 0; i < cards.Count; i++)
    {
        int cardToSwap = random.Next(cards.Count - 1);
    }
}
```

The final step is to swap the card at index *i* with the card at index *cardToSwap*. To do this, you must use a temporary local variable.

8. Add three statements to swap the card at index *i* with the card at index *cardToSwap*. Remember that the elements inside a collection class (such as *ArrayList*) are of type *object*. Also, notice that you can use regular array notation (square brackets and an index) to access existing elements in an *ArrayList*.

The *Shuffle* method should now look exactly like this (the new statements are shown in bold):

```
public void Shuffle()
{
    Random random = new Random();
    for (int i = 0; i < cards.Count; i++)
    {
        int cardToSwap = random.Next(cards.Count - 1);
        object temp = cards[i];
        cards[i] = cards[cardToSwap];
        cards[cardToSwap] = temp;
    }
}
```

9. On the *Debug* menu, click *Start Without Debugging*.

10. On the form, click *Deal*.

This time the pack is shuffled before dealing, as shown here. (Your screen will differ slightly each time, because the card order is now random.)

11. Close the form.

The final step is to add the code to return the cards to the pack so that they can be dealt again.

Return the cards to the pack

1. Display the Hand.cs file in the *Code and Text Editor* window.

The *Hand* class, which also contains an *ArrayList* named *cards*, represents the cards held by a player. The idea is that at any one time, each card is either in the pack or in a hand.

2. Locate the *ReturnCardsTo* method in the *Hand* class.

The *Pack* class has a method named *Accept* that takes a single parameter of type *PlayingCard*. You need to create a loop that goes through the cards in the hand and passes them back to the pack.

3. Complete the *ReturnCardsTo* method as shown here in bold:

```
public void ReturnCardsTo(Pack pack)
{
    foreach (PlayingCard card in cards)
    {
        pack.Accept(card);
    }
    cards.Clear();
}
```

A *foreach* statement is convenient here because you do not need write access to the element and you do not need to know the index of the element. The *Clear* method removes all elements from a collection. It is important to call *cards.Clear* after returning the cards to the pack so that the cards aren't in both the pack and the hand. The *Clear* method of the *ArrayList* class empties the *ArrayList* of its contents.

4. On the *Debug* menu, click *Start Without Debugging*.

5. On the form, click *Deal*.

   The shuffled cards are dealt to the four hands as before.

6. Click *Return to Pack*.

   The hands are cleared. The cards are now back in the pack.

7. Click *Deal* again.

   The shuffled cards are once again dealt to the four hands.

8. Close the form.

> **Note** If you click the *Deal* button twice without clicking *Return to Pack*, you lose all the cards. In the real world, you would disable the *Deal* button until the *Return to Pack* button was clicked. In Part IV, "Working with Windows Applications," we will look at using C# to write code that modifies the user interface.

In this chapter, you have learned how to create and use arrays to manipulate sets of data. You have also seen how to use some of the common collection classes to store and access data in memory in different ways.

- If you want to continue to the next chapter:

  Keep Visual Studio 2008 running, and turn to Chapter 11.

- If you want to exit Visual Studio 2008 now:

  On the *File* menu, click *Exit*. If you see a *Save* dialog box, click *Yes* (if you are using Visual Studio 2008) or *Save* (if you are using Visual C# 2008 Express Edition) and save the project.

Chapter 10 Quick Reference

To	Do this
Declare an array variable	Write the name of the element type, followed by square brackets, followed by the name of the variable, followed by a semicolon. For example: `bool[] flags;`
Create an instance of an array	Write the keyword *new*, followed by the name of the element type, followed by the size of the array enclosed in square brackets. For example: `bool[] flags = new bool[10];`

Initialize the elements of an array (or of a collection that supports the *Add* method) to specific values	For an array, write the specific values in a comma-separated list enclosed in braces. For example:
	
	`bool[] flags = { true, false, true, false };`
	
	For a collection, use the *new* operator and the collection type with the specific values in a comma-separated list enclosed in braces. For example:
	
	`ArrayList numbers = new ArrayList(){10, 9, 8, 7, 6, 5};`

Find the number of elements in an array	Use the *Length* property. For example:
	
	`int [] flags = ...;`
	`...`
	`int noOfElements = flags.Length;`

Find the number of elements in a collection	Use the *Count* property. For example:
	
	`ArrayList flags = new ArrayList();`
	`...`
	`int noOfElements = flags.Count;`

Access a single array element	Write the name of the array variable, followed by the integer index of the element enclosed in square brackets. Remember, array indexing starts at 0, not 1. For example: `bool initialElement = flags[0];`

Iterate through the elements of an array or a collection	Use a *for* statement or a *foreach* statement. For example:
	
	`bool[] flags = { true, false, true, false };`
	`for (int i = 0; i < flags.Length; i++)`
	`{`
	`    Console.WriteLine(flags[i]);`
	`}`
	
	`foreach (bool flag in flags)`
	`{`
	`    Console.WriteLine(flag);`
	`}`

Chapter 11
Understanding Parameter Arrays

After completing this chapter, you will be able to:

- Write a method that can accept any number of arguments by using the *params* keyword.

- Write a method that can accept any number of arguments of any type by using the *params* keyword in combination with the *object* type.

Parameter arrays are useful if you want to write methods that can take any number of arguments, possibly of different types, as parameters. If you are familiar with object-oriented concepts, you might well be grinding your teeth in frustration at this sentence. After all, the object-oriented approach to solving this problem is to define overloaded methods.

Overloading is the technical term for declaring two or more methods with the same name in the same scope. Being able to overload a method is very useful in cases where you want to perform the same action on arguments of different types. The classic example of overloading in Microsoft Visual C# is *Console.WriteLine*. The *WriteLine* method is overloaded numerous times so that you can pass any primitive type argument:

```
class Console
{
    public static void WriteLine(int parameter)
    ...
    public static void WriteLine(double parameter)
    ...
    public static void WriteLine(decimal parameter)
    ...
}
```

As useful as overloading is, it doesn't cover every case. In particular, overloading doesn't easily handle a situation in which the type of parameters doesn't vary but the number of parameters does. For example, what if you want to write many values to the console? Do you have to provide versions of *Console.WriteLine* that can take two parameters, other versions that can take three parameters, and so on? That would quickly get tedious. And doesn't the massive duplication of all these overloaded methods worry you? It should. Fortunately, there is a way to write a method that takes a variable number of arguments (a *variadic method*): you can use a parameter array (a parameter declared with the *params* keyword).

To understand how *params* arrays solve this problem, it helps to first understand the uses and shortcomings of plain arrays.

Using Array Arguments

Suppose you want to write a method to determine the minimum value in a set of values passed as parameters. One way would be to use an array. For example, to find the smallest of several *int* values, you could write a static method named *Min* with a single parameter representing an array of *int* values:

```
class Util
{
    public static int Min(int[] paramList)
    {
        if (paramList == null || paramList.Length == 0)
        {
            throw new ArgumentException("Util.Min: not enough arguments");
        }
        int currentMin = paramList [0];
        foreach (int i in paramList)
        {
            if (i < currentMin)
            {
                currentMin = i;
            }
        }
        return currentMin;
    }
}
```

> **Note** The *ArgumentException* class is specifically designed to be thrown by a method if the arguments supplied do not meet the requirements of the method.

To use the *Min* method to find the minimum of two *int* values, you would write this:

```
int[] array = new int[2];
array[0] = first;
array[1] = second;
int min = Util.Min(array);
```

And to use the *Min* method to find the minimum of three *int* values, you would write this:

```
int[] array = new int[3];
array[0] = first;
array[1] = second;
array[2] = third;
int min = Util.Min(array);
```

You can see that this solution avoids the need for a large number of overloads, but it does so at a price: you have to write additional code to populate the array that you pass in. However, you can get the compiler to write some of this code for you by using the *params* keyword to declare a *params* array.

Declaring a *params* Array

You use the *params* keyword as an array parameter modifier. For example, here's *Min* again, this time with its array parameter declared as a *params* array:

```
class Util
{
    public static int Min(params int[] paramList)
    {
        // code exactly as before
    }
}
```

The effect of the *params* keyword on the *Min* method is that it allows you to call it by using any number of integer arguments. For example, to find the minimum of two integer values, you would write this:

```
int min = Util.Min(first, second);
```

The compiler translates this call into code similar to this:

```
int[] array = new int[2];
array[0] = first;
array[1] = second;
int min = Util.Min(array);
```

To find the minimum of three integer values, you would write the code shown here, which is also converted by the compiler to the corresponding code that uses an array:

```
int min = Util.Min(first, second, third);
```

Both calls to *Min* (one call with two arguments and another with three arguments) resolve to the same *Min* method with the *params* keyword. And as you can probably guess, you can call this *Min* method with any number of *int* arguments. The compiler just counts the number of *int* arguments, creates an *int* array of that size, fills the array with the arguments, and then calls the method by passing the single array parameter.

> **Note** C and C++ programmers might recognize *params* as a type-safe equivalent of the *varargs* macros from the header file stdarg.h.

There are several points worth noting about *params* arrays:

- You can't use the *params* keyword on multidimensional arrays. The code in the following example will not compile:

  ```
  // compile-time error
  public static int Min(params int[,] table)
  ...
  ```

- You can't overload a method based solely on the *params* keyword. The *params* keyword does not form part of a method's signature, as shown in this example:

```
// compile-time error: duplicate declaration
public static int Min(int[] paramList)
...
public static int Min(params int[] paramList)
...
```

- You're not allowed to specify the *ref* or *out* modifier with *params* arrays, as shown in this example:

```
// compile-time errors
public static int Min(ref params int[] paramList)
...
public static int Min(out params int[] paramList)
...
```

- A *params* array must be the last parameter. (This means that you can have only one *params* array per method.) Consider this example:

```
// compile-time error
public static int Min(params int[] paramList, int i)
...
```

- A non-*params* method always takes priority over a *params* method. This means that if you want to, you can still create an overloaded version of a method for the common cases. For example:

```
public static int Min(int leftHandSide, int rightHandSide)
...
public static int Min(params int[] paramList)
...
```

The first version of the *Min* method is used when called using two *int* arguments. The second version is used if any other number of *int* arguments is supplied. This includes the case where the method is called with no arguments.

Adding the non-*params* array method might be a useful optimization technique because the compiler won't have to create and populate so many arrays.

- The compiler detects and rejects any potentially ambiguous overloads. For example, the following two *Min* methods are ambiguous; it's not clear which one should be called if you pass two *int* arguments:

```
// compile-time error
public static int Min(params int[] paramList)
...
public static int Min(int, params int[] paramList)
...
```

Using *params object[ ]*

A parameter array of type *int* is very useful because it enables you to pass any number of *int* arguments in a method call. However, what if not only the number of arguments varies but also the argument type? C# has a way to solve this problem, too. The technique is based on the facts that *object* is the root of all classes and that the compiler can generate code that converts value types (things that aren't classes) to objects by using boxing, as described in Chapter 8, "Understanding Values and References." You can use a parameters array of type *object* to declare a method that accepts any number of *object* arguments, allowing the arguments passed in to be of any type. Look at this example:

```
class Black
{
    public static void Hole(params object [] paramList)
    ...
}
```

I've called this method *Black.Hole*, not because it swallows every argument, but because no argument can escape from it:

- You can pass the method no arguments at all, in which case the compiler will pass an object array whose length is 0:

  ```
  Black.Hole();
  // converted to Black.Hole(new object[0]);
  ```

  > **Tip** It's perfectly safe to attempt to iterate through a zero-length array by using a *foreach* statement.

- You can call the *Black.Hole* method by passing *null* as the argument. An array is a reference type, so you're allowed to initialize an array with *null*:

  ```
  Black.Hole(null);
  ```

- You can pass the *Black.Hole* method an actual array. In other words, you can manually create the array normally created by the compiler:

  ```
  object[] array = new object[2];
  array[0] = "forty two";
  array[1] = 42;
  Black.Hole(array);
  ```

- You can pass the *Black.Hole* method any other arguments of different types, and these arguments will automatically be wrapped inside an *object* array:

  ```
  Black.Hole("forty two", 42);
  //converted to Black.Hole(new object[]{"forty two", 42});
  ```

The *Console.WriteLine* Method

The *Console* class contains many overloads for the *WriteLine* method. One of these overloads looks like this:

```
public static void WriteLine(string format, params object[] arg);
```

This overload enables the *WriteLine* method to support a format string argument that contains placeholders, each of which can be replaced at run time with a variable of any type. Here's an example of a call to this method that you have seen several times in earlier chapters:

```
Console.WriteLine("Name:{0}, Age:{1}", name, age);
```

The compiler resolves this call into the following:

```
Console.WriteLine("Name:{0}, Age:{1}", new object[2]{name, age});
```

Using a *params* Array

In the following exercise, you will implement and test a *static* method named *Util.Sum*. The purpose of this method is to calculate the sum of a variable number of *int* arguments passed to it, returning the result as an *int*. You will do this by writing *Util.Sum* to take a *params int[]* parameter. You will implement two checks on the *params* parameter to ensure that the *Util. Sum* method is completely robust. You will then call the *Util.Sum* method with a variety of different arguments to test it.

Write a *params* array method

1. Start Microsoft Visual Studio 2008 if it is not already running.

2. Open the *ParamsArray* project, located in the \Microsoft Press\Visual CSharp Step by Step\Chapter 11\ ParamArrays folder in your Documents folder.

3. Display the Util.cs file in the *Code and Text Editor* window.

   The Util.cs file contains an empty class named *Util* in the *ParamsArray* namespace.

4. Add a public static method named *Sum* to the *Util* class.

   The *Sum* method returns an *int* and accepts a *params* array of *int* values. The *Sum* method should look like this:

```
public static int Sum(params int[] paramList)
{
}
```

   The first step in implementing the *Sum* method is to check the *paramList* parameter. Apart from containing a valid set of integers, it could also be *null* or it could be an array of zero length. In both of these cases, it is difficult to calculate the sum, so the best option is to throw an *ArgumentException*. (You could argue that the sum of the integers in a zero-length array is 0, but we will treat this situation as an exception in this example.)

5. Add code to *Sum* that throws an *ArgumentException* if *paramList* is *null*.

The *Sum* method should now look like this:

```
public static int Sum(params int[] paramList)
{
    if (paramList == null)
    {
        throw new ArgumentException("Util.Sum: null parameter list");
    }
}
```

6. Add code to the *Sum* method that throws an *ArgumentException* if the length of *array* is 0, as shown in bold here:

```
public static int Sum(params int[] paramList)
{
    if (paramList == null)
    {
        throw new ArgumentException("Util.Sum: null parameter list");
    }

    if (paramList.Length == 0)
    {
        throw new ArgumentException("Util.Sum: empty parameter list");
    }
}
```

If the array passes these two tests, the next step is to add all the elements inside the array together.

7. You can use a *foreach* statement to add all the elements together. You will need a local variable to hold the running total. Declare an integer variable named *sumTotal* and initialize it to 0 following the code from the preceding step. Add a *foreach* statement to the *Sum* method to iterate through the *paramList* array. The body of this *foreach* loop should add each element in the array to *sumTotal*. At the end of the method, return the value of *sumTotal* by using a *return* statement.

```
class Util
{
    public static int Sum(params int[] paramList)
    {
        ...
        int sumTotal = 0;
        foreach (int i in paramList)
        {
            sumTotal += i;
        }
        return sumTotal;
    }
}
```

8. On the *Build* menu, click *Build Solution*. Confirm that your solution builds without any errors.

Test the *Util.Sum* method

1. Display the Program.cs file in the *Code and Text Editor* window.

2. In the *Code and Text Editor* window, locate the *Entrance* method in the *Program* class.

3. Add the following statement to the *Entrance* method:

```
Console.WriteLine(Util.Sum(null));
```

4. On the *Debug* menu, click *Start Without Debugging*.

   The program builds and runs, writing the following message to the console:

```
Exception: Util.Min: null parameter list
```

   This confirms that the first check in the method works.

5. Press the Enter key to close the program and return to Visual Studio 2008.

6. In the *Code and Text Editor* window, change the call to *Console.WriteLine* in *Entrance* as shown here:

```
Console.WriteLine(Util.Sum());
```

   This time, the method is being called without any arguments. The compiler will translate the empty argument list into an empty array.

7. On the *Debug* menu, click *Start Without Debugging*.

   The program builds and runs, writing the following message to the console:

```
Exception: Util.Min: empty parameter list
```

   This confirms that the second check in the method works.

8. Press the Enter key to close the program and return to Visual Studio 2008.

9. Change the call to *Console.WriteLine* in *Entrance* as follows:

```
Console.WriteLine(Util.Sum(10, 9, 8, 7, 6, 5, 4, 3, 2, 1));
```

10. On the *Debug* menu, click *Start Without Debugging*.

    The program builds, runs, and writes *55* to the console.

11. Press Enter to close the application.

In this chapter, you have learned how to use a *params* array to define a method that can take any number of arguments. You have also seen how to use a *params* array of *object* types to create a method that accepts any number of arguments of any type.

- If you want to continue to the next chapter:

  Keep Visual Studio 2008 running, and turn to Chapter 12.

- If you want to exit Visual Studio 2008 now:

  On the *File* menu, click *Exit*. If you see a *Save* dialog box, click *Yes* (if you are using Visual Studio 2008) or *Save* (if you are using Visual C# 2008 Express Edition) and save the project.

Chapter 11 Quick Reference

To	Do this

Write a method that accepts any number of arguments of a given type	Write a method whose parameter is a *params* array of the given type. For example, a method that accepts any number of *bool* arguments would be declared like this:
	
	`someType Method(params bool[] flags)`
	`{`
	`    ...`
	`}`

Write a method that accepts any number of arguments of any type	Write a method whose parameter is a *params* array whose elements are of type *object*. For example:
	
	`someType Method(params object[] paramList)`
	`{`
	`    ...`
	`}`

Chapter 12
Working with Inheritance

After completing this chapter, you will be able to:

- Create a derived class that inherits features from a base class.

- Control method hiding and overriding by using the new, virtual, and override keywords.

- Limit accessibility within an inheritance hierarchy by using the protected keyword.

- Define extension methods as an alternative mechanism to using inheritance.

Inheritance is a key concept in the world of object orientation. You can use inheritance as a tool to avoid repetition when defining different classes that have a number of features in common and are quite clearly related to each other. Perhaps they are different classes of the same type, each with its own distinguishing feature—for example, *managers*, *manual workers*, and *all employees* of a factory. If you were writing an application to simulate the factory, how would you specify that managers and manual workers have a number of features that are the same but also have other features that are different? For example, they all have an employee reference number, but managers have different responsibilities and perform different tasks than manual workers.

This is where inheritance proves useful.

What Is Inheritance?

If you ask several experienced programmers what they understand by the term *inheritance*, you will typically get different and conflicting answers. Part of the confusion stems from the fact that the word *inheritance* itself has several subtly different meanings. If someone bequeaths something to you in a will, you are said to inherit it. Similarly, we say that you inherit half of your genes from your mother and half of your genes from your father. Both of these uses of the word *inheritance* have very little to do with inheritance in programming.

Inheritance in programming is all about classification—it's a relationship between classes. For example, when you were at school, you probably learned about mammals, and you learned that horses and whales are examples of mammals. Each has every attribute that a mammal does (it breathes air, it suckles its young, it is warm-blooded, and so on), but each also has its own special features (a horse has hooves, unlike a whale).

How could you model a horse and a whale in a program? One way would be to create two distinct classes named *Horse* and *Whale*. Each class could implement the methods that are

217

unique to that type of mammal, such as *Trot* (for a horse) or *Swim* (for a whale) in its own way. How would you handle methods that are common to a horse and a whale, such as *Breathe* or *SuckleYoung*? You could add duplicate methods with these names to both classes, but this situation becomes a maintenance nightmare, especially if you also decide to start modeling other types of mammal, such as *Human* or *Aardvark*.

In C#, you can use class inheritance to address these issues. A horse, a whale, a human, and an aardvark are all types of mammal, so create a class named *Mammal* that provides the common functionality exhibited by these types. You can then declare that the *Horse*, *Whale*, *Human*, and *Aardvark* classes all inherit from *Mammal*. These classes would then automatically provide the functionality of the *Mammal* class (*Breathe*, *SuckleYoung*, and so on), but you could also add the functionality peculiar to a particular type of mammal to the corresponding class—the *Trot* method for the *Horse* class and the *Swim* method for the *Whale* class. If you need to modify the way in which a common method such as *Breathe* works, you need to change it in only one place, the *Mammal* class.

Using Inheritance

This section covers the essential inheritance-related syntax that you need to understand in order to create classes that inherit from other classes in C#.

Base Classes and Derived Classes

The syntax for declaring a class that inherits from another class is as follows:

```
class DerivedClass : BaseClass {
    ...
}
```

The derived class inherits from the base class, and the methods in the base class also become part of the derived class. In C#, a class is allowed to derive from, at most, one other class; a class is *not allowed* to derive from two or more classes. However, unless *DerivedClass* is declared as *sealed,* you can create further derived classes that inherit from *DerivedClass* using the same syntax. (You will learn about sealed classes in Chapter 13, "Creating Interfaces and Defining Abstract Classes.")

```
class DerivedSubClass : DerivedClass {
    ...
}
```

Important You cannot use inheritance with structures. A structure cannot inherit from a class or another structure.

In the example described earlier, you could declare the *Mammal* class as shown here. The methods *Breathe* and *SuckleYoung* are common to all mammals.

```
class Mammal
{
    public void Breathe()
    {
        ...
    }

    public void SuckleYoung()
    {
        ...
    }
    ...
}
```

You could then define classes for each different type of mammal, adding additional methods as necessary. For example:

```
class Horse : Mammal
{
    ...
    public void Trot()
    {
        ...
    }

}

class Horse : Whale
{
    ...
    public void Swim()
    {
        ...
    }

}
```

 Note C++ programmers should notice that you do not and cannot explicitly specify whether the inheritance is public, private, or protected. C# inheritance is always implicitly public. Java programmers should note the use of the colon and that there is no *extends* keyword.

Remember that the *System.Object* class is the root class of all classes. All classes implicitly derive from the *System.Object* class. Consequently, the C# compiler silently rewrites the *Mammal* class as the following code (which you can write explicitly if you really want to):

```
class Mammal : System.Object
{
    ...
}
```

Any methods in the *System.Object* class are automatically passed down the chain of inheritance to classes that derive from *Mammal*, such as *Horse* and *Whale*. What this means in practical terms is that all classes that you define automatically inherit all the features of the *System.Object* class. This includes methods such as *ToString* (first discussed in Chapter 2, "Working with Variables, Operators, and Expressions"), which is used to convert an object to a string, typically for display purposes.

Calling Base Class Constructors

A derived class automatically contains all fields from the base class. These fields usually require initialization when an object is created. You usually perform this kind of initialization in a constructor. Remember that all classes have at least one constructor. (If you don't provide one, the compiler generates a default constructor for you.) It is good practice for a constructor in a derived class to call the constructor for its base class as part of the initialization. You can specify the *base* keyword to call a base class constructor when you define a constructor for an inheriting class, as shown in this example:

```
class Mammal // base class
{
    public Mammal(string name)  // constructor for base class
    {
        ...
    }
    ...
}

class Horse : Mammal // derived class
{
    public Horse(string name)
            : base(name) // calls Mammal(name)
    {
        ...
    }
    ...
}
```

If you don't explicitly call a base class constructor in a derived class constructor, the compiler attempts to silently insert a call to the base class's default constructor before executing the code in the derived class constructor. Taking the earlier example, the compiler will rewrite this:

```
class Horse : Mammal
{
    public Horse(string name)
    {
        ...
    }
    ...
}
```

As this:

```
class Horse : Mammal
{
    public Horse(string name)
        : base()
    {
        ...
    }
    ...
}
```

This works if *Mammal* has a public default constructor. However, not all classes have a public default constructor (*Mammal* doesn't!), in which case forgetting to call the correct base class constructor results in a compile-time error.

Assigning Classes

In previous examples in this book, you have seen how to declare a variable by using a class type and then how to use the *new* keyword to create an object. You have also seen how the type-checking rules of C# prevent you from assigning an object of one type to a variable declared as a different type. For example, given the definitions of the *Mammal*, *Horse*, and *Whale* classes shown here, the code that follows these definitions is illegal:

```
class Mammal
{
    ...
}
class Horse : Mammal
{
    ...
}

class Whale : Mammal
{
    ...
}
...
Horse myHorse = new Horse("Neddy");   // constructor shown earlier expects a name!
Whale myWhale = myHorse;              // error - different types
```

However, it is possible to refer to an object from a variable of a different type as long as the type used is a class that is higher up the inheritance hierarchy. So the following statements are legal:

```
Horse myHorse = new Horse("Neddy");
Mammal myMammal = myHorse; // legal, Mammal is the base class of Horse
```

If you think about it in logical terms, all *Horse*s are *Mammal*s, so you can safely assign an object of type *Horse* to a variable of type *Mammal*. The inheritance hierarchy means that you can think of a *Horse* simply as a special type of *Mammal* (it has everything that a *Mammal* has) with a few extra bits (defined by any methods and fields you add to the *Horse* class). You can also make a *Mammal* variable refer to a *Whale* object. There is one significant limitation, however—when referring to a *Horse* or *Whale* object using a *Mammal* variable, you can access only methods and fields that are defined by the *Mammal* class. Any additional methods defined by the *Horse* or *Whale* class are not visible through the *Mammal* class:

```
Horse myHorse = new Horse("Neddy");
Mammal myMammal = myHorse;
myMammal.Breathe();        // OK - Breathe is part of the Mammal class
myMammal.Trot();           // error - Trot is not part of the Mammal class
```

Note This explains why you can assign almost anything to an *object* variable. Remember that *object* is an alias for *System.Object* and all classes inherit from *System.Object* either directly or indirectly.

Be warned that the converse situation is not true. You cannot unreservedly assign a *Mammal* object to a *Horse* variable:

```
Mammal myMammal = myMammal("Mammalia");
Horse myHorse = myMammal;   // error
```

This looks like a strange restriction, but remember that not all *Mammal* objects are *Horses*—some are *Whales*. You can assign a *Mammal* object to a *Horse* variable as long as you check that the *Mammal* is really a *Horse* first, by using the *as* or *is* operator or by using a cast. The following code example uses the *as* operator to check that *myMammal* refers to a *Horse*, and if it does, the assignment to *myHorseAgain* results in *myHorseAgain* referring to the same *Horse* object. If *myMammal* refers to some other type of *Mammal*, the *as* operator returns *null* instead.

```
Horse myHorse = new Horse("Neddy");
Mammal myMammal = myHorse;                   // myMammal refers to a Horse
...
Horse myHorseAgain = myMammal as Horse;  // OK - myMammal was a Horse
...
Whale myWhale = new Whale("Moby Dick");
myMammal = myWhale;
...
myHorseAgain = myMammal as Horse;          // returns null - myMammal was a Whale
```

Declaring *new* Methods

One of the hardest problems in the realm of computer programming is the task of thinking up unique and meaningful names for identifiers. If you are defining a method for a class and that class is part of an inheritance hierarchy, sooner or later you are going to try to reuse a name that is already in use by one of the classes higher up the hierarchy. If a base class and a derived class happen to declare two methods that have the same signature (the method signature is the name of the method and the number and types of its parameters), you will receive a warning when you compile the application. The method in the derived class masks (or hides) the method in the base class that has the same signature. For example, if you compile the following code, the compiler will generate a warning message telling you that *Horse.Talk* hides the inherited method *Mammal.Talk*:

```
class Mammal
{
    ...
    public void Talk() // all mammals talk
    {
        ...
    }
}

class Horse : Mammal
{
    ...
    public void Talk()  // horses talk in a different way from other mammals!
    {
        ...
    }
}
```

Although your code will compile and run, you should take this warning seriously. If another class derives from *Horse* and calls the *Talk* method, it might be expecting the method implemented in the *Mammal* class to be called. However, the *Talk* method in the *Horse* class hides the *Talk* method in the *Mammal* class, and the *Horse.Talk* method will be called instead. Most of the time, such a coincidence is at best a source of confusion, and you should consider renaming methods to avoid clashes. However, if you're sure that you want the two methods to have the same signature, thus hiding the *Mammal.Talk* method, you can silence the warning by using the *new* keyword as follows:

```
class Mammal
{
    ...
    public void Talk()
    {
        ...
    }
}

class Horse : Mammal
{
    ...
    new public void Talk()
    {
        ...
    }
}
```

Using the *new* keyword like this does not change the fact that the two methods are completely unrelated and that hiding still occurs. It just turns the warning off. In effect, the *new* keyword says, "I know what I'm doing, so stop showing me these warnings."

Declaring Virtual Methods

Sometimes you do want to hide the way in which a method is implemented in a base class. As an example, consider the *ToString* method in *System.Object*. The purpose of *ToString* is to convert an object to its string representation. Because this method is very useful, it is a member of *System.Object*, thereby automatically providing all classes with a *ToString* method. However, how does the version of *ToString* implemented by *System.Object* know how to convert an instance of a derived class to a string? A derived class might contain any number of fields with interesting values that should be part of the string. The answer is that the implementation of *ToString* in *System.Object* is actually a bit simplistic. All it can do is convert an object to a string that contains the name of its type, such as "Mammal" or "Horse." This is not very useful after all. So why provide a method that is so useless? The answer to this second question requires a bit of detailed thought.

Obviously, *ToString* is a fine idea in concept, and all classes should provide a method that can be used to convert objects to strings for display or debugging purposes. It is only the implementation that is problematic. In fact, you are not expected to call the *ToString* method defined by *System.Object*—it is simply a placeholder. Instead, you should provide your own version of the *ToString* method in each class you define, overriding the default implementation in *System.Object*. The version in *System.Object* is there only as a safety net, in case a class does not implement its own *ToString* method. In this way, you can be confident that you can call *ToString* on any object, and the method will return a string containing something.

A method that is intended to be overridden is called a *virtual* method. You should be clear on the difference between *overriding* a method and *hiding* a method. Overriding a method is a mechanism for providing different implementations of the same method—the methods are all related because they are intended to perform the same task, but in a class-specific manner. Hiding a method is a means of replacing one method with another—the methods are usually unrelated and might perform totally different tasks. Overriding a method is a useful programming concept; hiding a method is usually an error.

You can mark a method as a virtual method by using the *virtual* keyword. For example, the *ToString* method in the *System.Object* class is defined like this:

```
namespace System
{
    class Object
    {
        public virtual string ToString()
        {
            ...
```

```
        }
        ...
    }
    ...
}
```

 Note Java developers should note that C# methods are not virtual by default.

Declaring *override* Methods

If a base class declares that a method is *virtual*, a derived class can use the *override* keyword to declare another implementation of that method. For example:

```
class Horse : Mammal
{
    ...
    public override string ToString()
    {
        ...
    }
}
```

The new implementation of the method in the derived class can call the original implementation of the method in the base class by using the *base* keyword, like this:

```
    public override string ToString()
    {
        base.ToString();
        ...
    }
```

There are some important rules you must follow when declaring polymorphic methods (see the following sidebar, "Virtual Methods and Polymorphism") by using the *virtual* and *override* keywords:

- You're not allowed to declare a private method when using the virtual or override keyword. If you try, you'll get a compile-time error. Private really is private.

- The two method signatures must be identical—that is, they must have the same name, the same number and type of parameters, and the same return type.

- The two methods must have the same access. For example, if one of the two methods is public, the other must also be public. (Methods can also be protected, as you will find out in the next section.)

- You can override only a virtual method. If the base class method is not virtual and you try to override it, you'll get a compile-time error. This is sensible; it should be up to the designer of the base class to decide whether its methods can be overridden.

- If the derived class does not declare the method by using the override keyword, it does not override the base class method. In other words, it becomes an implementation of a completely different method that happens to have the same name. As before, this will cause a compile-time hiding warning, which you can silence by using the new keyword as previously described.

- An override method is implicitly virtual and can itself be overridden in a further derived class. However, you are not allowed to explicitly declare that an override method is virtual by using the virtual keyword.

Virtual Methods and Polymorphism

Virtual methods enable you to call different versions of the same method, based on the type of the object determined dynamically at run time. Consider the following example classes that define a variation on the *Mammal* hierarchy described earlier:

```
class Mammal
{
    ...
    public virtual string GetTypeName()
    {
        return "This is a mammal";
    }
}

class Horse : Mammal
{
    ...
    public override string GetTypeName()
    {
        return "This is a horse";
    }
}

class Whale : Mammal
{
    ...
    public override string GetTypeName ()
    {
        return "This is a whale";
    }
}

class Aardvark : Mammal
{
    ...
}
```

Notice two things: first, the *override* keyword used by the *GetTypeName* method (which will be described shortly) in the *Horse* and *Whale* classes, and second, the fact that the *Aardvark* class does not have a *GetTypeName* method.

Now examine the following block of code:

```
Mammal myMammal;
Horse myHorse = new Horse(...);
Whale myWhale = new Whale(...);
Aardvark myAardvark = new Aardvark(...);

myMammal = myHorse;
Console.WriteLine(myMammal.GetTypeName()); // Horse
myMammal = myWhale;
Console.WriteLine(myMammal.GetTypeName()); // Whale
myMammal = myAardvark;
Console.WriteLine(myMammal.GetTypeName()); // Aardvark
```

What will be output by the three different *Console.WriteLine* statements? At first glance, you would expect them all to print "This is a mammal," because each statement calls the *GetTypeName* method on the *myMammal* variable, which is a *Mammal*. However, in the first case, you can see that *myMammal* is actually a reference to a *Horse*. (Remember, you are allowed to assign a *Horse* to a *Mammal* variable because the *Horse* class is derived from the *Mammal* class.) Because the *GetTypeName* method is defined as *virtual*, the runtime works out that it should call the *Horse.GetTypeName* method, so the statement actually prints the message "This is a horse." The same logic applies to the second *Console.WriteLine* statement, which outputs the message "This is a whale." The third statement calls *Console.WriteLine* on an *Aardvark* object. However, the *Aardvark* class does not have a *GetTypeName* method, so the default method in the *Mammal* class is called, returning the string "This is a mammal."

This phenomenon of the same statement invoking a different method is called *polymorphism*, which literally means "many forms."

Understanding *protected* Access

The *public* and *private* access keywords create two extremes of accessibility: public fields and methods of a class are accessible to everyone, whereas private fields and methods of a class are accessible to only the class itself.

These two extremes are sufficient when considering classes in isolation. However, as all experienced object-oriented programmers know, isolated classes cannot solve complex problems. Inheritance is a powerful way of connecting classes, and there is clearly a special and close relationship between a derived class and its base class. Frequently it is useful for a base class to allow derived classes to access some of its members while hiding these same members from classes that are not part of the hierarchy. In this situation, you can use the *protected* keyword to tag members:

- If a class A is derived from another class B, it can access the protected class members of class B. In other words, inside the derived class A, a protected member of class B is effectively public.

- If a class A is not derived from another class B, it cannot access any protected members of class B. In other words, within class A, a protected member of class B is effectively private.

C# gives programmers complete freedom to declare methods and fields as protected. However, most object-oriented programming guidelines recommend keeping your fields strictly private. Public fields violate encapsulation because all users of the class have direct, unrestricted access to the fields. Protected fields maintain encapsulation for users of a class, for whom the protected fields are inaccessible. However, protected fields still allow encapsulation to be violated by classes that inherit from the class.

Note You can access a protected base class member not only in a derived class but also in classes derived from the derived class. A protected base class member retains its protected accessibility in a derived class and is accessible to further derived classes.

In the following exercise, you will define a simple class hierarchy for modeling different types of vehicles. You will define a base class named *Vehicle* and derived classes named *Airplane* and *Car*. You will define common methods named *StartEngine* and *StopEngine* in the *Vehicle* class, and you will add some additional methods to both of the derived classes that are specific to those classes. Last you will add a virtual method named *Drive* to the *Vehicle* class and override the default implementation of this method in both of the derived classes.

Create a hierarchy of classes

1. Start Microsoft Visual Studio 2008 if it is not already running.

2. Open the *Vehicles* project, located in the \Microsoft Press\Visual CSharp Step by Step\ Chapter 12\Vehicles folder in your Documents folder.

   The *Vehicles* project contains the file Program.cs, which defines the *Program* class with the *Main* and *Entrance* methods that you have seen in previous exercises.

3. In *Solution Explorer*, right-click the *Vehicles* project, point to *Add*, and then click *Class*.

   The *Add New Item—Vehicles* dialog box appears, enabling you to add a new file defining a class to the project.

4. In the *Add New Item—Vehicles* dialog box, in the *Name* box, type **Vehicle.cs**, and then click *Add*.

The file Vehicle.cs is created and added to the project and appears in the *Code and Text Editor* window. The file contains the definition of an empty class named *Vehicle*.

5. Add the *StartEngine* and *StopEngine* methods to the *Vehicle* class as shown here in bold:

```
class Vehicle
{
    public void StartEngine(string noiseToMakeWhenStarting)
    {
        Console.WriteLine("Starting engine: {0}", noiseToMakeWhenStarting);
    }

    public void StopEngine(string noiseToMakeWhenStopping)
    {
        Console.WriteLine("Stopping engine: {0}", noiseToMakeWhenStopping);
    }
}
```

All classes that derive from the *Vehicle* class will inherit these methods. The values for the *noiseToMakeWhenStarting* and *noiseToMakeWhenStopping* parameters will be different for each different type of vehicle and will help you to identify which vehicle is being started and stopped later.

6. On the *Project* menu, click *Add Class*.

The *Add New Item—Vehicles* dialog box appears again.

7. In the *Name* box, type **Airplane.cs**, and then click *Add*.

A new file containing a class named *Airplane* is added to the project and appears in the *Code and Text Editor* window.

8. In the *Code and Text Editor* window, modify the definition of the *Airplane* class so that it derives from the *Vehicle* class, as shown in bold here:

```
class Airplane : Vehicle
{
}
```

9. Add the *TakeOff* and *Land* methods to the *Airplane* class, as shown in bold here:

```
class Airplane : Vehicle
{
    public void TakeOff()
    {
        Console.WriteLine("Taking off");
    }

    public void Land()
    {
        Console.WriteLine("Landing");
    }
}
```

10. On the *Project* menu, click *Add Class*.

    The *Add New Item—Vehicles* dialog box appears again.

11. In the *Name* box, type **Car.cs**, and then click *Add*.

    A new file containing a class named *Car* is added to the project and appears in the *Code and Text Editor* window.

12. In the *Code and Text Editor* window, modify the definition of the *Car* class so that it derives from the *Vehicle* class, as shown here in bold:

    ```
    class Car : Vehicle
    {
    }
    ```

13. Add the *Accelerate* and *Brake* methods to the *Car* class, as shown in bold here:

    ```
    class Car : Vehicle
    {
        public void Accelerate()
        {
            Console.WriteLine("Accelerating");
        }

        public void Brake()
        {
            Console.WriteLine("Braking");
        }
    }
    ```

14. Display the Vehicle.cs file in the *Code and Text Editor* window.

15. Add the virtual *Drive* method to the *Vehicle* class, as shown here in bold:

    ```
    class Vehicle
    {
        ...
        public virtual void Drive()
        {
            Console.WriteLine("Default implementation of the Drive method");
        }
    }
    ```

16. Display the Program.cs file in the *Code and Text Editor* window.

17. In the *Entrance* method, create an instance of the *Airplane* class and exercise its methods by simulating a quick journey by airplane, as follows:

    ```
    static void Entrance()
    {
        Console.WriteLine("Journey by airplane:");
        Airplane myPlane = new Airplane();
        myPlane.StartEngine("Contact");
    ```

```
        myPlane.TakeOff();
        myPlane.Drive();
        myPlane.Land();
        myPlane.StopEngine("Whirr");
}
```

18. Add the following statements shown in bold to the *Entrance* method after the code you have just written. These statements create an instance of the *Car* class and test its methods.

```
static void Entrance()
{
    ...
    Console.WriteLine("\nJourney by car:");
    Car myCar = new Car();
    myCar.StartEngine("Brm brm");
    myCar.Accelerate();
    myCar.Drive();
    myCar.Brake();
    myCar.StopEngine("Phut phut");
}
```

19. On the *Debug* menu, click *Start Without Debugging*.

In the console window, verify that the program outputs messages simulating the different stages of performing a journey by airplane and by car, as shown in the following image:

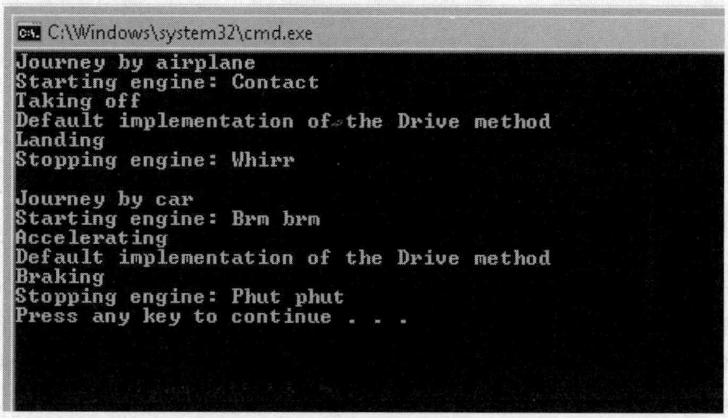

Notice that both modes of transport invoke the default implementation of the virtual *Drive* method because neither class currently overrides this method.

20. Press Enter to close the application and return to Visual Studio 2008.

21. Display the *Airplane* class in the *Code and Text Editor* window. Override the *Drive* method in the *Airplane* class, as follows:

```
public override void Drive()
{
    Console.WriteLine("Flying");
}
```

Note Notice that IntelliSense automatically displays a list of available virtual methods. If you select the *Drive* method from the IntelliSense list, Visual Studio automatically inserts into your code a statement that calls the *base.Drive* method. If this happens, delete the statement, as this exercise does not require it.

22. Display the *Car* class in the *Code and Text Editor* window. Override the *Drive* method in the *Car* class as follows:

```
public override void Drive()
{
    Console.WriteLine("Motoring");
}
```

23. On the *Debug* menu, click *Start Without Debugging*.

In the console window, notice that the *Airplane* object now displays the message *Flying* when the application calls the *Drive* method and the *Car* object displays the message *Motoring*.

24. Press Enter to close the application and return to Visual Studio 2008.

25. Display the Program.cs file in the *Code and Text Editor* window.

26. Add the statements shown here in bold to the end of the *Entrance* method:

```
static void Entrance()
{
    ...
    Console.WriteLine("\nTesting polymorphism");
    Vehicle v = myCar;
    v.Drive();
    v = myPlane;
    v.Drive();
}
```

This code tests the polymorphism provided by the virtual *Drive* method. The code creates a reference to the *Car* object using a *Vehicle* variable (this is safe, because all *Car* objects are *Vehicle* objects) and then calls the *Drive* method using this *Vehicle* variable. The final two statements refer the *Vehicle* variable to the *Airplane* object and call what seems to be the same *Drive* method again.

27. On the *Debug* menu, click *Start Without Debugging*.

In the console window, verify that the same messages appear as before, followed by this text:

```
Testing polymorphism
Motoring
Flying
```

The *Drive* method is virtual, so the runtime (not the compiler) works out which version of the *Drive* method to call when invoking it through a *Vehicle* variable based on the real type of the object referenced by this variable. In the first case, the *Vehicle* object refers to a *Car*, so the application calls the *Car.Drive* method. In the second case, the *Vehicle* object refers to an *Airplane*, so the application calls the *Airplane.Drive* method.

28. Press Enter to close the application and return to Visual Studio 2008.

Understanding Extension Methods

Inheritance is a very powerful feature, enabling you to extend the functionality of a class by creating a new class that derives from it. However, sometimes using inheritance is not the most appropriate mechanism for adding new behaviors, especially if you need to quickly extend a type without affecting existing code.

For example, suppose you want to add a new feature to the *int* type—a method named *Negate* that returns the negative equivalent value that an integer currently contains. (I know that you could simply use the unary minus operator [-] to perform the same task, but bear with me.) One way to achieve this would be to define a new type named *NegInt32* that inherits from *System.Int32* (*int* is an alias for *System.Int32*) and that adds the *Negate* method:

```
class NegInt32 : System.Int32  // don't try this!
{
    public int Negate()
    {
        ...
    }
}
```

The theory is that *NegInt32* will inherit all the functionality associated with the *System.Int32* type in addition to the *Negate* method. There are two reasons why you might not want to follow this approach:

- This method will apply only to the NegInt32 type, and if you want to use it with existing int variables in your code, you would have to change the definition of every int variable to the NegInt32 type.

- The System.Int32 type is actually a structure, not a class, and you cannot use inheritance with structures.

This is where extension methods become very useful.

An extension method enables you to extend an existing type (a class or a structure) with additional static methods. These static methods become immediately available to your code in any statements that reference data of the type being extended.

You define an extension method in a *static* class and specify the type that the method applies to as the first parameter to the method, along with the *this* keyword. Here's an example showing how you can implement the *Negate* extension method for the *int* type:

```
static class Util
{
    public static int Negate(this int i)
    {
        return -i;
    }
}
```

The syntax looks a little odd, but it is the *this* keyword prefixing the parameter to *Negate* that identifies it as an extension method, and the fact that the parameter that *this* prefixes is an *int* means that you are extending *the* int type.

To use the extension method, bring the *Util* class into scope (if necessary, add a *using* statement specifying the namespace to which the *Util* class belongs), and then you can simply use "." notation to reference the method, like this:

```
int x = 591;
Console.WriteLine("x.Negate {0}", x.Negate());
```

Notice that you do not need to reference the *Util* class anywhere in the statement that calls the *Negate* method. The C# compiler automatically detects all extension methods for a given type from all the static classes that are in scope. You can also invoke the *Utils.Negate* method passing an *int* as the parameter, using the regular syntax you have seen before, although this use obviates the purpose of defining the method as an extension method:

```
int x = 591;
Console.WriteLine("x.Negate {0}", Util.Negate(x));
```

In the following exercise, you will add an extension method to the *int* type. This extension method enables you to convert the value an *int* variable contains from base 10 to a representation of that value in a different number base.

Create an extension method

1. In Visual Studio 2008, open the *ExtensionMethod* project, located in the \Microsoft Press\Visual CSharp Step by Step\Chapter 12\ExtensionMethod folder in your Documents folder.

2. Display the Util.cs file in the *Code and Text Editor* window.

   This file contains a static class named *Util* in a namespace named *Extensions*. The class is empty apart from the *// to do* comment. Remember that you must define extension methods inside a static class.

3. Add a public static method to the *Util* class, named *ConvertToBase*. The method should take two parameters: an *int* parameter named *i*, prefixed with the *this* keyword to indicate that the method is an extension method for the *int* type, and another ordinary *int* parameter named *baseToConvertTo*. The method will convert the value in *i* to the base indicated by *baseToConvertTo*. The method should return an *int* containing the converted value.

The *ConvertToBase* method should look like this:

```
static class Util
{
    public static int ConvertToBase(this int i, int baseToConvertTo)
    {
    }
}
```

4. Add an *if* statement to the *ConvertToBase* method that checks that the value of the *baseToConvertTo* parameter is between 2 and 10. The algorithm used by this exercise does not work reliably outside this range of values. Throw an *ArgumentException* with a suitable message if the value of *baseToConvertTo* is outside this range.

The *ConvertToBase* method should look like this:

```
public static int ConvertToBase(this int i, int baseToConvertTo)
{
    if (baseToConvertTo < 2 || baseToConvertTo > 10)
        throw new ArgumentException("Value cannot be converted to base " +
                                    baseToConvertTo.ToString());
}
```

5. Add the following statements shown in bold to the *ConvertToBase* method, after the statement that throws the *ArgumentException*. This code implements a well-known algorithm that converts a number from base 10 to a different number base:

```
public static int ConvertToBase(this int i, int baseToConvertTo)
{
    ...
    int result = 0;
    int iterations = 0;
    do
    {
        int nextDigit = i % baseToConvertTo;
        result += nextDigit * (int)Math.Pow(10, iterations);
        iterations++;
        i /= baseToConvertTo;
    }
    while (i != 0);
    return result;
}
```

6. Display the Program.cs file in the *Code and Text Editor* window.

7. Add the following *using* statement after the *using System;* statement at the top of the file:

```
using Extensions;
```

This statement brings the namespace containing the *Util* class into scope. The *ConvertToBase* extension method will not be visible in the Program.cs file if you do not perform this task.

8. Add the following statements to the *Entrance* method of the *Program* class:

```
int x = 591;
for (int i = 2; i <= 10; i++)
{
    Console.WriteLine("{0} in base {1} is {2}", x, i, x.ConvertToBase(i));
}
```

This code creates an *int* named *x* and sets it to the value *591*. (You could pick any integer value you want.) The code then uses a loop to print out the value 591 in all number bases between 2 and 10. Notice that *ConvertToBase* appears as an extension method in IntelliSense when you type the period (.) after *x* in the *Console.WriteLine* statement.

9. On the *Debug* menu, click *Start Without Debugging*. Confirm that the program displays messages showing the value 591 in the different number bases to the console, like this:

```
C:\Windows\system32\cmd.exe
591 in base 2 is 1001001111
591 in base 3 is 210220
591 in base 4 is 21033
591 in base 5 is 4331
591 in base 6 is 2423
591 in base 7 is 1503
591 in base 8 is 1117
591 in base 9 is 726
591 in base 10 is 591
Press any key to continue . . .
```

10. Press Enter to close the program.

Congratulations. You have successfully used inheritance to define a hierarchy of classes, and you should now understand how to override inherited methods and implement virtual methods. You have also seen how to add an extension method to an existing type.

- If you want to continue to the next chapter:

  Keep Visual Studio 2008 running, and turn to Chapter 13.

- If you want to exit Visual Studio 2008 now:

  On the *File* menu, click *Exit*. If you see a *Save* dialog box, click *Yes* (if you are using Visual Studio 2008) or *Save* (if you are using Visual C# 2008 Express Edition) and save the project.

Chapter 12 Quick Reference

To	Do this

Create a derived class from a base class	Declare the new class name followed by a colon and the name of the base class. For example:
	
	`class Derived : Base`
	`{`
	`    ...`
	`}`

Call a base class constructor as part of the constructor for an inheriting class	Supply a constructor parameter list before the body of the derived class constructor. For example:
	
	`class Derived : Base`
	`{`
	`    ...`
	`    public Derived(int x) : Base(x)`
	`    {`
	`        ...`
	`    }`
	`    ...`
	`}`

Declare a virtual method	Use the *virtual* keyword when declaring the method. For example:
	
	`class Mammal`
	`{`
	`    public virtual void Breathe()`
	`    {`
	`        ...`
	`    }`
	`    ...`
	`}`

Implement a method in a derived class that overrides an inherited virtual method	Use the *override* keyword when declaring the method in the derived class. For example:

```
class Whale : Mammal
{
    public override void Breathe()
    {
        ...
    }
    ...
}
```

Define an extension method for a type	Add a static public method to a static class. The first parameter must be of the type being extended, preceded by the *this* keyword. For example:

```
static class Util
{
    public static int Negate(this int i)
    {
        return -i;
    }
}
```

Chapter 13
Creating Interfaces and Defining Abstract Classes

After completing this chapter, you will be able to:

- Define an interface identifying the names of methods.

- Implement an interface in a structure or class by writing the bodies of the methods.

- Capture common implementation details in an abstract class.

- Declare that a class cannot be used as a base class by using the sealed keyword.

Inheriting from a class is a powerful mechanism, but the real power of inheritance comes from inheriting from an interface. An interface does not contain any code or data; it just specifies the methods and properties that a class that inherits from the interface must provide. Using an interface enables you to completely separate the names and signatures of the methods of a class from the method's implementation.

Abstract classes are similar in many ways to interfaces except that they can contain code and data. However, you can specify that certain methods of an abstract class are virtual so that a class that inherits from the abstract class must provide its own implementation of these methods. You frequently use abstract classes with interfaces, and together they provide a key technique enabling you to build extensible programming frameworks, as you will discover in this chapter.

Understanding Interfaces

Suppose you want to define a new collection class that enables an application to store objects in a sequence that depends on the type of objects the collection contains. When you define the collection class, you do not want to restrict the types of objects that it can hold, and consequently you don't know how to order these objects. But you need to provide a way of sorting these unspecified objects. The question is, how do you provide a method that sorts objects whose types you do not know when you write the collection class? At first glance, this problem seems similar to the *ToString* problem described in Chapter 12, "Working with Inheritance," which could be resolved by declaring a virtual method that subclasses of your collection class can override. However, this is not the case. There is not usually any form of inheritance relationship between the collection class and the objects that it holds, so a virtual method would not be of much use. If you think for a moment, the problem is that the way in which the objects in the collection should be ordered is dependent on the type of the objects themselves, and not on the collection. The solution, therefore, is to require that all the objects provide a method that the collection can call, enabling the collection to compare these objects with one another. As an example, look at the *CompareTo* method shown here:

```
int CompareTo(object obj)
{
    // return 0 if this instance is equal to obj
    // return < 0 if this instance is less than obj
    // return > 0 if this instance is greater than obj
    ...
}
```

The collection class can make use of this method to sort its contents.

You can define an interface for collectable objects that includes the *CompareTo* method and specify that the collection class can collect only classes that implement this interface. In this way, an interface is similar to a contract. If a class implements an interface, the interface guarantees that the class contains all the methods specified in the interface. This mechanism ensures that you will be able to call the *CompareTo* method on all objects in the collection and sort them.

Interfaces enable you to truly separate the "what" from the "how." The interface tells you only the name, return type, and parameters of the method. Exactly how the method is implemented is not a concern of the interface. The interface represents how you want an object to be used, rather than how the usage happens to be implemented.

Interface Syntax

To declare an interface, you use the *interface* keyword instead of the *class* or *struct* keyword. Inside the interface, you declare methods exactly as in a class or a structure except that you never specify an access modifier (*public*, *private*, or *protected*), and you replace the method body with a semicolon. Here is an example:

```
interface IComparable
{
    int CompareTo(object obj);
}
```

> **Tip** The Microsoft .NET Framework documentation recommends that you preface the name of your interfaces with the capital letter *I*. This convention is the last vestige of Hungarian notation in C#. Incidentally, the *System* namespace already defines the *IComparable* interface as shown here.

Interface Restrictions

The essential idea to remember is that an interface never contains any implementation. The following restrictions are natural consequences of this:

- You're not allowed to define any fields in an interface, not even static ones. A field is an implementation detail of a class or structure.

- You're not allowed to define any constructors in an interface. A constructor is also considered to be an implementation detail of a class or structure.

- You're not allowed to define a destructor in an interface. A destructor contains the statements used to destroy an object instance. (Destructors are described in Chapter 14, "Using Garbage Collection and Resource Management.")

- You cannot specify an access modifier for any method. All methods in an interface are implicitly public.

- You cannot nest any types (such as enumerations, structures, classes, or interfaces) inside an interface.

- An interface is not allowed to inherit from a structure or a class, although an interface can inherit from another interface. Structures and classes contain implementation; if an interface were allowed to inherit from either, it would be inheriting some implementation.

Implementing an Interface

To implement an interface, you declare a class or structure that inherits from the interface and implements *all* the methods specified by the interface. For example, suppose you are defining the *Mammal* hierarchy shown in Chapter 12 but you need to specify that land-bound mammals provide a method named *NumberOfLegs* that returns as an *int* the number of legs that a mammal has. (Sea-bound mammals do not implement this interface.) You could define the *ILandBound* interface that contains this method as follows:

```
interface ILandBound
{
    int NumberOfLegs();
}
```

You could then implement this interface in the *Horse* class:

```
class Horse : ILandBound
{
    ...
    int ILandBound.NumberOfLegs()
    {
        return 4;
    }
}
```

When you implement an interface, you must ensure that each method matches its corresponding interface method exactly, according to the following rules:

- The method names and return types match exactly.

- Any parameters (including ref and out keyword modifiers) match exactly.

- The method name is prefaced by the name of the interface. This is known as explicit interface implementation and is a good habit to cultivate.

- All methods implementing an interface must be publicly accessible. However, if you are using explicit interface implementation, the method should not have an access qualifier.

If there is any difference between the interface definition and its declared implementation, the class will not compile.

The Advantages of Explicit Interface Implementations

Implementing an interface explicitly can seem a little verbose, but it does offer a number of advantages that help you to write clearer, more maintainable, and more predictable code.

You can implement a method without explicitly specifying the interface name, but this can lead to some differences in the way the implementation behaves. Some of these differences can cause confusion. For example, a method defined by using explicit interface implementation cannot be declared as *virtual*, whereas omitting the interface name allows this behavior.

It's possible for multiple interfaces to contain methods with the same names, return types, and parameters. If a class implements multiple interfaces with methods that have common signatures, you can use explicit interface implementation to disambiguate the method implementations. Explicit interface implementation identifies which methods in a class belong to which interface. Additionally, the methods for each interface are publicly accessible, but only through the interface itself. We will look at how to do this in the upcoming section "Referencing a Class Through Its Interface."

In this book, I recommend implementing an interface explicitly wherever possible.

A class can extend another class and implement an interface at the same time. In this case, C# does not denote the base class and the interface by using keywords as, for example, Java does. Instead, C# uses a positional notation. The base class is named first, followed by a comma, followed by the interface. The following example defines *Horse* as a class that is a *Mammal* but that additionally implements the *ILandBound* interface:

```
interface ILandBound
{
    ...
}

class Mammal
{
    ...
}

class Horse : Mammal , ILandBound
{
    ...
}
```

Referencing a Class Through Its Interface

In the same way that you can reference an object by using a variable defined as a class that is higher up the hierarchy, you can reference an object by using a variable defined as an interface that its class implements. Taking the preceding example, you can reference a *Horse* object by using an *ILandBound* variable, as follows:

```
Horse myHorse = new Horse(...);
ILandBound iMyHorse = myHorse; // legal
```

This works because all horses are land-bound mammals, although the converse is not true, and you cannot assign an *ILandBound* object to a *Horse* variable without casting it first.

The technique of referencing an object through an interface is useful because it enables you to define methods that can take different types as parameters, as long as the types implement a specified interface. For example, the *FindLandSpeed* method shown here can take any argument that implements the *ILandBound* interface:

```
int FindLandSpeed(ILandBound landBoundMammal)
{
    ...
}
```

Note that when referencing an object through an interface, you can invoke only methods that are visible through the interface.

Working with Multiple Interfaces

A class can have at most one base class, but it is allowed to implement an unlimited number of interfaces. A class must still implement all the methods it inherits from all its interfaces.

If an interface, a structure, or a class inherits from more than one interface, you write the interfaces in a comma-separated list. If a class also has a base class, the interfaces are listed *after* the base class. For example, suppose you define another interface named *IGrazable* that contains the *ChewGrass* method for all grazing animals. You can define the *Horse* class like this:

```
class Horse : Mammal, ILandBound, IGrazable
{
    ...
}
```

Abstract Classes

The *ILandBound* and *IGrazable* interfaces could be implemented by many different classes, depending on how many different types of mammals you want to model in your C# application. In situations such as this, it's quite common for parts of the derived classes to share common implementations. For example, the duplication in the following two classes is obvious:

```
class Horse : Mammal, ILandBound, IGrazable
{
    ...
    void IGrazable.ChewGrass()
    {
        Console.WriteLine("Chewing grass");
        // code for chewing grass
    };
}

class Sheep : Mammal, ILandBound, IGrazable
{
    ...
    void IGrazable.ChewGrass()
    {
        Console.WriteLine("Chewing grass");
        // same code as horse for chewing grass
    };
}
```

Duplication in code is a warning sign. You should refactor the code to avoid the duplication and reduce any maintenance costs. The way to achieve this refactoring is to put the common implementation into a new class created specifically for this purpose. In effect, you can insert a new class into the class hierarchy. For example:

```
class GrazingMammal : Mammal, IGrazable
{
    ...
    void IGrazable.ChewGrass()
    {
        Console.WriteLine("Chewing grass");
        // common code for chewing grass
    }
}

class Horse : GrazingMammal, ILandBound
{
    ...
}

class Sheep : GrazingMammal, ILandBound
{
    ...
}
```

This is a good solution, but there is one thing that is still not quite right: You can actually create instances of the *GrazingMammal* class (and the *Mammal* class for that matter). This doesn't really make sense. The *GrazingMammal* class exists to provide a common default implementation. Its sole purpose is to be inherited from. The *GrazingMammal* class is an abstraction of common functionality rather than an entity in its own right.

To declare that creating instances of a class is not allowed, you must explicitly declare that the class is abstract, by using the *abstract* keyword. For example:

```
abstract class GrazingMammal : Mammal, IGrazable
{
    ...
}
```

If you try to instantiate a *GrazingMammal* object, the code will not compile:

```
GrazingMammal myGrazingMammal = new GrazingMammal(...);  // illegal
```

Abstract Methods

An abstract class can contain abstract methods. An abstract method is similar in principle to a virtual method (you met virtual methods in Chapter 12) except that it does not contain a method body. A derived class *must* override this method. The following example defines the *DigestGrass* method in the *GrazingMammal* class as an abstract method; grazing mammals might use the same code for chewing grass, but they must provide their own implementation of the *DigestGrass* method. An abstract method is useful if it does not make sense to provide a default implementation in the abstract class and you want to ensure that an inheriting class provides its own implementation of that method.

```
abstract class GrazingMammal : Mammal, IGrazable
{
    abstract void DigestGrass();
    ...
}
```

Sealed Classes

Using inheritance is not always easy and requires forethought. If you create an interface or an abstract class, you are knowingly writing something that will be inherited from in the future. The trouble is that predicting the future is a difficult business. With practice and experience, you can develop the skills to craft a flexible, easy-to-use hierarchy of interfaces, abstract classes, and classes, but it takes effort and you also need a solid understanding of the problem you are modeling. To put it another way, unless you consciously design a class with the intention of using it as a base class, it's extremely unlikely that it will function very well as a base class. C# allows you to use the *sealed* keyword to prevent a class from being used as a base class if you decide that it should not be. For example:

```
sealed class Horse : GrazingMammal, ILandBound
{
    ...
}
```

If any class attempts to use *Horse* as a base class, a compile-time error will be generated. Note that a sealed class cannot declare any virtual methods and that an abstract class cannot be sealed.

 Note A structure is implicitly sealed. You can never derive from a structure.

Sealed Methods

You can also use the *sealed* keyword to declare that an individual method in an unsealed class is sealed. This means that a derived class cannot then override the sealed method. You can seal only an *override* method. (You declare the method as *sealed override*.) You can think of the *interface*, *virtual*, *override*, and *sealed* keywords as follows:

- An interface introduces the name of a method.

- A virtual method is the first implementation of a method.

- An override method is another implementation of a method.

- A sealed method is the last implementation of a method.

Implementing an Extensible Framework

In the following exercise, you will familiarize yourself with a hierarchy of interfaces and classes that together implement a simple framework for reading a C# source file and classifying its contents into *tokens* (identifiers, keywords, operators, and so on). This framework performs some of the tasks that a typical compiler might perform. The framework provides a mechanism for "visiting" each token in turn, to perform specific tasks. For example, you could create:

- A displaying visitor class that displays the source file in a rich text box.

- A printing visitor class that converts tabs to spaces and aligns braces correctly.

- A spelling visitor class that checks the spelling of each identifier.

- A guideline visitor class that checks that public identifiers start with a capital letter and that interfaces start with the capital letter I.

- A complexity visitor class that monitors the depth of the brace nesting in the code.

- A counting visitor class that counts the number of lines in each method, the number of members in each class, and the number of lines in each source file.

 Note This framework implements the Visitor pattern, first documented by Erich Gamma, Richard Helm, Ralph Johnson, and John Vlissides in *Design Patterns: Elements of Reusable Object-Oriented Software* (Addison Wesley Longman, 1995).

Understand the inheritance hierarchy and its purpose

1. Start Microsoft Visual Studio 2008 if it is not already running.

2. Open the *Tokenizer* project, located in the \Microsoft Press\Visual CSharp Step by Step\ Chapter 13\Tokenizer folder in your Documents folder.

3. Display the SourceFile.cs file in the *Code and Text Editor* window.

   The *SourceFile* class contains a private array field named *tokens* that looks like this and is essentially a hard-coded version of a source file that has already been parsed and tokenized:

   ```
   private IVisitableToken[] tokens =
   {
       new KeywordToken("using"),
       new WhitespaceToken(" "),
       new IdentifierToken("System"),
       new PunctuatorToken(";"),
       ...
   };
   ```

The *tokens* array contains a sequence of objects that all implement the *IVisitableToken* interface (which is explained shortly). Together, these tokens simulate the tokens of a simple "hello, world" source file. (A complete compiler would parse a source file, identify the type of each token, and dynamically create the *tokens* array. Each token would be created using the appropriate class type, typically through a *switch* statement.) The *SourceFile* class also contains a public method named *Accept*. The *SourceFile.Accept* method has a single parameter of type *ITokenVisitor*. The body of the *SourceFile.Accept* method iterates through the tokens, calling their *Accept* methods. The *Token.Accept* method will process the current token in some way, according to the type of the token:

```
public void Accept(ITokenVisitor visitor)
{
    foreach (IVisitableToken token in tokens)
    {
        token.Accept(visitor);
    }
}
```

In this way, the *visitor* parameter "visits" each token in sequence. The *visitor* parameter is an instance of some visitor class that processes the token that the *visitor* object visits. When the *visitor* object processes the token, the token's own class methods come into play.

4. Display the IVisitableToken.cs file in the *Code and Text Editor* window.

   This file defines the *IVisitableToken* interface. The *IVisitableToken* interface inherits from two other interfaces, the *IVisitable* interface and the *IToken* interface, but does not define any methods of its own:

```
interface IVisitableToken : IVisitable, IToken
{
}
```

5. Display the IVisitable.cs file in the *Code and Text Editor* window.

   This file defines the *IVisitable* interface. The *IVisitable* interface declares a single method named *Accept*:

```
interface IVisitable
{
    void Accept(ITokenVisitor visitor);
}
```

   Each object in the array of tokens inside the *SourceFile* class is accessed using the *IVisitableToken* interface. The *IVisitableToken* interface inherits the *Accept* method, and each token implements the *Accept* method. (Recall that each token *must* implement the *Accept* method because any class that inherits from an interface must implement all the methods in the interface.)

6. On the *View* menu, click *Class View*.

The *Class View* window appears in the pane used by *Solution Explorer*. This window displays the namespaces, classes, and interfaces defined by the project.

7. In the *Class View* window, expand the *Tokenizer* project, and then expand the *{} Tokenizer* namespace. The classes and interfaces in this namespace are listed. Notice the different icons used to distinguish interfaces from classes.

   Expand the *IVisitableToken* interface, and then expand the *Base Types* node. The interfaces that the *IVisitableToken* interface extends (*IToken* and *IVisitable*) are displayed, like this:

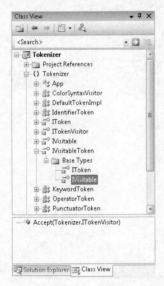

8. In the *Class View* window, right-click the *IdentifierToken* class, and then click *Go To Definition* to display this class in the *Code and Text Editor* window. (It is actually located in SourceFile.cs.)

   The *IdentifierToken* class inherits from the *DefaultTokenImpl* abstract class and the *IVisitableToken* interface. It implements the *Accept* method as follows:

```
void IVisitable.Accept(ITokenVisitor visitor)
{
    visitor.VisitIdentifier(this.ToString());
}
```

> **Note** The *VisitIdentifier* method processes the token passed to it as a parameter in whatever way the *visitor* object sees fit. In the following exercise, you will provide an implementation of the *VisitIdentifier* method that simply renders the token in a particular color.

The other token classes in this file follow a similar pattern.

9. In the *Class View* window, right-click the *ITokenVisitor* interface, and then click *Go To Definition*. This action displays the ITokenVisitor.cs source file in the *Code and Text Editor* window.

The *ITokenVisitor* interface contains one method for each type of token. The result of this hierarchy of interfaces, abstract classes, and classes is that you can create a class that implements the *ITokenVisitor* interface, create an instance of this class, and pass this instance as the parameter to the *Accept* method of a *SourceFile* object. For example:

```
class MyVisitor : ITokenVisitor
{
    public void VisitIdentifier(string token)
    {
        ...
    }

    public void VisitKeyword(string token)
    {
        ...
    }
}

...

class Program
{
    static void Main()
    {
        SourceFile source = new SourceFile();
        MyVisitor visitor = new MyVisitor();
        source.Accept(visitor);
    }
}
```

The code in the *Main* method will result in each token in the source file calling the matching method in the visitor object.

In the following exercise, you will create a class that derives from the *ITokenVisitor* interface and whose implementation displays the tokens from our hard-coded source file in a rich text box in color syntax (for example, keywords in blue) by using the "visitor" mechanism.

Write the *ColorSyntaxVisitor* class

1. In Solution Explorer (click the *Solution Explorer* tab below the *Class View* window), double-click Window1.xaml to display the *Color Syntax* form in the *Design View* window.

You will use this form to test the framework. This form contains a button for opening a file to be tokenized and a rich text box for displaying the tokens:

The rich text box in the middle of the form is named *codeText*, and the button is named *Open*.

Note A *rich text box* is like an ordinary text box except that it can display formatted content rather than simple, unformatted text.

2. Right-click the form, and then click *View Code* to display the code for the form in the *Code and Text Editor* window.

3. Locate the *openClick* method.

This method is called when the user clicks the *Open* button. You must implement this method so that it displays the tokens defined in the *SourceFile* class in the rich text box, by using a *ColorSyntaxVisitor* object. Add the code shown here in bold to the *openClick* method:

```
private void openClick(object sender, RoutedEventArgs e)
{
    SourceFile source = new SourceFile();
    ColorSyntaxVisitor visitor = new ColorSyntaxVisitor(codeText);
    source.Accept(visitor);
}
```

Remember that the *Accept* method of the *SourceFile* class iterates through all the tokens, processing each one by using the specified visitor. In this case, the visitor is the *ColorSyntaxVisitor* object, which will render each token in color.

Note In the current implementation, the *Open* button uses just data that is hard-coded in the *SourceFile* class. In a fully functional implementation, the *Open* button would prompt the user for the name of a text file and then parse and tokenize it into the format shown in the *SourceFile* class before calling the *Accept* method.

4. Open the ColorSyntaxVisitor.cs file in the *Code and Text Editor* window.

The *ColorSyntaxVisitor* class has been partially written. This class implements the *ITokenVisitor* interface and already contains two fields and a constructor to initialize a reference to the rich text box, named *target*, used to display tokens. Your task is to implement the methods inherited from the *ITokenVisitor* interface and also create a method that will write the tokens to the rich text box.

5. In the *Code and Text Editor* window, add the *Write* method to the *ColorSyntaxVisitor* class exactly as follows:

```
private void Write(string token, SolidColorBrush color)
{
    target.AppendText(token);
    int offsetToStartOfToken = -1 * token.Length - 2;
    int offsetToEndOfToken = -2;
    TextPointer start =
      target.Document.ContentEnd.GetPositionAtOffset(offsetToStartOfToken);
    TextPointer end =
      target.Document.ContentEnd.GetPositionAtOffset(offsetToEndOfToken);
    TextRange text = new TextRange(start, end);
    text.ApplyPropertyValue(TextElement.ForegroundProperty, color);
}
```

This code appends each token to the rich text box identified by the *target* variable using the specified color. The two *TextPointer* variables, *start* and *end*, indicate where the new token starts and ends in the rich text box control. (Don't worry about how these positions are calculated. If you're wondering, they are negative values because they are offset from the *ContentEnd* property.) The *TextRange* variable *text* obtains a reference to the portion of the text in the rich text box control displaying the newly appended token. The *ApplyPropertyValue* method sets the color of this text to the color specified as the second parameter.

Each of the various "visit" methods in the *ColorSyntaxVisitor* class will call this *Write* method with an appropriate color to display color-coded results.

6. In the *Code and Text Editor* window, add the following methods that implement the *ITokenVisitor* interface to the *ColorSyntaxVisitor* class. Specify *Brushes.Blue* for keywords, *Brushes.Green* for *StringLiterals*, and *Brushes.Black* for all other methods. (*Brushes* is a class defined in the *System.Windows.Media* namespace.) Notice that this code implements the interface explicitly; it qualifies each method with the interface name.

```
void ITokenVisitor.VisitComment(string token)
{
    Write(token, Brushes.Black);
}

void ITokenVisitor.VisitIdentifier(string token)
{
    Write(token, Brushes.Black);
}
```

```
void ITokenVisitor.VisitKeyword(string token)
{
    Write(token, Brushes.Blue);
}

void ITokenVisitor.VisitOperator(string token)
{
    Write(token, Brushes.Black);
}

void ITokenVisitor.VisitPunctuator(string token)
{
    Write(token, Brushes.Black);
}

void ITokenVisitor.VisitStringLiteral(string token)
{
    Write(token, Brushes.Green);
}

void ITokenVisitor.VisitWhitespace(string token)
{
    Write(token, Brushes.Black);
}
```

It is the class type of the token in the token array that determines which of these methods is called through the token's override of the *Token.Accept* method.

> **Tip** You can either type these methods into the *Code and Text Editor* window directly or use Visual Studio 2008 to generate default implementations for each one and then modify the method bodies with the appropriate code. To do this, right-click the *ITokenVisitor* identifier in the class definition sealed class, *ColorSyntaxVisitor : ITokenVisitor.* On the shortcut menu, point to *Implement Interface* and then click *Implement Interface Explicitly.* Each method will contain a statement that throws a *NotImplementedException.* Replace this code with that shown here.

7. On the *Build* menu, click *Build Solution.* Correct any errors, and rebuild if necessary.

8. On the *Debug* menu, click *Start Without Debugging.*

The *Color Syntax* form appears.

9. On the form, click *Open.*

The dummy code is displayed in the rich text box, with keywords in blue and string literals in green.

10. Close the form, and return to Visual Studio 2008.

Generating a Class Diagram

The *Class View* window is useful for displaying and navigating the hierarchy of classes and interfaces in a project. Visual Studio 2008 also enables you to generate class diagrams that depict this same information graphically. (You can also use a class diagram to add new classes and interfaces and to define methods, properties, and other class members.)

> **Note** This feature is not available in Visual C# 2008 Express Edition.

To generate a new class diagram, on the *Project* menu, click *Add New Item*. In the *Add New Item* dialog box, select the *Class Diagram* template, and then click *Add*. This action will generate an empty diagram, and you can create new types by dragging items from the *Class Designer* category in the *Toolbox*. You can generate a diagram of all existing classes by dragging them individually from the *Class View* window or by dragging the namespace to which they belong. The diagram shows the relationships between the classes and interfaces, and you can expand the definition of each class to show its contents. You can drag the classes and interfaces around to make the diagram more readable, as shown in the image on the following page.

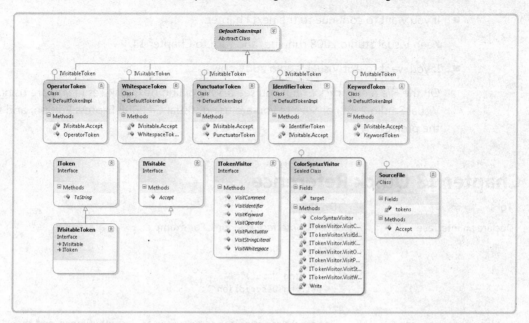

Summarizing Keyword Combinations

The following table summarizes the various valid (yes), invalid (no), and mandatory (required) keyword combinations when creating classes and interfaces.

Keyword	Interface	Abstract class	Class	Sealed class	Structure
abstract	no	yes	no	no	no
new	yes[1]	yes	yes	yes	no[2]
override	no	yes	yes	yes	no[3]
private	no	yes	yes	yes	yes
protected	no	yes	yes	yes	no[4]
public	no	yes	yes	yes	yes
sealed	no	yes	yes	required	no
virtual	no	yes	yes	no	no

[1] An interface can extend another interface and introduce a new method with the same signature.

[2] A structure implicitly derives from *System.Object*, which contains methods that the structure can hide.

[3] A structure implicitly derives from *System.Object*, which contains no virtual methods.

[4] A structure is implicitly sealed and cannot be derived from.

- If you want to continue to the next chapter:

  Keep Visual Studio 2008 running, and turn to Chapter 14.

- If you want to exit Visual Studio 2008 now:

  On the *File* menu, click *Exit*. If you see a *Save* dialog box, click *Yes* (if you are using Visual Studio 2008) or *Save* (if you are using Visual C# 2008 Express Edition) and save the project.

Chapter 13 Quick Reference

To	Do this
| Declare an interface | Use the *interface* keyword. For example:<br><br>```
interface IDemo
{
    string Name();
    string Description();
}
``` |
| Implement an interface | Declare a class using the same syntax as class inheritance, and then implement all the member functions of the interface. For example:<br><br>```
class Test : IDemo
{
    public string IDemo.Name()
    {
    ...
    }

    public string IDemo.Description()
    {
    ...
    }
}
``` |
| Create an abstract class that can be used only as a base class, containing abstract methods | Declare the class using the *abstract* keyword. For each abstract method, declare the method with the *abstract* keyword and without a method body. For example:<br><br>```
abstract class GrazingMammal
{
    abstract void DigestGrass();
    ...
}
``` |
| Create a sealed class that cannot be used as a base class | Declare the class using the *sealed* keyword. For example:<br><br>```
sealed class Horse

{
    ...
}
``` |

Chapter 14
Using Garbage Collection and Resource Management

After completing this chapter, you will be able to:

- Manage system resources by using garbage collection.

- Write code that runs when an object is finalized by using a destructor.

- Release a resource at a known point in time in an exception-safe manner by writing a *try/finally* statement.

- Release a resource at a known point in time in an exception-safe manner by writing a *using* statement.

You have seen in earlier chapters how to create variables and objects, and you should understand how memory is allocated when you create variables and objects. (In case you don't remember, value types are created on the stack, and reference types are given memory from the heap.) Computers do not have infinite amounts of memory, so memory must be reclaimed when a variable or an object no longer needs it. Value types are destroyed and their memory reclaimed when they go out of scope. That's the easy bit. How about reference types? You create an object by using the *new* keyword, but how and when is an object destroyed? That's what this chapter is all about.

The Life and Times of an Object

First, let's recap what happens when you create an object.

You create an object by using the *new* operator. The following example creates a new instance of the *TextBox* class. (This class is provided as part of the Microsoft .NET Framework.)

```
TextBox message = new TextBox(); // TextBox is a reference type
```

From your point of view, the *new* operation is atomic, but underneath, object creation is really a two-phase process:

1. The *new* operation allocates a chunk of *raw* memory from the heap. You have no control over this phase of an object's creation.

2. The *new* operation converts the chunk of raw memory to an object; it has to initialize the object. You can control this phase by using a constructor.

Note C++ programmers should note that in C#, you cannot overload *new* to control allocation.

After you have created an object, you can access its members by using the dot operator (.). For example, the *TextBox* class includes a member named *Text* that you can access like this:

```
message.Text = "People of Earth, your attention please";
```

You can make other reference variables refer to the same object:

```
TextBox messageRef = message;
```

How many references can you create to an object? As many as you want! This has an impact on the lifetime of an object. The runtime has to keep track of all these references. If the variable *message* disappears (by going out of scope), other variables (such as *messageRef*) might still exist. The lifetime of an object cannot be tied to a particular reference variable. An object can be destroyed and its memory reclaimed only when *all* the references to it have disappeared.

Note C++ programmers should note that C# does not have a delete operator. The runtime controls when an object is destroyed.

Like object creation, object destruction is a two-phase process. The two phases of destruction exactly mirror the two phases of creation:

1. The runtime has to perform some tidying up. You can control this by writing a *destructor*.

2. The runtime has to return the memory previously belonging to the object back to the heap; the memory that the object lived in has to be deallocated. You have no control over this phase.

The process of destroying an object and returning memory back to the heap is known as *garbage collection*.

Writing Destructors

You can use a destructor to perform any tidying up required when an object is garbage collected. A destructor is a special method, a little like a constructor, except that the runtime calls it after the last reference to an object has disappeared. The syntax for writing a destructor is a tilde (~) followed by the name of the class. For example, here's a simple class that counts the number of existing instances by incrementing a static variable in the constructor and decrementing the same static variable in the destructor:

```
class Tally
{
    public Tally()
    {
        this.instanceCount++;
    }

    ~Tally()
    {
        this.instanceCount--;
    }

    public static int InstanceCount()
    {
        return this.instanceCount;
    }
    ...
    private static int instanceCount = 0;
}
```

There are some very important restrictions that apply to destructors:

- Destructors apply only to reference types. You cannot declare a destructor in a value type, such as a *struct*.

  ```
  struct Tally
  {
      ~Tally() { ... } // compile-time error
  }
  ```

- You cannot specify an access modifier (such as *public*) for a destructor. You never call the destructor in your own code—part of the the runtime called the *garbage collector* does this for you.

  ```
  public ~Tally() { ... } // compile-time error
  ```

- You never declare a destructor with parameters, and the destructor cannot take any parameters. Again, this is because you never call the destructor yourself.

  ```
  ~Tally(int parameter) { ... } // compile-time error
  ```

The compiler automatically translates a destructor into an override of the *Object.Finalize* method. The compiler translates the following destructor:

```
class Tally
{
    ~Tally() { ... }
}
```

into this:

```
class Tally
{
    protected override void Finalize()
    {
        try { ... }
        finally { base.Finalize(); }
    }
}
```

The compiler-generated *Finalize* method contains the destructor body inside a *try* block, followed by a *finally* block that calls the *Finalize* method in the base class. (The *try* and *finally* keywords are described in Chapter 6, "Managing Errors and Exceptions.") This ensures that a destructor always calls its base class destructor. It's important to realize that only the compiler can make this translation. You can't override *Finalize* yourself, and you can't call *Finalize* yourself.

Why Use the Garbage Collector?

You should now understand that you can never destroy an object yourself by using C# code. There just isn't any syntax to do it, and there are good reasons why the designers of C# decided to forbid you from doing it. If it were *your* responsibility to destroy objects, sooner or later one of the following situations would arise:

- You'd forget to destroy the object. This would mean that the object's destructor (if it had one) would not be run, tidying up would not occur, and memory would not be deallocated back to the heap. You could quite easily run out of memory.

- You'd try to destroy an active object. Remember, objects are accessed by reference. If a class held a reference to a destroyed object, it would be a *dangling reference*. The dangling reference would end up referring either to unused memory or possibly to a completely different object in the same piece of memory. Either way, the outcome of using a dangling reference would be undefined at best or a security risk at worst. All bets would be off.

- You'd try and destroy the same object more than once. This might or might not be disastrous, depending on the code in the destructor.

These problems are unacceptable in a language like C#, which places robustness and security high on its list of design goals. Instead, the garbage collector is responsible for destroying objects for you. The garbage collector makes the following guarantees:

- Every object will be destroyed and its destructors run. When a program ends, all outstanding objects will be destroyed.

- Every object will be destroyed exactly once.

■ Every object will be destroyed only when it becomes unreachable—that is, when no references refer to the object.

These guarantees are tremendously useful and free you, the programmer, from tedious housekeeping chores that are easy to get wrong. They allow you to concentrate on the logic of the program itself and be more productive.

When does garbage collection occur? This might seem like a strange question. After all, surely garbage collection occurs when an object is no longer needed. Well, it does, but not necessarily immediately. Garbage collection can be an expensive process, so the runtime collects garbage only when it needs to (when it thinks available memory is starting to run low), and then it collects as much as it can. Performing a few large sweeps of memory is more efficient than performing lots of little dustings!

> **Note** You can invoke the garbage collector in a program by calling the static method *System. GC.Collect*. However, except in a few cases, this is not recommended. The *System.GC.Collect* method starts the garbage collector, but the process runs asynchronously, and when the method call is complete, you still don't know whether your objects have been destroyed. Let the runtime decide when it is best to collect garbage!

One feature of the garbage collector is that you don't know, and should not rely upon, the order in which objects will be destroyed. The final point to understand is arguably the most important: destructors do not run until objects are garbage collected. If you write a destructor, you know it will be executed, but you just don't know when.

How Does the Garbage Collector Work?

The garbage collector runs in its own thread and can execute only at certain times—typically, when your application reaches the end of a method. While it runs, other threads running in your application will temporarily halt. This is because the garbage collector might need to move objects around and update object references; it cannot do this while objects are in use. The steps that the garbage collector takes are as follows:

1. It builds a map of all reachable objects. It does this by repeatedly following reference fields inside objects. The garbage collector builds this map very carefully and makes sure that circular references do not cause an infinite recursion. Any object *not* in this map is deemed to be unreachable.

2. It checks whether any of the unreachable objects has a destructor that needs to be run (a process called *finalization*). Any unreachable object that requires finalization is placed in a special queue called the *freachable queue* (pronounced "F-reachable").

3. It deallocates the remaining unreachable objects (those that don't require finalization) by moving the *reachable* objects down the heap, thus defragmenting the heap and freeing memory at the top of the heap. When the garbage collector moves a reachable object, it also updates any references to the object.

4. At this point, it allows other threads to resume.

5. It finalizes the unreachable objects that require finalization (now in the freachable queue) by its own thread.

Recommendations

Writing classes that contain destructors adds complexity to your code and to the garbage collection process and makes your program run more slowly. If your program does not contain any destructors, the garbage collector does not need to place unreachable objects in the freachable queue and finalize them. Clearly, not doing something is faster than doing it. Therefore, try to avoid using destructors except when you really need them. For example, consider a *using* statement instead. (See the section "The *using* Statement" later in this chapter.)

You need to be very careful when you write a destructor. In particular, you need to be aware that, if your destructor calls other objects, those other objects might have *already* had their destructor called by the garbage collector. Remember that the order of finalization is not guaranteed. Therefore, ensure that destructors do not depend on one another or overlap with one another. (Don't have two destructors that try to release the same resource, for example.)

Resource Management

Sometimes it's inadvisable to release a resource in a destructor; some resources are just too valuable to lie around waiting for an arbitrary length of time until the garbage collector actually releases them. Scarce resources need to be released, and they need to be released as soon as possible. In these situations, your only option is to release the resource yourself. You can achieve this by creating a *disposal* method. A disposal method is a method that explicitly disposes of a resource. If a class has a disposal method, you can call it and control when the resource is released.

 Note The term *disposal method* refers to the purpose of the method rather than its name. A disposal method can be named using any valid C# identifier.

Disposal Methods

An example of a class that implements a disposal method is the *TextReader* class from the *System.IO* namespace. This class provides a mechanism to read characters from a sequential stream of input. The *TextReader* class contains a virtual method named *Close*, which closes the stream. The *StreamReader* class (which reads characters from a stream, such as an open file) and the *StringReader* class (which reads characters from a string) both derive from *TextReader*, and both override the *Close* method. Here's an example that reads lines of text from a file by using the *StreamReader* class and then displays them on the screen:

```
TextReader reader = new StreamReader(filename);
string line;
while ((line = reader.ReadLine()) != null)
{
    Console.WriteLine(line);
}
reader.Close();
```

The *ReadLine* method reads the next line of text from the stream into a string. The *ReadLine* method returns *null* if there is nothing left in the stream. It's important to call *Close* when you have finished with *reader* to release the file handle and associated resources. However, there is a problem with this example: it's not exception-safe. If the call to *ReadLine* or *WriteLine* throws an exception, the call to *Close* will not happen; it will be bypassed. If this happens often enough, you will run out of file handles and be unable to open any more files.

Exception-Safe Disposal

One way to ensure that a disposal method (such as *Close*) is always called, regardless of whether there is an exception, is to call the disposal method inside a *finally* block. Here's the preceding example coded using this technique:

```
TextReader reader = new StreamReader(filename);
try
{
    string line;
    while ((line = reader.ReadLine()) != null)
    {
        Console.WriteLine(line);
    }
}
finally
{
    reader.Close();
}
```

Using a *finally* block like this works, but it has several drawbacks that make it a less than ideal solution:

- It quickly gets unwieldy if you have to dispose of more than one resource. (You end up with nested *try* and *finally* blocks.)

- In some cases, you might have to modify the code. (For example, you might need to reorder the declaration of the resource reference, remember to initialize the reference to *null*, and remember to check that the reference isn't *null* in the *finally* block.)

- It fails to create an abstraction of the solution. This means that the solution is hard to understand and you must repeat the code everywhere you need this functionality.

- The reference to the resource remains in scope after the *finally* block. This means that you can accidentally try to use the resource after it has been released.

The *using* statement is designed to solve all these problems.

The *using* Statement

The *using* statement provides a clean mechanism for controlling the lifetimes of resources. You can create an object, and this object will be destroyed when the *using* statement block finishes.

Important Do not confuse the *using* statement shown in this section with the *using* directive that brings a namespace into scope. It is unfortunate that the same keyword has two different meanings.

The syntax for a *using* statement is as follows:

```
using ( type variable = initialization )
{
    StatementBlock
}
```

Here is the best way to ensure that your code always calls *Close* on a *TextReader*:

```
using (TextReader reader = new StreamReader(filename))
{
    string line;
    while ((line = reader.ReadLine()) != null)
    {
        Console.WriteLine(line);
    }
}
```

This *using* statement is precisely equivalent to the following transformation:

```
{
    TextReader reader = new StreamReader(filename);
    try
    {
        string line;
        while ((line = reader.ReadLine()) != null)
        {
            Console.WriteLine(line);
        }
    }
    finally
    {
        if (reader != null)
        {
            ((IDisposable)reader).Dispose();
        }
    }
}
```

> **Note** The *using* statement introduces its own block for scoping purposes. This arrangement means that the variable you declare in a *using* statement automatically goes out of scope at the end of the embedded statement and you cannot accidentally attempt to access a disposed resource.

The variable you declare in a *using* statement must be of a type that implements the *IDisposable* interface. The *IDisposable* interface lives in the *System* namespace and contains just one method, named *Dispose*:

```
namespace System
{
    interface IDisposable
    {
        void Dispose();
    }
}
```

It just so happens that the *StreamReader* class implements the *IDisposable* interface, and its *Dispose* method calls *Close* to close the stream. You can employ a *using* statement as a clean, exception-safe, and robust way to ensure that a resource is always released. This approach solves all of the problems that existed in the manual *try/finally* solution. You now have a solution that:

- Scales well if you need to dispose of multiple resources.

- Doesn't distort the logic of the program code.

- Abstracts away the problem and avoids repetition.

- Is robust. You can't use the variable declared inside the *using* statement (in this case, *reader*) after the *using* statement has ended because it's not in scope anymore—you'll get a compile-time error.

Calling the *Dispose* Method from a Destructor

When writing a class, should you write a destructor or implement the *IDisposable* interface? A call to a destructor *will* happen, but you just don't know when. On the other hand, you know exactly when a call to the *Dispose* method happens, but you just can't be sure that it will actually happen, because it relies on the programmer remembering to write a *using* statement. However, it is possible to ensure that the *Dispose* method always runs by calling it from the destructor. This acts as a useful backup. You might forget to call the *Dispose* method, but at least you can be sure that it will be called, even if it's only when the program shuts down. Here's an example of how to do this:

```
class Example : IDisposable
{
    ...
    ~Example()
    {
        Dispose();
    }

    public virtual void Dispose()
    {
        if (!this.disposed)
        {
            try {
                // release scarce resource here
            }
            finally {
                this.disposed = true;
                GC.SuppressFinalize(this);
            }
        }
    }

    public void SomeBehavior() // example method
    {
        checkIfDisposed();
        ...
    }
    ...
    private void checkIfDisposed()
    {
        if (this.disposed)
        {
            throw new ObjectDisposedException("Example: object has been disposed");
        }
    }

    private Resource scarce;
    private bool disposed = false;
}
```

Notice the following features of the *Example* class:

- The class implements the *IDisposable* interface.

- The destructor calls *Dispose*.

- The *Dispose* method is public and can be called at any time.

- The *Dispose* method can safely be called multiple times. The variable *disposed* indicates whether the method has already been run. The scarce resource is released only the first time the method runs.

- The *Dispose* method calls the static *GC.SuppressFinalize* method. This method stops the garbage collector from calling the destructor on this object, because the object has now been finalized.

- All the regular methods of the class (such as *SomeBehavior*) check to see whether the object has already been disposed. If it has, they throw an exception.

Making Code Exception-Safe

In the following exercise, you will rewrite a small piece of code to make the code exception-safe. The code opens a text file, reads its contents one line at a time, writes these lines to a text box on a form on the screen, and then closes the text file. However, if an exception arises as the file is read or as the lines are written to the text box, the call to close the text file will be bypassed. You will rewrite the code to use a *using* statement instead, ensuring that the code is exception-safe.

Write a *using* statement

1. Start Microsoft Visual Studio 2008 if it is not already running.

2. Open the *UsingStatement* project, located in the \Microsoft Press\Visual CSharp Step by Step\Chapter 14\UsingStatement folder in your Documents folder.

3. On the *Debug* menu, click *Start Without Debugging*.

   A Windows Presentation Foundation (WPF) form appears.

4. On the form, click *Open File*.

5. In the *Open* dialog box, move to the \Microsoft Press\Visual CSharp Step by Step\Chapter 14\UsingStatement\UsingStatement folder in your Documents folder, and select the Window1.xaml.cs source file.

   This is the source file for the application itself.

6. Click *Open*.

The contents of the file are displayed in the form, as shown here:

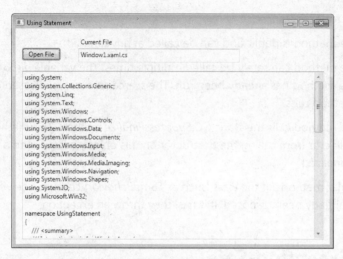

7. Close the form to return to Visual Studio 2008.

8. Open the Window1.xaml.cs file in the *Code and Text Editor* window, and then locate the *openFileDialogFileOk* method.

The method looks like this:

```
private void openFileDialogFileOk(object sender,
System.ComponentModel.CancelEventArgs e)
{
    string fullPathname = openFileDialog.FileName;
    FileInfo src = new FileInfo(fullPathname);
    fileName.Text = src.Name;
    source.Clear();

    TextReader reader = new StreamReader(fullPathname);
    string line;
    while ((line = reader.ReadLine()) != null)
    {
        source.Text += line + "\n";
    }
    reader.Close();
}
```

The variables *fileName*, *openFileDialog*, and *source* are three private fields of the *Window1* class. The problem with this code is that the call to *reader.Close* is not guaranteed to execute. If an exception occurs after opening the file, the method will terminate with an exception, but the file will remain open until the application finishes.

9. Modify the *openFileDialogFileOk* method, and wrap the code that processes the file in a *using* statement (including opening and closing braces), as shown in bold here. Remove the statement that closes the *TextReader* object.

```
private void openFileDialogFileOk(object sender,
System.ComponentModel.CancelEventArgs e)
{
    string fullPathname = openFileDialog.FileName;
    FileInfo src = new FileInfo(fullPathname);
    fileName.Text = src.Name;
    source.Clear();
    using (TextReader reader = new StreamReader(fullPathname))
    {
        string line;
        while ((line = reader.ReadLine()) != null)
        {
            source.Text += line + "\n";
        }
    }
}
```

You no longer need to call *reader.Close* because it will be invoked automatically by the *Dispose* method of the *StreamReader* class when the *using* statement completes. This applies whether the *using* statement finishes naturally or terminates because of an exception.

10. On the *Debug* menu, click *Start Without Debugging*.

11. Verify that the application works as before, and then close the form.

■ If you want to continue to the next chapter:

Keep Visual Studio 2008 running, and turn to Chapter 15.

■ If you want to exit Visual Studio 2008 now:

On the *File* menu, click *Exit*. If you see a *Save* dialog box, click *Yes* (if you are using Visual Studio 2008) or *Save* (if you are using Visual C# 2008 Express Edition) and save the project.

Chapter 14 Quick Reference

To	Do this
| Write a destructor | Write a method whose name is the same as the name of the class and is prefixed with a tilde (~). The method must not have an access modifier (such as *public*) and cannot have any parameters or return a value. For example:<br><br>```csharp
class Example
{
    ~Example()
    {
        ...
    }
}
``` |
| Call a destructor | You can't call a destructor. Only the garbage collector can call a destructor. |
| Force garbage collection (not recommended) | Call *System.GC.Collect*. |
| Release a resource at a known point in time (but at the risk of memory leaks if an exception interrupts the execution) | Write a disposal method (a method that disposes of a resource) and call it explicitly from the program. For example:<br><br>```csharp
class TextReader
{
    ...
    public virtual void Close()
    {
        ...
    }
}

class Example
{
    void Use()
    {
        TextReader reader = ...;
        // use reader
        reader.Close();
    }
}
``` |

Release a resource at a known point in time in an exception-safe manner (the recommended approach)	Release the resource with a *using* statement. For example:

```
class TextReader : IDisposable
{
    ...
    public virtual void Dispose()
    {
        // calls Close
    }
    public virtual void Close()
    {
        ...
    }
}

class Example
{
    void Use()
    {
        using (TextReader reader = ...)
        {
            // use reader
        }
    }
}
```

Part III

Creating Components

Chapter 15
Implementing Properties to Access Fields

After completing this chapter, you will be able to:

- Encapsulate logical fields by using properties.

- Control read access to properties by declaring *get* accessors.

- Control write access to properties by declaring *set* accessors.

- Create interfaces that declare properties.

- Implement interfaces containing properties by using structures and classes.

- Generate properties automatically based on field definitions.

- Use properties to initialize objects.

The first two parts of this book have introduced the core syntax of the C# language and have shown you how to use C# to build new types using structures, enumerations, and classes. You have also seen how the runtime manages the memory used by variables and objects when a program runs, and you should now understand the life cycle of C# objects. The chapters in Part III, "Creating Components," build on this information, showing you how to use C# to create reusable components—functional classes that you can reuse in many different applications.

This chapter looks at how to define and use properties to hide fields in a class. Previous chapters have emphasized that you should make the fields in a class private and provide methods to store values in them and to retrieve their values. This approach provides safe and controlled access to fields and enables you to encapsulate additional logic and rules concerning the values that are permitted. However, the syntax for accessing a field in this way is unnatural. When you want to read or write a variable, you normally use an assignment statement, so calling a method to achieve the same effect on a field (which is, after all, just a variable) feels a little clumsy. Properties are designed to alleviate this awkwardness.

Implementing Encapsulation by Using Methods

First let's recap the original motivation for using methods to hide fields.

Consider the following structure that represents a position on a computer screen as a pair of coordinates, *x* and *y*. Assume that the range of valid values for the *x*-coordinate lies between 0 and 1280 and the range of valid values for the *y*-coordinate lies between 0 and 1024:

```
struct ScreenPosition
{
    public ScreenPosition(int x, int y)
    {
        this.X = rangeCheckedX(x);
        this.Y = rangeCheckedY(y);
    }

    public int X;
    public int Y;

    private static int rangeCheckedX(int x)
    {
        if (x < 0 || x > 1280)
        {
            throw new ArgumentOutOfRangeException("X");
        }
        return x;
    }

    private static int rangeCheckedY(int y)
    {
        if (y < 0 || y > 1024)
        {
            throw new ArgumentOutOfRangeException("Y");
        }
        return y;
    }
}
```

One problem with this structure is that it does not follow the golden rule of encapsulation—that is, it does not keep its data private. Public data is a bad idea because its use cannot be checked and controlled. For example, the *ScreenPosition* constructor range checks its parameters, but no such check can be done on the "raw" access to the public fields. Sooner or later (probably sooner), either X or Y will stray out of its acceptable range, possibly as the result of a programming error:

```
ScreenPosition origin = new ScreenPosition(0, 0);
...
int xpos = origin.X;
origin.Y = -100; // oops
```

The common way to solve this problem is to make the fields private and add an accessor method and a modifier method to respectively read and write the value of each private field. The modifier methods can then range-check new field values. For example, the following code contains an accessor (*GetX*) and a modifier (*SetX*) for the *X* field. Notice that *SetX* checks its parameter value.

```
struct ScreenPosition
{
    ...
    public int GetX()
    {
        return this.x;
    }

    public void SetX(int newX)
    {
        this.x = rangeCheckedX(newX);
    }
    ...
    private static int rangeCheckedX(int x) { ... }
    private static int rangeCheckedY(int y) { ... }
    private int x, y;
}
```

The code now successfully enforces the range constraints, which is good. However, there is a price to pay for this valuable guarantee—*ScreenPosition* no longer has a natural fieldlike syntax; it uses awkward method-based syntax instead. The following example increases the value of *X* by 10. To do so, it has to read the value of *X* by using the *GetX* accessor method and then write the value of *X* by using the *SetX* modifier method.

```
int xpos = origin.GetX();
origin.SetX(xpos + 10);
```

Compare this with the equivalent code if the *X* field were public:

```
origin.X += 10;
```

There is no doubt that, in this case, using public fields is cleaner, shorter, and easier. Unfortunately, using public fields breaks encapsulation. Properties enable you to combine the best of both examples: to retain encapsulation while allowing a fieldlike syntax.

What Are Properties?

A *property* is a cross between a field and a method—it looks like a field but acts like a method. You access a property using exactly the same syntax that you use to access a field. However, the compiler automatically translates this fieldlike syntax into calls to accessor methods. A property declaration looks like this:

```
AccessModifier Type PropertyName
{
    get
    {
        // read accessor code
    }

    set
    {
        // write accessor code
    }
}
```

A property can contain two blocks of code, starting with the *get* and *set* keywords. The *get* block contains statements that execute when the property is read, and the *set* block contains statements that run when the property is written to. The type of the property specifies the type of data read and written by the *get* and *set* accessors.

The next code example shows the *ScreenPosition* structure rewritten by using properties. When reading this code, notice the following:

- Lowercase *x* and *y* are *private* fields.

- Uppercase *X* and *Y* are *public* properties.

- All *set* accessors are passed the data to be written by using a hidden, built-in parameter named *value*.

> **Tip** The fields and properties follow the standard Microsoft Visual C# *public/private* naming convention. Public fields and properties should start with an uppercase letter, but private fields and properties should start with a lowercase letter.

```
struct ScreenPosition
{
    public ScreenPosition(int X, int Y)
    {
        this.x = rangeCheckedX(X);
        this.y = rangeCheckedY(Y);
    }
```

```
        public int X
        {
            get { return this.x; }
            set { this.x = rangeCheckedX(value); }
        }

        public int Y
        {
            get { return this.y; }
            set { this.y = rangeCheckedY(value); }
        }

        private static int rangeCheckedX(int x) { ... }
        private static int rangeCheckedY(int y) { ... }
        private int x, y;
}
```

In this example, a private field directly implements each property, but this is only one way to implement a property. All that is required is that a *get* accessor return a value of the specified type. Such a value could easily be calculated dynamically rather than being simply retrieved from stored data, in which case there would be no need for a physical field.

> **Note** Although the examples in this chapter show how to define properties for a structure, they are equally applicable to classes; the syntax is the same.

Using Properties

When you use a property in an expression, you can use it in a read context (when you are reading its value) and in a write context (when you are modifying its value). The following example shows how to read values from the *X* and *Y* properties of a *ScreenPosition* structure:

```
ScreenPosition origin = new ScreenPosition(0, 0);
int xpos = origin.X;    // calls origin.X.get
int ypos = origin.Y;    // calls origin.Y.get
```

Notice that you access properties and fields by using the same syntax. When you use a property in a read context, the compiler automatically translates your fieldlike code into a call to the *get* accessor of that property. Similarly, if you use a property in a write context, the compiler automatically translates your fieldlike code into a call to the *set* accessor of that property:

```
origin.X = 40;      // calls origin.X.set, with value set to 40
origin.Y = 100;     // calls origin.Y.Set, with value set to 100
```

The values being assigned are passed in to the *set* accessors by using the *value* variable, as described in the preceding section. The runtime does this automatically.

It's also possible to use a property in a read/write context. In this case, both the *get* accessor and the *set* accessor are used. For example, the compiler automatically translates statements such as the following into calls to the *get* and *set* accessors:

```
origin.X += 10;
```

> **Tip** You can declare *static* properties in the same way that you can declare *static* fields and methods. Static properties are accessed by using the name of the class or structure rather than an instance of the class or structure.

Read-Only Properties

You're allowed to declare a property that contains only a *get* accessor. In this case, you can use the property only in a read context. For example, here's the *X* property of the *ScreenPosition* structure declared as a read-only property:

```
struct ScreenPosition
{
    ...
    public int X
    {
        get { return this.x; }
    }
}
```

The *X* property does not contain a *set* accessor; therefore, any attempt to use *X* in a write context will fail. For example:

```
origin.X = 140; // compile-time error
```

Write-Only Properties

Similarly, you can declare a property that contains only a *set* accessor. In this case, you can use the property only in a write context. For example, here's the *X* property of the *ScreenPosition* structure declared as a write-only property:

```
struct ScreenPosition
{
    ...
    public int X
    {
        set { this.x = rangeCheckedX(value); }
    }
}
```

The *X* property does not contain a *get* accessor; any attempt to use *X* in a read context will fail. For example:

```
Console.WriteLine(origin.X);    // compile-time error
origin.X = 200;                 // compiles OK
origin.X += 10;                 // compile-time error
```

> **Note** Write-only properties are useful for secure data such as passwords. Ideally, an application that implements security should allow you to set your password but should never allow you to read it back. A login method should compare a user-supplied string with the stored password and return only an indication of whether they match.

Property Accessibility

You can specify the accessibility of a property (*public*, *private*, or *protected*) when you declare it. However, it is possible within the property declaration to override the property accessibility for the *get* and *set* accessors. For example, the version of the *ScreenPosition* structure shown here defines the *set* accessors of the *X* and *Y* properties as *private*. (The *get* accessors are *public*, because the properties are *public*.)

```
struct ScreenPosition
{
    ...
    public int X
    {
        get { return this.x; }
        private set { this.x = rangeCheckedX(value); }
    }

    public int Y
    {
        get { return this.y; }
        private set { this.y = rangeCheckedY(value); }
    }
    ...
    private int x, y;
}
```

You must observe some rules when defining accessors with different accessibility from one another:

- You can change the accessibility of only one of the accessors when you define it. It wouldn't make much sense to define a property as *public* only to change the accessibility of both accessors to *private* anyway!

- The modifier must not specify an accessibility that is less restrictive than that of the property. For example, if the property is declared as *private*, you cannot specify the read accessor as *public*. (Instead, you would make the property *public* and make the read accessor *private*.)

Properties and Field Names: A Warning

Although it is a commonly accepted practice to give properties and private fields the same name that differs only in the case of the initial letter, you should be aware of one drawback. Examine the following code, which implements a class named *Employee*. The *employeeID* field is private, but the *EmployeeID* property provides pubic access to this field.

```
class Employee
{
    private int employeeID;

    public int EmployeeID;
    {
        get { return this.EmployeeID; }
        set { this.EmployeeID = value; }
    }
}
```

This code will compile perfectly well, but it results in a program raising a *StackOverflowException* whenever the *EmployeeID* property is accessed. This is because the *get* and *set* accessors reference the property (uppercase *E*) rather than the private field (lowercase *e*), which causes an endless recursive loop that eventually causes the process to exhaust the available memory. This sort of bug is very difficult to spot!

Understanding the Property Restrictions

Properties look, act, and feel like fields. However, they are not true fields, and certain restrictions apply to them:

- You can assign a value through a property of a structure or class only after the structure or class has been initialized. The following code example is illegal, as the location variable has not been initialized (by using *new*):

```
ScreenPosition location;
location.X = 40; // compile-time error, location not assigned
```

Note This might seem trivial, but if *X* were a field rather than a property, the code would be legal. What this really means is that there are some differences between fields and properties. You should define structures and classes by using properties from the start, rather than by using fields that you later migrate to properties—code that uses your classes and structures might no longer work after you change fields into properties. We will return to this matter in the section "Generating Automatic Properties" later in this chapter.

- You can't use a property as a *ref* or an *out* argument to a method (although you can use a writable field as a *ref* or an *out* argument). This makes sense because the property doesn't really point to a memory location but rather to an accessor method. For example:

```
MyMethod(ref location.X); // compile-time error
```

- A property can contain at most one *get* accessor and one *set* accessor. A property cannot contain other methods, fields, or properties.

- The *get* and *set* accessors cannot take any parameters. The data being assigned is passed to the *set* accessor automatically by using the *value* variable.

- You can't declare *const* properties. For example:

```
const int X { get { ... } set { ... } } // compile-time error
```

Using Properties Appropriately

Properties are a powerful feature with a clean, fieldlike syntax. Used in the correct manner, properties help to make code easier to understand and maintain. However, they are no substitute for careful object-oriented design that focuses on the behavior of objects rather than on the properties of objects. Accessing private fields through regular methods or through properties does not, by itself, make your code well-designed. For example, a bank account holds a balance. You might therefore be tempted to create a *Balance* property on a *BankAccount* class, like this:

```
class BankAccount
{
    ...
    public money Balance
    {
        get { ... }
        set { ... }
    }

    private money balance;
}
```

This would be a poor design. It fails to represent the functionality required when withdrawing money from and depositing money into an account. (If you know of a bank that allows you to set the balance of your account directly without depositing money, please let me know!) When you're programming, try to express the problem you are solving in the solution and don't get lost in a mass of low-level syntax:

```
class BankAccount
{
    ...
    public money Balance { get { ... } }
    public void Deposit(money amount) { ... }
    public bool Withdraw(money amount) { ... }
    private money balance;
}
```

Declaring Interface Properties

You encountered interfaces in Chapter 13, "Creating Interfaces and Defining Abstract Classes." Interfaces can define properties as well as methods. To do this, you specify the *get* or *set* keyword, or both, but replace the body of the *get* or *set* accessor with a semicolon. For example:

```
interface IScreenPosition
{
    int X { get; set; }
    int Y { get; set; }
}
```

Any class or structure that implements this interface must implement the *X* and *Y* properties with *get* and *set* accessor methods. For example:

```
struct ScreenPosition : IScreenPosition
{
    ...
    public int X
    {
        get { ... }
        set { ... }
    }

    public int Y
    {
        get { ... }
        set { ... }
    }
    ...
}
```

If you implement the interface properties in a class, you can declare the property implementations as *virtual*, which enables derived classes to override the implementations. For example:

```
class ScreenPosition : IScreenPosition
{
    ...
    public virtual int X
    {
        get { ... }
        set { ... }
    }

    public virtual int Y
    {
        get { ... }
        set { ... }
    }
    ...
}
```

Note This example shows a *class*. Remember that the *virtual* keyword is not valid when creating a *struct* because structures are implicitly sealed.

You can also choose to implement a property by using the explicit interface implementation syntax covered in Chapter 13. An explicit implementation of a property is nonpublic and nonvirtual (and cannot be overridden). For example:

```
struct ScreenPosition : IScreenPosition
{
    ...
    int IScreenPosition.X
    {
        get { ... }
        set { ... }
    }

    int IScreenPosition.Y
    {
        get { ... }
        set { ... }
    }
    ...
    private int x, y;
}
```

Using Properties in a Windows Application

When you set property values of objects such as *TextBox* controls, *Windows*, and *Button* controls by using the Properties window in Microsoft Visual Studio 2008, you are actually generating code that sets the values of these properties at run time. Some components have a large number of properties, although some properties are more commonly used than others. You can write your own code to modify many of these properties at run time by using the same syntax you have seen throughout this chapter.

In the following exercise, you will use some predefined properties of the *TextBox* controls and the *Window* class to create a simple application that continually displays the size of its main window, even when the window is resized.

Use properties

1. Start Visual Studio 2008 if it is not already running.

2. Open the *WindowProperties* project, located in the \Microsoft Press\Visual CSharp Step by Step\Chapter 15\WindowProperties folder in your Documents folder.

3. On the *Debug* menu, click *Start Without Debugging*.

The project builds and runs. A window (a Windows Presentation Foundation [WPF] form) appears, displaying two empty text boxes labeled *Width* and *Height*.

In the program, the text box controls are named *width* and *height*. They are currently empty. You will add code to the application that displays the current size of the window.

4. Close the form, and return to the Visual Studio 2008 programming environment.

5. Display the Window1.xaml.cs file in the *Code and Text Editor* window, and locate the *sizeChanged* method.

This method is called by the *Window1* constructor. You will use it to display the current size of the form in the *width* and *height* text boxes. You will make use of the *ActualWidth* and *ActualHeight* properties of the *Window* class. These properties return the current width and height of the form as *double* values.

6. Add two statements to the *sizeChanged* method to display the size of the form. The first statement should read the value of the *ActualWidth* property of the form, convert it to a string, and assign this value to the *Text* property of the *width* text box. The second statement should read the value of the *ActualHeight* property of the form, convert it to a string, and assign this value to the *Text* property of the *height* text box.

The *sizeChanged* method should look exactly like this:

```
private void sizeChanged()
{
    width.Text = this.ActualWidth.ToString();
    height.Text = this.ActualHeight.ToString();
}
```

7. Locate the *window1SizeChanged* method.

This method runs whenever the size of the window changes when the application is running. Notice that this method calls the *sizeChanged* method to display the new size of the window in the text boxes.

8. On the *Debug* menu, click *Start Without Debugging* to build and run the project.

The form displays the two text boxes containing the values *305* and *155*. These are the default dimensions of the form, specified when the form was designed.

9. Resize the form. Notice that the text in the text boxes changes to reflect the new size.

10. Close the form, and return to the Visual Studio 2008 programming environment.

Generating Automatic Properties

This chapter mentioned earlier that the principal purpose of properties is to hide the implementation of fields from the outside world. This is fine if your properties actually perform some useful work, but if the *get* and *set* accessors simply wrap operations that just read or assign a value to a field, you might be questioning the value of this approach. There are at least two good reasons why you should define properties rather than exposing data as public fields:

- **Compatibility with applications**   Fields and properties expose themselves by using different metadata in assemblies. If you develop a class and decide to use public fields, any applications that use this class will reference these items as fields. Although you use the same C# syntax for reading and writing a field that you use when reading and writing a property, the compiled code is actually quite different—the C# compiler just hides the differences from you. If you later decide that you really do need to change these fields to properties (maybe the business requirements have changed, and you need to perform additional logic when assigning values), existing applications will not be able to use the updated version of the class without being recompiled. This is awkward if you have deployed the application on a large number of users' desktops throughout an organization. There are ways around this, but it is generally better to avoid getting into this situation in the first place.

- **Compatibility with interfaces**   If you are implementing an interface and the interface defines an item as a property, you must write a property that matches the specification in the interface, even if the property just reads and writes data in a private field. You cannot implement a property simply by exposing a public field with the same name.

The designers of the C# language recognized that programmers are busy people who should not have to waste their time writing more code than they need to. To this end, the C# compiler can generate the code for properties for you automatically, like this:

```
class Circle
{
    public int Radius{ get; set; }
    ...
}
```

In this example, the *Circle* class contains a property named *Radius*. Apart from the type of this property, you have not specified how this property works—the *get* and *set* accessors are empty. The C# compiler converts this definition to a private field and a default implementation that looks similar to this:

```
class Circle
{
    private int _radius;
    public int Radius{
        get
        {
            return this._radius;
        }
        set
        {
            this._radius = value;
        }
    }
    ...
}
```

So for very little effort, you can implement a simple property by using automatically generated code, and if you need to include additional logic later, you can do so without breaking any existing applications. You should note, however, that you must specify both a *get* and a *set* accessor with an automatically generated property—an automatic property cannot be read-only or write-only.

 Note The syntax for defining an automatic property is almost identical to the syntax for defining a property in an interface. The exception is that an automatic property can specify an access modifier, such as *private*, *public*, or *protected*.

Initializing Objects by Using Properties

In Chapter 7, "Creating and Managing Classes and Objects," you learned how to define constructors to initialize an object. An object can have multiple constructors, and you can define constructors with varying parameters to initialize different elements in an object. For example, you could define a class that models a triangle like this:

```
public class Triangle
{
    private int side1Length;
    private int side2Length;
    private int side3Length;

    // default constructor - default values for all sides
    public Triangle()
    {
        this.side1Length = this.side2Length = this.side3Length = 10;
    }

    // specify length for side1Length, default values for the others
    public Triangle(int length1)
```

```
    {
        this.side1Length = length1;
        this.side2Length = this.side3Length = 10;
    }

    // specify length for side1Length and side2Length,
    // default value for side3Length
    public Triangle(int length1, int length2)
    {
        this.side1Length = length1;
        this.side2Length = length2;
        this.side3Length = 10;
    }

    // specify length for all sides
    public Triangle(int length1, int length2, int length3)
    {
        this.side1Length = length1;
        this.side2Length = length2;
        this.side3Length = length3;
    }
}
```

Depending on how many fields a class contains and the various combinations you want to enable for initializing the fields, you could end up writing a lot of constructors. There are also potential problems if many of the fields have the same type: you might not be able to write a unique constructor for all combinations of fields. For example, in the preceding *Triangle* class, you could not easily add a constructor that initializes only the *side1Length* and *side3Length* fields because it would not have a unique signature; it would take two *int* parameters, and the constructor that initializes *side1Length* and *side2Length* already has this signature. The solution is to initialize the private fields to their default values and to define properties, like this:

```
public class Triangle
{
    private int side1Length = 10;
    private int side2Length = 10;
    private int side3Length = 10;

    public int Side1Length
    {
        set { this.side1Length = value; }
    }

    public int Side2Length
    {
        set { this.side2Length = value; }
    }

    public int Side3Length
    {
        set { this.side3Length = value; }
    }
}
```

When you create an instance of a class, you can initialize it by specifying values for any public properties that have *set* accessors. This means that you can create *Triangle* objects and initialize any combination of the three sides, like this:

```
Triangle tri1 = new Triangle { Side3Length = 15 };
Triangle tri2 = new Triangle { Side1Length = 15, Side3Length = 20 };
Triangle tri3 = new Triangle { Side2Length = 12, Side3Length = 17 };
Triangle tri4 = new Triangle { Side1Length = 9, Side2Length = 12,
                               Side3Length = 15 };
```

This syntax is known as an *object initializer*. When you invoke an object initializer in this way, the C# compiler generates code that calls the default constructor and then calls the *set* accessor of each named property to initialize it with the value specified. You can specify object initializers in combination with nondefault constructors as well. For example, if the *Triangle* class also provided a constructor that took a single string parameter describing the type of triangle, you could invoke this constructor and initialize the other properties like this:

```
Triangle tri5 = new Triangle("Equilateral triangle") { Side1Length = 3,
                                                       Side2Length = 3,
                                                       Side3Length = 3 };
```

The important point to remember is that the constructor runs first and the properties are set afterward. Understanding this sequencing is important if the constructor sets fields in an object to specific values and the properties that you specify change these values.

You can also use object initializers with automatic properties, as you will see in the next exercise. In this exercise, you will define a class for modeling regular polygons, containing automatic properties for providing access to the number of sides the polygon contains and the length of these sides.

Define automatic properties and use object initializers

1. In Visual Studio 2008, open the *AutomaticProperties* project, located in the \Microsoft Press\Visual CSharp Step by Step\Chapter 15\AutomaticProperties folder in your Documents folder.

   The *AutomaticProperties* project contains the Program.cs file, defining the *Program* class with the *Main* and *Entrance* methods that you have seen in previous exercises.

2. In *Solution Explorer*, right-click the *AutomaticProperties* project, point to *Add*, and then click *Class*. In the *Add New Item—AutomaticProperties* dialog box, in the *Name* text box, type **Polygon.cs**, and then click *Add*.

   The Polygon.cs file, holding the *Polygon* class, is created and added to the project and appears in the *Code and Text Editor* window.

3. Add the automatic properties *NumSides* and *SideLength*, shown here in bold, to the *Polygon* class:

```
class Polygon
{
    public int NumSides { get; set; }
    public double SideLength { get; set; }
}
```

4. Add the following default constructor to the *Polygon* class:

```
class Polygon
{
    public Polygon()
    {
        this.NumSides = 4;
        this.SideLength = 10.0;
    }
    ...
}
```

In this exercise, the default polygon is a square with sides 10 units long.

5. Display the Program.cs file in the *Code and Text Editor* window.

6. Add the statements shown here in bold to the *Entrance* method:

```
static void Entrance()
{
    Polygon square = new Polygon();
    Polygon triangle = new Polygon { NumSides = 3 };
    Polygon pentagon = new Polygon { SideLength = 15.5, NumSides = 5 };
}
```

These statements create *Polygon* objects. The *square* variable is initialized by using the default constructor. The *triangle* and *pentagon* variables are also initialized by using the default constructor, and then this code changes the value of the properties exposed by the *Polygon* class. In the case of the *triangle* variable, the *NumSides* property is set to 3, but the *SideLength* property is left at its default value of *10.0*. For the *pentagon* variable, the code changes the values of the *SideLength* and *NumSides* properties.

7. Add the following code to the end of the *Entrance* method:

```
static void Entrance()
{
    ...
    Console.WriteLine("Square: number of sides is {0}, length of each side is {1}",
        square.NumSides, square.SideLength);
    Console.WriteLine("Triangle: number of sides is {0}, length of each side is {1}",
        triangle.NumSides, triangle.SideLength);
    Console.WriteLine("Pentagon: number of sides is {0}, length of each side is {1}",
        pentagon.NumSides, pentagon.SideLength);
}
```

These statements display the values of the *NumSides* and *SideLength* properties for each *Polygon* object.

8. On the *Debug* menu, click *Start Without Debugging*.

Verify that the program builds and runs, writing the message shown here to the console:

9. Press the Enter key to close the application and return to Visual Studio 2008.

You have now seen how to create automatic properties and how to use properties when initializing objects.

- If you want to continue to the next chapter:

  Keep Visual Studio 2008 running, and turn to Chapter 16.

- If you want to exit Visual Studio 2008 now:

  On the *File* menu, click *Exit*. If you see a *Save* dialog box, click *Yes* (if you are using Visual Studio 2008) or *Save* (if you are using Visual C# 2008 Express Edition) and save the project.

Chapter 15 Quick Reference

To	Do this
Declare a read/write property for a structure or class	Declare the type of the property, its name, a *get* accessor, and a *set* accessor. For example:

```
struct ScreenPosition
{
    ...
    public int X
    {
        get { ... }
        set { ... }
    }
    ...
}
```

Declare a read-only property for a structure or class	Declare a property with only a *get* accessor. For example:
	
	``
	struct ScreenPosition
	{
	    ...
	    public int X
	    {
	        get { ... }
	    }
	    ...
	}
	``

Declare a write-only property for a structure or class	Declare a property with only a *set* accessor. For example:
	
	``
	struct ScreenPosition
	{
	    ...
	    public int X
	    {
	        set { ... }
	    }
	    ...
	}
	``

Declare a property in an interface	Declare a property with just the *get* or *set* keyword, or both. For example:
	
	``
	interface IScreenPosition
	{
	    int X { get; set; } // no body
	    int Y { get; set; } // no body
	}
	``

Implement an interface property in a structure or class	In the class or structure that implements the interface, declare the property and implement the accessors. For example:
	
	``
	struct ScreenPosition : IScreenPosition
	{
	    public int X
	    {
	        get { ... }
	        set { ... }
	    }
	    public int Y
	    {
	        get { ... }
	        set { ... }
	    }
	}
	``

Create an automatic property	In the class or structure that contains the property, define the property with empty *get* and *set* accessors. For example:

```
class Polygon
{
    public int NumSides { get; set; }
}
```

Use properties to initialize an object	Specify the properties and their values as a list enclosed in braces when constructing the object. For example:

```
Triangle tri3 = new Triangle { Side2Length = 12,

                               Side3Length = 17 };
```

Chapter 16
Using Indexers

After completing this chapter, you will be able to:

- Encapsulate logical arraylike access to an object by using indexers.

- Control read access to indexers by declaring *get* accessors.

- Control write access to indexers by declaring *set* accessors.

- Create interfaces that declare indexers.

- Implement indexers in structures and classes that inherit from interfaces.

The preceding chapter described how to implement and use properties as a means of providing controlled access to the fields in a class. Properties are useful for mirroring fields that contain a single value. However, indexers are invaluable if you want to provide access to items that contain multiple values by using a natural and familiar syntax.

What Is an Indexer?

An *indexer* is a smart array in exactly the same way that a property is a smart field. The syntax that you use for an indexer is exactly the same as the syntax that you use for an array. The best way to understand indexers is to work through an example. First we'll examine a problem and examine a weak solution that doesn't use indexers. Then we'll work through the same problem and look at a better solution that does use indexers. The problem concerns integers, or more precisely, the *int* type.

An Example That Doesn't Use Indexers

You normally use an *int* to hold an integer value. Internally, an *int* stores its value as a sequence of 32 bits, where each bit can be either 0 or 1. Most of the time, you don't care about this internal binary representation; you just use an *int* type as a bucket to hold an integer value. However, sometimes programmers use the *int* type for other purposes: some programs manipulate the individual bits within an *int*. In other words, occasionally a program might use an *int* because it holds 32 bits and not because it can represent an integer value. (If you are an old C hack like I am, what follows should have a very familiar feel!)

Note Some older programs used *int* types to try to save memory. Such programs typically date back to when the size of computer memory was measured in kilobytes rather than the gigabytes available these days and memory was at an absolute premium. A single *int* holds 32 bits, each of which can be 1 or 0. In some cases, programmers assigned 1 to indicate the value *true* and 0 to indicate *false* and then employed an *int* as a set of Boolean values.

As an example, the following expression uses the left-shift (<<) and bitwise AND (&) operators to determine whether the sixth bit of the *int* named *bits* is set to *0* or to *1*:

```
(bits & (1 << 6)) != 0
```

If the bit at position 6 is *0*, this expression evaluates to *false*; if the bit at position 6 is *1*, this expression evaluates to *true*. This is a fairly complicated expression, but it's trivial in comparison with the following expression, which uses the compound assignment operator &= to set the bit at position 6 to *0*:

```
bits &= ~(1 << 6)
```

Note The bitwise operators count the positions of bits from right to left, so bit 0 is the rightmost bit, and the bit at position 6 is the bit six places from the right.

Similarly, if you want to set the bit at position 6 to *1*, you can use a bitwise OR (|) operator. The following complicated expression is based on the compound assignment operator |=:

```
bits |= (1 << 6)
```

The trouble with these examples is that although they work, it's not clear why or how they work. They're complicated, and the solution is a very low-level one: it fails to create an abstraction of the problem that it solves.

The Bitwise and Shift Operators

You might have noticed some unfamiliar symbols in the expressions shown in these examples—in particular, ~, <<, |, and &. These are some of the bitwise and shift operators, and they are used to manipulate the individual bits held in *int* and *long* data types.

- The NOT (~) operator is a unary operator that performs a bitwise complement. For example, if you take the 8-bit value *11001100* (*204* decimal) and apply the ~ operator to it, you obtain the result *00110011* (*51* decimal)—all the 1s in the original value become 0s, and all the 0s become 1s.

- The left-shift (<<) operator is a binary operator that performs a left shift. The expression *204 << 2* returns the value *48*. (In binary, *204* decimal is *11001100*, and left-shifting it by two places yields *00110000*, or *48* decimal.) The far-left bits are discarded, and zeros are introduced from the right. There is a corresponding right-shift operator >>.

- The OR (|) operator is a binary operator that performs a bitwise OR operation, returning a value containing a 1 in each position in which either of the operands has a 1. For example, the expression *204 | 24* has the value *220* (*204* is *11001100*, *24* is *00011000*, and *220* is *11011100*).

- The AND (&) operator performs a bitwise AND operation. AND is similar to the bitwise OR operator, except that it returns a value containing a 1 in each position where both of the operands have a 1. So *204 & 20* is *8* (*204* is *11001100*, *24* is *00011000*, and *8* is *00001000*).

- The XOR (^) operator performs a bitwise exclusive OR operation, returning a 1 in each bit where there is a 1 in one operand or the other but not both. (Two 1s yield a 0—this is the "exclusive" part of the operator.) So *204 ^ 24* is *212* (*11001100 ^ 00011000* is *11010100*).

The Same Example Using Indexers

Let's pull back from the preceding low-level solution for a moment and stop to remind ourselves what the problem is. We'd like to use an *int* not as an *int* but as an array of 32 bits. Therefore, the best way to solve this problem is to use an *int* as if it were an array of 32 bits! In other words, what we'd like to be able to write to access the bit at index 6 of the *bits* variable is something like this:

```
bits[6]
```

And, for example, to set the bit at index 6 to *true*, we'd like to be able to write:

```
bits[6] = true
```

Unfortunately, you can't use the square bracket notation on an *int*—it works only on an array or on a type that behaves like an array. So the solution to the problem is to create a new type that acts like, feels like, and is used like an array of *bool* variables but is implemented by using an *int*. You can achieve this feat by defining an indexer. Let's call this new type *IntBits*. *IntBits* will contain an *int* value (initialized in its constructor), but the idea is that we'll use *IntBits* as an array of *bool* variables.

> **Tip** The *IntBits* type is small and lightweight, so it makes sense to create it as a structure rather than as a class.

```
struct IntBits
{
    public IntBits(int initialBitValue)
    {
        bits = initialBitValue;
    }

    // indexer to be written here

    private int bits;
}
```

To define the indexer, you use a notation that is a cross between a property and an array. The indexer for the *IntBits* struct looks like this:

```
struct IntBits
{
    ...
    public bool this [ int index ]
    {
        get
        {
            return (bits & (1 << index)) != 0;
        }

        set
        {
            if (value)  // turn the bit on if value is true; otherwise, turn it off
                bits |=  (1 << index);
            else
                bits &= ~(1 << index);
        }
    }
    ...
}
```

Notice the following points:

- An indexer is not a method—there are no parentheses containing a parameter, but there are square brackets that specify an index. This index is used to specify which element is being accessed.

- All indexers use the *this* keyword in place of the method name. A class or structure can define at most one indexer, and it is always named *this*.

- Indexers contain *get* and *set* accessors just like properties. In this example, the *get* and *set* accessors contain the complicated bitwise expressions previously discussed.

- The index specified in the indexer declaration is populated with the index value specified when the indexer is called. The *get* and *set* accessor methods can read this argument to determine which element should be accessed.

> **Note** You should perform a range check on the index value in the indexer to prevent any unexpected exceptions from occurring in your indexer code.

After you have declared the indexer, you can use a variable of type *IntBits* instead of an *int* and apply the square bracket notation, as shown in the next example:

```
int adapted = 62;      // 62 has the binary representation 111110
IntBits bits = new IntBits(adapted);
bool peek = bits[6];   // retrieve bool at index 6; should be true (1)
bits[0] = true;        // set the bit at index 0 to true (1)
bits[3] = false;       // set the bit at index 3 to false (0)
                       // the value in adapted is now 111011, or 59 in decimal
```

This syntax is certainly much easier to understand. It directly and succinctly captures the essence of the problem.

> **Note** Indexers and properties are similar in that both use *get* and *set* accessors. An indexer is like a property with multiple values. However, although you're allowed to declare *static* properties, *static* indexers are illegal.

Understanding Indexer Accessors

When you read an indexer, the compiler automatically translates your arraylike code into a call to the *get* accessor of that indexer. Consider the following example:

```
bool peek = bits[6];
```

This statement is converted to a call to the *get* accessor for *bits*, and the *index* argument is set to *6*.

Similarly, if you write to an indexer, the compiler automatically translates your arraylike code into a call to the *set* accessor of that indexer, setting the *index* argument to the value enclosed in the square brackets. For example:

```
bits[6] = true;
```

This statement is converted to a call to the *set* accessor for *bits* where *index* is *6*. As with ordinary properties, the data you are writing to the indexer (in this case, *true*) is made available inside the *set* accessor by using the *value* keyword. The type of *value* is the same as the type of indexer itself (in this case, *bool*).

It's also possible to use an indexer in a combined read/write context. In this case, the *get* and *set* accessors are both used. Look at the following statement:

```
bits[6] ^= true;
```

This code is automatically translated into:

```
bits[6] = bits[6] ^ true;
```

This code works because the indexer declares both a *get* and a *set* accessor.

> **Note** You can declare an indexer that contains only a *get* accessor (a read-only indexer) or only a *set* accessor (a write-only accessor).

Comparing Indexers and Arrays

When you use an indexer, the syntax is deliberately very arraylike. However, there are some important differences between indexers and arrays:

- Indexers can use non-numeric subscripts, whereas arrays can use only integer subscripts:

  ```
  public int this [ string name ] { ... } // OK
  ```

> **Tip** Many collection classes, such as *Hashtable*, that implement an associative lookup based on key/value pairs implement indexers to provide a convenient alternative to using the *Add* method to add a new value and as an alternative to iterating through the *Values* property to locate a value in your code. For example, instead of this:
>
> ```
> Hashtable ages = new Hashtable();
> ages.Add("John", 42);
> ```
>
> you can use this:
>
> ```
> Hashtable ages = new Hashtable();
> ages["John"] = 42;
> ```

- Indexers can be overloaded (just like methods), whereas arrays cannot:

  ```
  public Name        this [ PhoneNumber number ] { ... }
  public PhoneNumber this [ Name name ] { ... }
  ```

- Indexers cannot be used as *ref* or *out* parameters, whereas array elements can:

  ```
  IntBits bits;        // bits contains an indexer
  Method(ref bits[1]); // compile-time error
  ```

Properties, Arrays, and Indexers

It is possible for a property to return an array, but remember that arrays are reference types, so exposing an array as a property makes it possible to accidentally overwrite a lot of data. Look at the following structure that exposes an array property named *Data*:

```
struct Wrapper
{
    private int[] data;
    ...
    public int[] Data
    {
        get { return this.data; }
        set { this.data = value; }
    }
}
```

Now consider the following code that uses this property:

```
Wrapper wrap = new Wrapper();
...
int[] myData = wrap.Data;
myData[0]++;
myData[1]++;
```

This looks pretty innocuous. However, because arrays are reference types, the variable *myData* refers to the same object as the private *data* variable in the *Wrapper* structure. Any changes you make to elements in *myData* are made to the *data* array; the expression *myData[0]++* has exactly the same effect as *data[0]++*. If this is not the intention, you should use the *Clone* method in the *get* and *set* accessors of the *Data* property to return a copy of the data array, or make a copy of the value being set, as shown here. (The *Clone* method returns an object, which you must cast to an integer array.)

```
struct Wrapper
{
    private int[] data;
    ...
    public int[] Data
    {
        get { return this.data.Clone() as int[]; }
        set { this.data = value.Clone() as int[]; }
    }
}
```

However, this approach can become very messy and expensive in terms of memory use. Indexers provide a natural solution to this problem—don't expose the entire array as a property; just make its individual elements available through an indexer:

```
struct Wrapper
{
    private int[] data;
    ...
```

```
        public int this [int i]
        {
            get { return this.data[i]; }
            set { this.data[i] = value; }
        }
    }
```

The following code uses the indexer in a similar manner to the property shown earlier:

```
Wrapper wrap = new Wrapper();
...
int[] myData = new int[2];
myData[0] = wrap[0];
myData[1] = wrap[1];
myData[0]++;
myData[1]++;
```

This time, incrementing the values in the *MyData* array has no effect on the original array in the *Wrapper* object. If you really want to modify the data in the *Wrapper* object, you must write statements such as this:

```
wrap[0]++;
```

This is much clearer, and safer!

Indexers in Interfaces

You can declare indexers in an interface. To do this, specify the *get* keyword, the *set* keyword, or both, but replace the body of the *get* or *set* accessor with a semicolon. Any class or structure that implements the interface must implement the *indexer* accessors declared in the interface. For example:

```
interface IRawInt
{
    bool this [ int index ] { get; set; }
}

struct RawInt : IRawInt
{
    ...
    public bool this [ int index ]
    {
        get { ... }
        set { ... }
    }
    ...
}
```

If you implement the interface indexer in a class, you can declare the indexer implementations as virtual. This allows further derived classes to override the *get* and *set* accessors. For example:

```
class RawInt : IRawInt
{
    ...
    public virtual bool this [ int index ]
    {
        get { ... }
        set { ... }
    }
    ...
}
```

You can also choose to implement an indexer by using the explicit interface implementation syntax covered in Chapter 12, "Working with Inheritance." An explicit implementation of an indexer is nonpublic and nonvirtual (and so cannot be overridden). For example:

```
struct RawInt : IRawInt
{
    ...
    bool IRawInt.this [ int index ]
    {
        get { ... }
        set { ... }
    }
    ...
}
```

Using Indexers in a Windows Application

In the following exercise, you will examine a simple phone book application and complete its implementation. You will write two indexers in the *PhoneBook* class: one that accepts a *Name* parameter and returns a *PhoneNumber* and another that accepts a *PhoneNumber* parameter and returns a *Name*. (The *Name* and *PhoneNumber* structures have already been written.) You will also need to call these indexers from the correct places in the program.

Familiarize yourself with the application

1. Start Microsoft Visual Studio 2008 if it is not already running.

2. Open the *Indexers* project, located in the \Microsoft Press\Visual CSharp Step by Step\ Chapter 16\Indexers folder in your Documents folder.

   This is a Windows Presentation Foundation (WPF) application that enables a user to search for the telephone number for a contact and also find the name of a contact that matches a given telephone number.

3. On the *Debug* menu, click *Start Without Debugging*.

   The project builds and runs. A form appears, displaying two empty text boxes labeled *Name* and *Phone Number*. The form also contains three buttons—one to add a name/phone number pair to a list of names and phone numbers held by the application, one to find a phone number when given a name, and one to find a name when given a phone number. These buttons currently do nothing. Your task is to complete the application so that these buttons work.

4. Close the form, and return to Visual Studio 2008.

5. Display the Name.cs file in the *Code and Text Editor* window. Examine the *Name* structure. Its purpose is to act as a holder for names.

   The name is provided as a string to the constructor. The name can be retrieved by using the read-only string property named *Text*. (The *Equals* and *GetHashCode* methods are used for comparing *Name*s when searching through an array of *Name* values—you can ignore them for now.)

6. Display the PhoneNumber.cs file in the *Code and Text Editor* window, and examine the *PhoneNumber* structure. It is similar to the *Name* structure.

7. Display the PhoneBook.cs file in the *Code and Text Editor* window, and examine the *PhoneBook* class.

   This class contains two private arrays: an array of *Name* values named *names*, and an array of *PhoneNumber* values named *phoneNumbers*. The *PhoneBook* class also contains an *Add* method that adds a phone number and name to the phone book. This method is called when the user clicks the *Add* button on the form. The *enlargeIfFull* method is called by *Add* to check whether the arrays are full when the user adds another entry. This method creates two new bigger arrays, copies the contents of the existing arrays to them, and then discards the old arrays.

Write the indexers

1. In the PhoneBook.cs file, add a *public* read-only indexer to the *PhoneBook* class, as shown in bold in the following code. The indexer should return a *Name* and take a *PhoneNumber* item as its index. Leave the body of the *get* accessor blank.

   The indexer should look like this:

```
sealed class PhoneBook
{
    ...
    public Name this [PhoneNumber number]
    {
        get
        {
        }
    }
    ...
}
```

2. Implement the *get* accessor as shown in bold in the following code. The purpose of the accessor is to find the name that matches the specifed phone number. To do this, you will need to call the static *IndexOf* method of the *Array* class. The *IndexOf* method performs a search through an array, returning the index of the first item in the array that matches the specified value. The first argument to *IndexOf* is the array to search through (*phoneNumbers*). The second argument to *IndexOf* is the item you are searching for. *IndexOf* returns the integer index of the element if it finds it; otherwise, *IndexOf* returns *–1*. If the indexer finds the phone number, it should return it; otherwise, it should return an empty *Name* value. (Note that *Name* is a structure and will always have a default constructor that sets its *private* field to *null*.)

The indexer with its completed *get* accessor should look like this:

```
sealed class PhoneBook
{
    ...
    public Name this [PhoneNumber number]
    {
        get
        {
            int i = Array.IndexOf(this.phoneNumbers, number);
            if (i != -1)
                return this.names[i];
            else
                return new Name();
        }
    }
    ...
}
```

3. Add a second *public* read-only indexer to the *PhoneBook* class that returns a *PhoneNumber* and accepts a single *Name* parameter. Implement this indexer in the same way as the first one. (Again note that *PhoneNumber* is a structure and therefore always has a default constructor.)

The second indexer should look like this:

```
sealed class PhoneBook
{
    ...
    public PhoneNumber this [Name name]
    {
        get
        {
            int i = Array.IndexOf(this.names, name);
            if (i != -1)
                return this.phoneNumbers[i];
            else
                return new PhoneNumber();
        }
    }
    ...
}
```

Notice that these overloaded indexers can coexist because they return different types, which means that their signatures are different. If the *Name* and *PhoneNumber* structures were replaced by simple strings (which they wrap), the overloads would have the same signature and the class would not compile.

4. On the *Build* menu, click *Build Solution*. Correct any syntax errors, and then rebuild if necessary.

Call the indexers

1. Display the Window1.xaml.cs file in the *Code and Text Editor* window, and then locate the *findPhoneClick* method.

This method is called when the *Search by Name* button is clicked. This method is currently empty. Add the code shown in bold in the following example to perform these tasks:

1.1. Read the value of the *Text* property from the *name* text box on the form. This is a string containing the contact name that the user has typed in.

1.2. If the string is not empty, search for the phone number corresponding to that name in the *PhoneBook* by using the indexer. (Notice that the *Window1* class contains a private *PhoneBook* field named *phoneBook*.) Construct a *Name* object from the string, and pass it as the parameter to the *PhoneBook* indexer.

1.3. Write the *Text* property of the *PhoneNumber* structure returned by the indexer to the *phoneNumber* text box on the form.

The *findPhoneClick* method should look like this:

```
private void findPhoneClick(object sender, RoutedEventArgs e)
{
    string text = name.Text;
    if (!String.IsNullOrEmpty(text))
    {
        phoneNumber.Text = phoneBook[new Name(text)].Text;
    }
}
```

 Tip Notice the use of the static *String* method *IsNullOrEmpty* to determine whether a string is empty or contains a null value. This is the preferred method for testing whether a string contains a value. It returns *true* if the string has a value and *false* otherwise.

2. Locate the *findNameClick* method in the Window1.xaml.cs file. It is below the *findPhoneClick* method.

The *findName_Click* method is called when the *Search by Phone* button is clicked. This method is currently empty, so you need to implement it as follows. (The code is shown in bold in the following example.)

2.1. Read the value of the *Text* property from the *phoneNumber* text box on the form. This is a string containing the phone number that the user has typed.

2.2. If the string is not empty, search for the name corresponding to that phone number in the *PhoneBook* by using the indexer.

2.3. Write the *Text* property of the *Name* structure returned by the indexer to the *name* text box on the form.

The completed method should look like this:

```
private void findNameClick(object sender, RoutedEventArgs e)
{
    string text = phoneNumber.Text;
    if (!String.IsNullOrEmpty(text))
    {
        name.Text = phoneBook[new PhoneNumber(text)].Text;
    }
}
```

3. On the *Build* menu, click *Build Solution*. Correct any errors that occur.

Run the application

1. On the *Debug* menu, click *Start Without Debugging*.

2. Type your name and phone number in the text boxes, and then click *Add*.

When you click the *Add* button, the *Add* method stores the information in the phone book and clears the text boxes so that they are ready to perform a search.

3. Repeat step 2 several times with some different names and phone numbers so that the phone book contains a selection of entries.

> **Note** The application performs no checking of the names and telephone numbers that you enter, and you can input the same name and telephone number more than once. To avoid confusion, please make sure that you provide different names and telephone numbers.

4. Type a name that you used in step 2 into the *Name* text box, and then click *Search by Name*.

The phone number you added for this contact in step 2 is retrieved from the phone book and is displayed in the *Phone Number* text box.

5. Type a phone number for a different contact in the *Phone Number* text box, and then click *Search by Phone*.

The contact name is retrieved from the phone book and is displayed in the *Name* text box.

6. Type a name that you did not enter in the phone book into the *Name* text box, and then click *Search by Name*.

This time the *Phone Number* text box is empty, indicating that the name could not be found in the phone book.

7. Close the form, and return to Visual Studio 2008.

■ If you want to continue to the next chapter:

Keep Visual Studio 2008 running, and turn to Chapter 17.

■ If you want to exit Visual Studio 2008 now:

On the *File* menu, click *Exit*. If you see a *Save* dialog box, click *Yes* (if you are using Visual Studio 2008) or *Save* (if you are using Visual C# 2008 Express Edition) and save the project.

Chapter 16 Quick Reference

To	Do this
Create an indexer for a class or structure	Declare the type of the indexer, followed by the keyword *this* and then the indexer arguments in square brackets. The body of the indexer can contain a *get* and/or *set* accessor. For example:

```
struct RawInt
{
    ...
    public bool this [ int index ]
    {
        get { ... }
        set { ... }
    }
    ...
}
```

Define an indexer in an interface	Define an indexer with the *get* and/or *set* keywords. For example:

```
interface IRawInt
{
    bool this [ int index ] { get;  set; }
}
```

| Implement an interface indexer in a class or structure | In the class or structure that implements the interface, define the indexer and implement the accessors. For example: |

```
struct RawInt : IRawInt
{
    ...
    public bool this [ int index  ]
    {
        get { ... }
        set { ... }
    }
    ...
}
```

| Implement an interface indexer by using explicit interface implementation in a class or structure | In the class or structure that implements the interface, explicitly name the interface, but do not specify the indexer accessibility. For example: |

```
struct RawInt : IRawInt
{
    ...
    bool IRawInt.this [ int index  ]
    {
        get { ... }
        set { ... }
    }
    ...
}
```

Chapter 17
Interrupting Program Flow and Handling Events

After completing this chapter, you will be able to:

- Declare a delegate type to create an abstraction of a method signature.

- Create an instance of a delegate to refer to a specific method.

- Call a method through a delegate.

- Define a lambda expression to specify the code for a delegate.

- Declare an event field.

- Handle an event by using a delegate.

- Raise an event.

Much of the code you have written in the various exercises in this book has assumed that statements execute sequentially. Although this is a common scenario, you will find that it is sometimes necessary to interrupt the current flow of execution and perform another, more important, task. When the task has completed, the program can continue where it left off. The classic example of this style of program is the Microsoft Windows Presentation Foundation (WPF) form. A WPF form displays controls such as buttons and text boxes. When you click a button or type text in a text box, you expect the form to respond immediately. The application has to temporarily stop what it is doing and handle your input. This style of operation applies not just to graphical user interfaces but to any application where an operation must be performed urgently—shutting down the reactor in a nuclear power plant if it is getting too hot, for example.

To handle this type of application, the runtime has to provide two things: a means of indicating that something urgent has happened and a way of indicating the code that should be run when it happens. This is the purpose of events and delegates. We start by looking at delegates.

Declaring and Using Delegates

A delegate is a pointer to a method, and you can call it in the same way as you would call a method. When you invoke a delegate, the runtime actually executes the method to which the delegate refers. You can dynamically change the method that a delegate references so that code that calls a delegate might actually run a different method each time it executes. The best way to understand delegates is to see them in action, so let's work through an example.

> **Note** If you are familiar with C++, a delegate is similar to a function pointer. However, unlike function pointers, delegates are type-safe; you can make a delegate refer to only a method that matches the signature of the delegate, and you cannot call a delegate that does not refer to a valid method.

The Automated Factory Scenario

Suppose you are writing the control systems for an automated factory. The factory contains a large number of different machines, each performing distinct tasks in the production of the articles manufactured by the factory—shaping and folding metal sheets, welding sheets together, painting sheets, and so on. Each machine was built and installed by a specialist vendor. The machines are all computer-controlled, and each vendor has provided a set of APIs that you can use to control its machine. Your task is to integrate the different systems used by the machines into a single control program. One aspect on which you have decided to concentrate is to provide a means of shutting down all the machines, quickly if needed!

> **Note** The term *API* stands for application programming interface. It is a method, or set of methods, exposed by a piece of software that you can use to control that software. You can think of the Microsoft .NET Framework as a set of APIs because it provides methods that you can use to control the .NET common language runtime and the Microsoft Windows operating system.

Each machine has its own unique computer-controlled process (and API) for shutting down safely. These are summarized here:

```
StopFolding();      // Folding and shaping machine
FinishWelding();    // Welding machine
PaintOff();         // Painting machine
```

Implementing the Factory Without Using Delegates

A simple approach to implementing the shutdown functionality in the control program is as follows:

```
class Controller
{
    // Fields representing the different machines
    private FoldingMachine folder;
    private WeldingMachine welder;
    private PaintingMachine painter;
    ...
```

```
    public void ShutDown()
    {
        folder.StopFolding();
        welder.FinishWelding();
        painter.PaintOff();
    }
    ...
}
```

Although this approach works, it is not very extensible or flexible. If the factory buys a new machine, you must modify this code; the *Controller* class and the machines are tightly coupled.

Implementing the Factory by Using a Delegate

Although the names of each method are different, they all have the same "shape": They take no parameters, and they do not return a value. (We consider what happens if this isn't the case later, so bear with me!) The general format of each method is, therefore:

```
void methodName();
```

This is where a delegate is useful. A delegate that matches this shape can be used to refer to any of the machinery shutdown methods. You declare a delegate like this:

```
delegate void stopMachineryDelegate();
```

Note the following points:

- Use the *delegate* keyword when declaring a delegate.

- A delegate defines the shape of the methods it can refer to. You specify the return type (*void* in this example), a name for the delegate (*stopMachineryDelegate*), and any parameters (there are none in this case).

After you have defined the delegate, you can create an instance and make it refer to a matching method by using the += compound assignment operator. You can do this in the constructor of the controller class like this:

```
class Controller
{
    delegate void stopMachineryDelegate();
    private stopMachineryDelegate stopMachinery; // an instance of the delegate
    ...
    public Controller()
    {
        this.stopMachinery += folder.StopFolding;
    }
    ...
}
```

This syntax takes a bit of getting used to. You *add* the method to the delegate; you are not actually calling the method at this point. The + operator is overloaded to have this new meaning when used with delegates. (You will learn more about operator overloading in Chapter 21, "Operator Overloading.") Notice that you simply specify the method name and should not include any parentheses or parameters.

It is safe to use the += operator on an uninitialized delegate. It will be initialized automatically. You can also use the *new* keyword to initialize a delegate explicitly with a single specific method, like this:

```
this.stopMachinery = new stopMachineryDelegate(folder.StopFolding);
```

You can call the method by invoking the delegate, like this:

```
public void ShutDown()
{
    this.stopMachinery();
    ...
}
```

You use the same syntax to invoke a delegate as you use to make a method call. If the method that the delegate refers to takes any parameters, you should specify them at this time, between parentheses.

> **Note** If you attempt to invoke a delegate that is uninitialized and does not refer to any methods, you will get a *NullReferenceException*.

The principal advantage of using a delegate is that it can refer to more than one method; you simply use the += operator to add methods to the delegate, like this:

```
public Controller()
{
    this.stopMachinery += folder.StopFolding;
    this.stopMachinery += welder.FinishWelding;
    this.stopMachinery += painter.PaintOff;
}
```

Invoking *this.stopMachinery()* in the *Shutdown* method of the *Controller* class automatically calls each of the methods in turn. The *Shutdown* method does not need to know how many machines there are or what the method names are.

You can remove a method from a delegate by using the -= compound assignment operator:

```
this.stopMachinery -= folder.StopFolding;
```

The current scheme adds the machine methods to the delegate in the *Controller* constructor. To make the *Controller* class totally independent of the various machines, you need to make *stopMachineryDelegate* type public and supply a means of enabling classes outside *Controller* to add methods to the delegate. You have several options:

- Make the delegate variable, *stopMachinery*, public:

```
public stopMachineryDelegate stopMachinery;
```

- Keep the *stopMachinery* delegate variable private, but provide a read/write property to provide access to it:

```
public delegate void stopMachineryDelegate();
...
public stopMachineryDelegate StopMachinery
{
    get
    {
        return this.stopMachinery;
    }

    set
    {
        this.stopMachinery = value;
    }
}
```

- Provide complete encapsulation by implementing separate *Add* and *Remove* methods. The *Add* method takes a method as a parameter and adds it to the delegate, while the *Remove* method removes the specified method from the delegate (notice that you specify a method as a parameter by using a delegate type):

```
public void Add(stopMachineryDelegate stopMethod)
{
    this.stopMachinery += stopMethod;
}

public void Remove(stopMachineryDelegate stopMethod)
{
    this.stopMachinery -= stopMethod;
}
```

If you are an object-oriented purist, you will probably opt for the *Add/Remove* approach. However, the others are viable alternatives that are frequently used, which is why they are shown here.

Whichever technique you choose, you should remove the code that adds the machine methods to the delegate from the *Controller* constructor. You can then instantiate a

Controller and objects representing the other machines like this (this example uses the *Add/Remove* approach):

```
Controller control = new Controller();
FoldingMachine folder = new FoldingMachine();
WeldingMachine welder = new WeldingMachine();
PaintingMachine painter = new PaintingMachine();
...
control.Add(folder.StopFolding);
control.Add(welder.FinishWelding);
control.Add(painter.PaintOff);
...
control.ShutDown();
...
```

Using Delegates

In the following exercise, you will create a delegate to encapsulate a method that displays the time in a text box acting as a digital clock on a WPF form. You will attach the delegate object to a class called *Ticker* that invokes the delegate every second.

Complete the digital clock application

1. Start Microsoft Visual Studio 2008 if it is not already running.

2. Open the Delegates project located in the \Microsoft Press\Visual CSharp Step by Step \Chapter 17\Delegates folder in your Documents folder.

3. On the *Debug* menu, click *Start Without Debugging*.

   The project builds and runs. A form appears, displaying a digital clock. The clock displays the current time as "00:00:00," which is probably wrong unless you happen to be reading this chapter at midnight.

4. Click *Start* to start the clock, and then click *Stop* to stop it again.

   Nothing happens. The *Start* and *Stop* methods have not been written yet. Your task is to implement these methods.

5. Close the form, and return to the Visual Studio 2008 environment.

6. Open the Ticker.cs file, and display it in the *Code and Text Editor* window. This file contains a class called *Ticker* that models the inner workings of a clock. Scroll to the bottom of the file. The class contains a *DispatcherTimer* object called *ticking* to arrange for a pulse to be sent at regular intervals. The constructor for the class sets this interval to 1 second. The class catches the pulse by using an event (you will learn how events work shortly) and then arranges for the display to be updated by invoking a delegate.

Note The .NET Framework provides another timer class called *System.Timers.Timer*. This class offers similar functionality to the *DispatcherTimer* class, but it is not suitable for use in a WPF application.

7. In the *Code and Text Editor* window, find the declaration of the *Tick* delegate. It is located near the top of the file and looks like this:

```
public delegate void Tick(int hh, int mm, int ss);
```

The *Tick* delegate can be used to refer to a method that takes three integer parameters and that does not return a value. A delegate variable called *tickers* at the bottom of the file is based on this type. By using the *Add* and *Remove* methods in this class (shown in the following code example), you can add methods with matching signatures to (and remove them from) the *tickers* delegate variable:

```
class Ticker
{
    ...
    public void Add(Tick newMethod)
    {
        this.tickers += newMethod;
    }

    public void Remove(Tick oldMethod)
    {
        this.tickers -= oldMethod;
    }
    ...
    private Tick tickers;
}
```

8. Open the Clock.cs file, and display it in the *Code and Text Editor* window. The *Clock* class models the clock display. It has methods called *Start* and *Stop* that are used to start and stop the clock running (after you have implemented them) and a method called *RefreshTime* that formats a string to depict the time specified by its three parameters (hours, minutes, and seconds) and then displays it in the *TextBox* field called *display*. This *TextBox* field is initialized in the constructor. The class also contains a private *Ticker* field called *pulsed* that tells the clock when to update its display:

```
class Clock
{
    ...

    public Clock(TextBox displayBox)
    {
        this.display = displayBox;
    }
    ...
```

```
        private void RefreshTime(int hh, int mm, int ss )
        {
            this.display.Text = string.Format("{0:D2}:{1:D2}:{2:D2}", hh, mm, ss);
        }

        private Ticker pulsed = new Ticker();
        private TextBox display;
    }
```

9. Display the code for the Window1.xaml.cs file in the *Code and Text Editor* window. Notice that the constructor creates a new instance of the *Clock* class, passing in the *TextBox* field called *digital* as its parameter:

```
public Window1()
{
    ...
    clock = new Clock(digital);
}
```

The *digital* field is the *TextBox* control displayed on the form. The clock will display its output in this *TextBox* control.

10. Return to the Clock.cs file. Implement the *Clock.Start* method so that it adds the *Clock. RefreshTime* method to the delegate in the *pulsed* object by using the *Ticker.Add* method, as follows in bold type. The *pulsed* delegate is invoked every time a pulse occurs, and this statement causes the *RefreshTime* method to execute when this happens.

The *Start* method should look like this:

```
public void Start()
{
    pulsed.Add(this.RefreshTime);
}
```

11. Implement the *Clock.Stop* method so that it removes the *Clock.RefreshTime* method from the *pulsed* delegate by using the *Ticker.Remove* method, as follows in bold type.

The *Stop* method should look like this:

```
public void Stop()
{
    pulsed.Remove(this.RefreshTime);
}
```

12. On the *Debug* menu, click *Start Without Debugging*.

13. On the WPF form, click *Start*.

The form now displays the correct time and updates every second.

14. Click *Stop*.

The display stops responding, or "freezes." This is because the *Stop* button calls the *Clock.Stop* method, which removes the *RefreshTime* method from the *Ticker* delegate; *RefreshTime* is no longer being called every second, although the timer continues to pulse.

Note If you click *Start* more than one time, you must click *Stop* the same number of times. Each time you click *Start* you add a reference to the *RefreshTime* method to the delegate. You must remove them all before the clock will stop.

15. Click *Start* again.

The display resumes processing, corrects the time, and updates the time every second. This is because the *Start* button calls the *Clock.Start* method, which attaches the *RefreshTime* method to the *Ticker* delegate again.

16. Close the form, and return to Visual Studio 2008.

Lambda Expressions and Delegates

All the examples of adding a method to a delegate that you have seen so far use the method's name. For example, returning to the automated factory scenario described earlier, you add the *StopFolding* method of the *folder* object to the *stopMachinery* delegate like this:

```
this.stopMachinery += folder.StopFolding;
```

This approach is very useful if there is a convenient method that matches the signature of the delegate, but what if this is not the case? Suppose that the *StopFolding* method actually had the following signature:

```
void StopFolding(int shutDownTime); // Shut down in the specified number of seconds
```

This signature is now different from that of the *FinishWelding* and *PaintOff* methods, and therefore you cannot use the same delegate to handle all three methods.

Creating a Method Adapter

One way around this problem is to create another method that calls *StopFolding* but that takes no parameters itself, like this:

```
void FinishFolding()
{
    folder.StopFolding(0); // Shut down immediately
}
```

You can then add the *FinishFolding* method to the *stopMachinery* delegate in place of the *StopFolding* method, using the same syntax as before:

```
this.stopMachinery += folder.FinishFolding;
```

When the *stopMachinery* delegate is invoked, it calls *FinishFolding*, which in turn calls the *StopFolding* method, passing in the parameter of 0.

> **Note** The *FinishFolding* method is a classic example of an adapter, a method that converts (or adapts) a method to give it a different signature. This pattern is very common and is one of the set of patterns documented in the book *Design Patterns: Elements of Reusable Object-Oriented Architecture* by Gamma, Helm, Johnson, and Vlissides (Addison-Wesley Professional, 1994).

In many cases, adapter methods such as this are small, and it is easy to lose them in a sea of methods, especially in a large class. Furthermore, apart from using it to adapt the *StopFolding* method for use by the delegate, it is unlikely to be called elsewhere. C# provides lambda expressions for situations such as this.

Using a Lambda Expression as an Adapter

A lambda expression is an expression that returns a method. This sounds rather odd because most expressions that you have met so far in C# actually return a value. If you are familiar with functional programming languages such as Haskell, you are probably comfortable with this concept. For the rest of you, fear not: lambda expressions are not particularly complicated, and after you have gotten used to a new bit of syntax, you will see that they are very useful.

You saw in Chapter 3, "Writing Methods and Applying Scope," that a typical method consists of four elements: a return type, a method name, a list of parameters, and a method body. A lambda expression contains two of these elements: a list of parameters and a method body. Lambda expressions do not define a method name, and the return type (if any) is inferred from the context in which the lambda expression is used. In the *StopFolding* method of the *FoldingMachine* class, the problem is that this method now takes a parameter, so you need to create an adapter that takes no parameters that you can add to the *stopMachinery* delegate. You can use the following statement to do this:

```
this.stopMachinery += () => { folder.StopFolding(0); };
```

All of the text to the right of the += operator is a lambda expression, which defines the method to be added to the *stopMachinery* delegate. It has the following syntactic items:

- A list of parameters enclosed in parentheses. As with a regular method, if the method you are defining (as in the preceding example) takes no parameters, you must still provide the parentheses.

- The => operator, which indicates to the C# compiler that this is a lambda expression.

- The body of the method. The example shown here is very simple, containing a single statement. However, a lambda expression can contain multiple statements, and you can format it in whatever way you feel is most readable. Just remember to add a semicolon after each statement as you would in an ordinary method.

Strictly speaking, the body of a lambda expression can be a method body containing multiple statements, or it can actually be a single expression. If the body of a lambda expression contains only a single expression, you can omit the braces and the semicolon (you still need a semicolon to complete the entire statement), like this:

```
this.stopMachinery += () => folder.StopFolding(0) ;
```

When you invoke the *stopMachinery* delegate, it will run the code defined by the lambda expression.

The Form of Lambda Expressions

Lambda expressions can take a number of subtly different forms. Lambda expressions were originally part of a mathematical notation called the Lambda Calculus that provides a notation for describing functions. (You can think of a function as a method that returns a value.) Although the C# language has extended the syntax and semantics of the Lambda Calculus in its implementation of lambda expressions, many of the original principles still apply. Here are some examples showing the different forms of lambda expression available in C#:

```
x => x * x  // A simple expression that returns the square of its parameter
            // The type of parameter x is inferred from the context.

x => { return x * x ; } // Semantically the same as the preceding
                        // expression, but using a C# statement block as
                        // a body rather than a simple expression

(int x) => x / 2  // A simple expression that returns the value of the
                  // parameter divided by 2
                  // The type of parameter x is stated explicitly.

() => folder.StopFolding(0) // Calling a method
                            // The expression takes no parameters.
                            // The expression might or might not
                            // return a value.

(x, y) => { x++; return x / y; } // Multiple parameters; the compiler
                                 // infers the parameter types.
                                 // The parameter x is passed by value, so
                                 // the effect of the ++ operation is
                                 // local to the expression.

(ref int x, int y) { x++; return x / y; } // Multiple parameters
                                          // with explicit types
                                          // Parameter x is passed by
                                          // reference, so the effect of
                                          // the ++ operation is permanent.
```

To summarize, here are some features of lambda expressions that you should be aware of:

- If a lambda expression takes parameters, you specify them in the parentheses to the left of the => operator. You can omit the types of parameters, and the C# compiler will infer their types from the context of the lambda expression. You can pass parameters by reference (by using the *ref* keyword) if you want the lambda expression to be able to change their values other than locally, but this is not recommended.

- Lambda expressions can return values, but the return type must match that of the delegate they are being added to.

- The body of a lambda expression can be a simple expression or a block of C# code made up of multiple statements, method calls, variable definitions, and so on.

- Variables defined in a lambda expression method go out of scope when the method finishes.

- A lambda expression can access and modify all variables outside the lambda expression that are in scope when the lambda expression is defined. Be very careful with this feature!

You will learn more about lambda expressions and see further examples that take parameters and return values in later chapters in this book.

Lambda Expressions and Anonymous Methods

Lambda expressions are a new addition to the C# language in version 3.0. C# version 2.0 introduced anonymous methods that can perform a similar task but that are not as flexible. Anonymous methods were added primarily so that you can define delegates without having to create a named method; you simply provide the definition of the method body in place of the method name, like this:

```
this.stopMachinery += delegate { folder.StopFolding(0); };
```

You can also pass an anonymous method as a parameter in place of a delegate, like this:

```
control.Add(delegate { folder.StopFolding(0); } );
```

Notice that whenever you introduce an anonymous method, you must prefix it with the *delegate* keyword. Also, any parameters needed are specified in braces following the *delegate* keyword. For example:

```
control.Add(delegate(int param1, string param2) { /* code that uses param1 and param2 */ ... });
```

After you are used to them, you will notice that lambda expressions provide a more succinct syntax than anonymous methods do and they pervade many of the more advanced aspects of C#, as you will see later in this book. Generally speaking, you should use lambda expressions rather than anonymous methods in your code.

Enabling Notifications with Events

You have now seen how to declare a delegate type, call a delegate, and create delegate instances. However, this is only half the story. Although by using delegates you can invoke any number of methods indirectly, you still have to invoke the delegate explicitly. In many cases, it would be useful to have the delegate run automatically when something significant happens. For example, in the automated factory scenario, it could be vital to be able to invoke the *stopMachinery* delegate and halt the equipment if the system detects that a machine is overheating.

The .NET Framework provides *events*, which you can use to define and trap significant actions and arrange for a delegate to be called to handle the situation. Many classes in the .NET Framework expose events. Most of the controls that you can place on a WPF form, and the *Windows* class itself, use events so that you can run code when, for example, the user clicks a button or types something in a field. You can also define your own events.

Declaring an Event

You declare an event in a class intended to act as an event source. An event source is usually a class that monitors its environment and raises an event when something significant happens. In the automated factory, an event source could be a class that monitors the temperature of each machine. The temperature-monitoring class would raise a "machine overheating" event if it detects that a machine has exceeded its thermal radiation boundary (that is, it has become too hot). An event maintains a list of methods to call when it is raised. These methods are sometimes referred to as subscribers. These methods should be prepared to handle the "machine overheating" event and take the necessary corrective action: shut down the machines.

You declare an event similarly to how you declare a field. However, because events are intended to be used with delegates, the type of an event must be a delegate, and you must prefix the declaration with the *event* keyword. Use the following syntax to declare an event:

event *delegateTypeName eventName*

As an example, here's the *StopMachineryDelegate* delegate from the automated factory. It has been relocated to a new class called *TemperatureMonitor*, which provides an interface to the various electronic probes monitoring the temperature of the equipment (this is a more logical place for the event than the *Controller* class is):

```
class TemperatureMonitor
{
    public delegate void StopMachineryDelegate();
    ...
}
```

You can define the *MachineOverheating* event, which will invoke the *stopMachineryDelegate*, like this:

```
class TemperatureMonitor
{
    public delegate void StopMachineryDelegate();
    public event StopMachineryDelegate MachineOverheating;
    ...
}
```

The logic (not shown) in the *TemperatureMonitor* class raises the *MachineOverheating* event as necessary. You will see how to raise an event in the upcoming section titled "Raising an Event." Also, you add methods to an event (a process known as *subscribing* to the event) rather than adding them to the delegate that the event is based on. You will look at this aspect of events next.

Subscribing to an Event

Like delegates, events come ready-made with a += operator. You subscribe to an event by using this += operator. In the automated factory, the software controlling each machine can arrange for the shutdown methods to be called when the *MachineOverheating* event is raised, like this:

```
class TemperatureMonitor
{
    public delegate void StopMachineryDelegate();
    public event StopMachineryDelegate MachineOverheating;
    ...
}
...
TemperatureMonitor tempMonitor = new TemperatureMonitor();
...
tempMonitor.MachineOverheating += () => { folder.StopFolding(0); };
tempMonitor.MachineOverheating += welder.FinishWelding;
tempMonitor.MachineOverheating += painter.PaintOff;
```

Notice that the syntax is the same as for adding a method to a delegate. You can even subscribe by using a lambda expression. When the *tempMonitor.MachineOverheating* event runs, it will call all the subscribing methods and shut down the machines.

Unsubscribing from an Event

Knowing that you use the += operator to attach a delegate to an event, you can probably guess that you use the −= operator to detach a delegate from an event. Calling the −= operator removes the method from the event's internal delegate collection. This action is often referred to as unsubscribing from the event.

Raising an Event

An event can be raised, just like a delegate, by calling it like a method. When you raise an event, all the attached delegates are called in sequence. For example, here's the *TemperatureMonitor* class with a private *Notify* method that raises the *MachineOverheating* event:

```
class TemperatureMonitor
{
    public delegate void StopMachineryDelegate;
    public event StopMachineryDelegate MachineOverheating;
    ...
    private void Notify()
    {
        if (this.MachineOverheating != null)
        {
            this.MachineOverheating();
        }
    }
    ...
}
```

This is a common idiom. The *null* check is necessary because an event field is implicitly *null* and only becomes non-*null* when a method subscribes to it by using the += operator. If you try to raise a *null* event, you will get a *NullReferenceException*. If the delegate defining the event expects any parameters, the appropriate arguments must be provided when you raise the event. You will see some examples of this later.

> **Important** Events have a very useful built-in security feature. A public event (such as *MachineOverheating*) can be raised only by methods in the class that defines it (the *TemperatureMonitor* class). Any attempt to raise the method outside the class results in a compiler error.

Understanding WPF User Interface Events

As mentioned earlier, the .NET Framework classes and controls used for building graphical user interfaces (GUIs) employ events extensively. You'll see and use GUI events on many occasions in the second half of this book. For example, the WPF *Button* class derives from the *ButtonBase* class, inheriting a public event called *Click* of type *RoutedEventHandler*. The *RoutedEventHandler* delegate expects two parameters: a reference to the object that caused the event to be raised and a *RoutedEventArgs* object that contains additional information about the event:

```
public delegate void RoutedEventHandler(Object sender, RoutedEventArgs e);
```

The *Button* class looks like this:

```
public class ButtonBase: ...
{
    public event RoutedEventHandler Click;
    ...
}

public class Button: ButtonBase
{
    ...
}
```

The *Button* class automatically raises the *Click* event when you click the button on-screen. (How this actually happens is beyond the scope of this book.) This arrangement makes it easy to create a delegate for a chosen method and attach that delegate to the required event. The following example shows the code for a WPF form that contains a button called *okay* and the code to connect the *Click* event of the *okay* button to the *okayClick* method:

```
public partial class Example : System.Windows.Window, System.Windows.Markup.
IComponentConnector
{
    internal System.Windows.Controls.Button okay;
    ...
    void System.Windows.Markup.IComponentConnector.Connect(...)
    {
        ...
        this.okay.Click += new System.Windows.RoutedEventHandler(this.okayClick);
        ...
    }
    ...
}
```

This code is usually hidden from you. When you use the *Design View* window in Visual Studio 2008 and set the *Click* property of the *okay* button to *okayClick* in the Extensible Application Markup Language (XAML) description of the form, Visual Studio 2008 generates this code for you. All you have to do is write your application logic in the event handling method, *okayClick*, in the part of the code that you do have access to, in the Example.xaml.cs file in this case:

```
public partial class Example : System.Windows.Window
{
    ...
    private void okayClick(object sender, RoutedEventArgs args)
    {
        // your code to handle the Click event
    }
}
```

The events that the various GUI controls generate always follow the same pattern. The events are of a delegate type whose signature has a *void* return type and two arguments. The first argument is always the sender (the source) of the event, and the second argument is always an *EventArgs* argument (or a class derived from *EventArgs*).

With the *sender* argument, you can reuse a single method for multiple events. The delegated method can examine the *sender* argument and respond accordingly. For example, you can use the same method to subscribe to the *Click* event for two buttons (you add the same method to two different events). When the event is raised, the code in the method can examine the *sender* argument to ascertain which button was clicked.

You learn more about how to handle events for WPF controls in Chapter 22, "Introducing Windows Presentation Foundation."

Using Events

In the following exercise, you will use events to simplify the program you completed in the first exercise. You will add an event field to the *Ticker* class and delete its *Add* and *Remove* methods. You will then modify the *Clock.Start* and *Clock.Stop* methods to subscribe to the event. You will also examine the *Timer* object, used by the *Ticker* class to obtain a pulse once each second.

Rework the digital clock application

1. Return to the Visual Studio 2008 window displaying the Delegates project.

2. Display the Ticker.cs file in the *Code and Text Editor* window.

   This file contains the declaration of the *Tick* delegate type in the *Ticker* class:

   ```
   public delegate void Tick(int hh, int mm, int ss);
   ```

3. Add a public event called *tick* of type *Tick* to the *Ticker* class, as shown in bold type in the following code:

   ```
   class Ticker
   {
       public delegate void Tick(int hh, int mm, int ss);
       public event Tick tick;
       ...
   }
   ```

4. Comment out the following delegate variable *tickers* near the bottom of the *Ticker* class definition because it is now obsolete:

   ```
   // private Tick tickers;
   ```

5. Comment out the *Add* and *Remove* methods from the *Ticker* class.

   The add and remove functionality is automatically provided by the += and –= operators of the event object.

6. Locate the *Ticker.Notify* method. This method previously invoked an instance of the *Tick* delegate called *tickers*. Modify it so that it calls the *tick* event instead. Don't forget to check whether *tick* is *null* before calling the event.

The *Notify* method should look like this:

```
private void Notify(int hours, int minutes, int seconds)
{
    if (this.tick != null)
        this.tick(hours, minutes, seconds);
}
```

Notice that the *Tick* delegate specifies parameters, so the statement that raises the *tick* event must specify arguments for each of these parameters.

7. Examine the definition of the *ticking* variable at the end of the class:

```
private DispatcherTimer ticking = new DispatcherTimer();
```

The *DispatcherTimer* class can be programmed to raise an event repeatedly at a specified interval.

8. Examine the constructor for the *Ticker* class:

```
public Ticker()
{
    this.ticking.Tick += new EventHandler(this.OnTimedEvent);
    this.ticking.Interval = new TimeSpan(0, 0, 1); // 1 second
    this.ticking.Start();
}
```

The *DispatcherTimer* class exposes the *Tick* event, which can be raised at regular intervals according to the value of the *Interval* property. Setting *Interval* to the *TimeSpan* shown causes the *Tick* event to be raised once a second. The timer starts when you invoke the *Start* method. Methods that subscribe to the *Tick* event must match the signature of the *EventHandler* delegate. The *EventHandler* delegate has the same signature as the *RoutedEventHandler* delegate described earlier. The *Ticker* constructor creates an instance of this delegate referring to the *OnTimedEvent* method and subscribes to the *Tick* event.

The *OnTimedEvent* method in the *Ticker* class obtains the current time by examining the static *DateTime.Now* property. The *DateTime* structure is part of the .NET Framework class library. The *Now* property returns a *DateTime* structure. This structure has several fields, including those used by the *OnTimedEvent* method shown in the following code and called *Hour*, *Minute*, and *Second*. The *OnTimedEvent* method uses this information in turn to raise the *tick* event through the *Notify* method:

```
private void OnTimedEvent(object source, EventArgs args)
{
    DateTime now = DateTime.Now;
    int hh = now.Hour;
    int mm = now.Minutes;
```

```
        int ss = now.Seconds;
        Notify(hh, mm, ss);
    }
```

9. Display the Clock.cs file in the *Code and Text Editor* window.

10. Modify the *Clock.Start* method so that the delegate is attached to the *tick* event of the *pulsed* field by using the += operator, like this:

```
public void Start()
{
    pulsed.tick += this.RefreshTime;
}
```

11. Modify the *Clock.Stop* method so that the delegate is detached from the *tick* event of the *pulsed* field by using the −= operator, like this:

```
public void Stop()
{
    pulsed.tick -= this.RefreshTime;
}
```

12. On the *Debug* menu, click *Start Without Debugging*.

13. Click *Start*.

    The digital clock form displays the correct time and updates the display every second.

14. Click *Stop*, and verify that the clock stops.

15. Close the form, and return to Visual Studio 2008.

- If you want to continue to the next chapter

    Keep Visual Studio 2008 running and turn to Chapter 18.

- If you want to exit Visual Studio 2008 now

    On the *File* menu, click *Exit*. If you see a *Save* dialog box, click *Yes* (if you are using Visual Studio 2008) or *Save* (if you are using Microsoft Visual C# 2008 Express Edition) and save the project.

Chapter 17 Quick Reference

To	Do this
Declare a delegate type	Write the keyword *delegate*, followed by the return type, followed by the name of the delegate type, followed by any parameter types. For example: `delegate void myDelegate();`

Create an instance of a delegate initialized with a single specific method	Use the same syntax you use for a class or structure: Write the keyword *new*, followed by the name of the type (the name of the delegate), followed by the argument between parentheses. The argument must be a method whose signature exactly matches the signature of the delegate. For example:

```
delegate void myDelegate();
private void myMethod() { ... }
...
myDelegate del = new myDelegate(this.myMethod);
```

Invoke a delegate	Use the same syntax as a method call. For example:

```
myDelegate del;
...
del();
```

Declare an event	Write the keyword *event*, followed by the name of the type (the type must be a delegate type), followed by the name of the event. For example:

```
delegate void myEvent();

class MyClass
{
    public event myDelegate MyEvent;
}
```

Subscribe to an event	Create a delegate instance (of the same type as the event), and attach the delegate instance to the event by using the += operator. For example:

```
class MyEventHandlingClass
{
    ...
    public void Start()
    {
        myClass.MyEvent += new myDelegate
            (this.eventHandlingMethod);
    }

    private void eventHandlingMethod()
    {
        ...
    }

    private MyClass myClass = new MyClass();
}
```

You can also get the compiler to generate the new delegate automatically simply by specifying the subscribing method:

```
public void Start()
{
    myClass.MyEvent += this.eventHandlingMethod;
}
```

Unsubscribe from an event	Create a delegate instance (of the same type as the event), and detach the delegate instance from the event by using the −= operator. For example:

```
class MyEventHandlingClass
{
    ...
    public void Stop()
    {
        myClass.MyEvent -= new myDelegate
            (this.eventHandlingMethod);
    }

    private void eventHandlingMethod()
    {
        ...
    }
    private MyClass myClass = new MyClass();
}
```

Or:

```
public void Stop()
{
    myClass.MyEvent -= this.eventHandlingMethod;
}
```

Raise an event	Use parentheses exactly as if the event were a method. You must supply arguments to match the type of the event. Don't forget to check whether the event is *null*. For example:

```
class MyClass
{
    public event myDelegate MyEvent;
    ...
    private void RaiseEvent()
    {
        if (this.MyEvent != null)
        {
            this.MyEvent();
        }
    }
    ...
}
```

Chapter 18
Introducing Generics

After completing this chapter, you will be able to:

- Define a type-safe class by using generics.

- Create instances of a generic class based on types specified as type parameters.

- Implement a generic interface.

- Define a generic method that implements an algorithm independent of the type of data on which it operates.

In Chapter 8, "Understanding Values and References," you learned how to use the *object* type to refer to an instance of any class. You can use the *object* type to store a value of any type, and you can define parameters by using the *object* type when you need to pass values of any type into a method. A method can also return values of any type by specifying *object* as the return type. Although this practice is very flexible, it puts the onus on the programmer to remember what sort of data is actually being used and can lead to run-time errors if the programmer makes a mistake. In this chapter, you will learn about generics, a feature that has been designed to help you prevent that kind of mistake.

The Problem with *objects*

To understand generics, it is worth looking in detail at the problems they are designed to solve, specifically when using the *object* type.

You can use the *object* type to refer to a value or variable of any type. All reference types automatically inherit (either directly or indirectly) from the *System.Object* class in the Microsoft .NET Framework. You can use this information to create highly generalized classes and methods. For example, many of the classes in the *System.Collections* namespace exploit this fact, so you can create collections holding almost any type of data. (You have already been introduced to the collection classes in Chapter 10, "Using Arrays and Collections.") By homing in on one particular collection class as a detailed example, you will also notice in the

System.Collections.Queue class that you can create queues containing practically anything. The following code example shows how to create and manipulate a queue of *Circle* objects:

```
using System.Collections;
...
Queue myQueue = new Queue();
Circle myCircle = new Circle();
myQueue.Enqueue(myCircle);
...
myCircle = (Circle)myQueue.Dequeue();
```

The *Enqueue* method adds an *object* to the head of a queue, and the *Dequeue* method removes the *object* at the other end of the queue. These methods are defined like this:

```
public void Enqueue( object item );
public object Dequeue();
```

Because the *Enqueue* and *Dequeue* methods manipulate *object*s, you can operate on queues of *Circle*s, *PhoneBook*s, *Clock*s, or any of the other classes you have seen in earlier exercises in this book. However, it is important to notice that you have to cast the value returned by the *Dequeue* method to the appropriate type because the compiler will not perform the conversion from the *object* type automatically. If you don't cast the returned value, you will get the compiler error "Cannot implicitly convert type 'object' to 'Circle'. "

This need to perform an explicit cast denigrates much of the flexibility afforded by the *object* type. It is very easy to write code such as this:

```
Queue myQueue = new Queue();
Circle myCircle = new Circle();
myQueue.Enqueue(myCircle);
...
Clock myClock = (Clock)myQueue.Dequeue(); // run-time error
```

Although this code will compile, it is not valid and throws a *System.InvalidCastException* at run time. The error is caused by trying to store a reference to a *Circle* in a *Clock* variable, and the two types are not compatible. This error is not spotted until run time because the compiler does not have enough information to perform this check at compile time. The real type of the object being dequeued becomes apparent only when the code runs.

Another disadvantage of using the *object* approach to create generalized classes and methods is that it can use additional memory and processor time if the runtime needs to convert an *object* to a value type and back again. Consider the following piece of code that manipulates a queue of *int* variables:

```
Queue myQueue = new Queue();
int myInt = 99;
myQueue.Enqueue(myInt);          // box the int to an object
...
myInt = (int)myQueue.Dequeue(); // unbox the object to an int
```

The *Queue* data type expects the items it holds to be reference types. Enqueueing a value type, such as an *int*, requires it to be boxed to convert it to a reference type. Similarly, dequeueing into an *int* requires the item to be unboxed to convert it back to a value type. See the sections titled "Boxing" and "Unboxing" in Chapter 8 for more details. Although boxing and unboxing happen transparently, they add a performance overhead because they involve dynamic memory allocations. This overhead is small for each item, but it adds up when a program creates queues of large numbers of value types.

The Generics Solution

C# provides generics to remove the need for casting, improve type safety, reduce the amount of boxing required, and make it easier to create generalized classes and methods. Generic classes and methods accept *type parameters*, which specify the type of objects that they operate on. The .NET Framework class library includes generic versions of many of the collection classes and interfaces in the *System.Collections.Generic* namespace. The following code example shows how to use the generic *Queue* class found in this namespace to create a queue of *Circle* objects:

```
using System.Collections.Generic;
...
Queue<Circle> myQueue = new Queue<Circle>();
Circle myCircle = new Circle();
myQueue.Enqueue(myCircle);
...
myCircle = myQueue.Dequeue();
```

There are two new things to note about the code in the preceding example:

- The use of the type parameter between the angle brackets, *<Circle>*, when declaring the *myQueue* variable
- The lack of a cast when executing the *Dequeue* method

The type parameter in angle brackets specifies the type of objects accepted by the queue. All references to methods in this queue will automatically expect to use this type rather than *object*, rendering unnecessary the cast to the *Circle* type when invoking the *Dequeue* method. The compiler will check to ensure that types are not accidentally mixed and will generate an error at compile time rather than at run time if you try to dequeue an item from *circleQueue* into a *Clock* object, for example.

If you examine the description of the generic *Queue* class in the Microsoft Visual Studio 2008 documentation, you will notice that it is defined as follows:

```
public class Queue<T> : ...
```

The *T* identifies the type parameter and acts as a placeholder for a real type at compile time. When you write code to instantiate a generic *Queue*, you provide the type that should be substituted for *T* (*Circle* in the preceding example). Furthermore, if you then look at the methods of the *Queue<T>* class, you will observe that some of them, such as *Enqueue* and *Dequeue*, specify *T* as a parameter type or return value:

```
public void Enqueue( T item );
public T Dequeue();
```

The type parameter, *T*, will be replaced with the type you specified when you declared the queue. What is more, the compiler now has enough information to perform strict type checking when you build the application and can trap any type mismatch errors early.

You should also be aware that this substitution of *T* for a specified type is not simply a textual replacement mechanism. Instead, the compiler performs a complete semantic substitution so that you can specify any valid type for *T*. Here are more examples:

```
struct Person
{
    ...
}
...
Queue<int> intQueue = new Queue<int>();
Queue<Person> personQueue = new Queue<Person>();
Queue<Queue<int>> queueQueue = new Queue<Queue<int>>();
```

The first two examples create queues of value types, while the third creates a queue of queues (of *ints*). For example, for the *intQueue* variable the compiler will also generate the following versions of the *Enqueue* and *Dequeue* methods:

```
public void Enqueue( int item );
public int Dequeue();
```

Contrast these definitions with those of the nongeneric *Queue* class shown in the preceding section. In the methods derived from the generic class, the *item* parameter to *Enqueue* is passed as a value type that does not require boxing. Similarly, the value returned by *Dequeue* is also a value type that does not need to be unboxed.

It is also possible for a generic class to have multiple type parameters. For example, the generic *System.Collections.Generic.Dictionary* class expects two type parameters: one type for keys and another for the values. The following definition shows how to specify multiple type parameters:

```
public class Dictionary<TKey, TValue>
```

A dictionary provides a collection of key/value pairs. You store values (type *TValue*) with an associated key (type *TKey*) and then retrieve them by specifying the key to look up. The

Dictionary class provides an indexer that allows you to access items by using array notation. It is defined like this:

```
public virtual TValue this[ TKey key ] { get; set; }
```

Notice that the indexer accesses values of type *TValue* by using a key of type *TKey*. To create and use a dictionary called *directory* containing *Person* values identified by *string* keys, you could use the following code:

```
struct Person
{
    ...
}
...
Dictionary<string, Person> directory = new Dictionary<string, Person>();
Person john = new Person();
directory["John"] = john;
...
Person author = directory["John"];
```

As with the generic *Queue* class, the compiler will detect attempts to store values other than *Person* structures in the directory, as well as ensure that the key is always a *string* value. For more information about the *Dictionary* class, you should read the Visual Studio 2008 documentation.

 Note You can also define generic structures and interfaces by using the same type–parameter syntax as generic classes.

Generics vs. Generalized Classes

It is important to be aware that a generic class that uses type parameters is different from a *generalized* class designed to take parameters that can be cast to different types. For example, the *System.Collections.Queue* class is a generalized class. There is a *single* implementation of this class, and its methods take *object* parameters and return *object* types. You can use this class with *ints*, *strings*, and many other types; in each case, you are using instances of the same class.

Compare this with the *System.Collections.Generic.Queue<T>* class. Each time you use this class with a type parameter (such as *Queue<int>* or *Queue<string>*) you actually cause the compiler to generate an entirely new class that happens to have functionality defined by the generic class. You can think of a generic class as one that defines a template that is then used by the compiler to generate new type-specific classes on demand. The type-specific versions of a generic class (*Queue<int>*, *Queue<string>*, and so on) are referred to as *constructed types*, and you should treat them as distinctly different types (albeit ones that have a similar set of methods and properties).

Generics and Constraints

Occasionally, you will want to ensure that the type parameter used by a generic class identifies a type that provides certain methods. For example, if you are defining a *PrintableCollection* class, you might want to ensure that all objects stored in the class have a *Print* method. You can specify this condition by using a *constraint*.

By using a constraint, you can limit the type parameters of a generic class to those that implement a particular set of interfaces, and therefore provide the methods defined by those interfaces. For example, if the *IPrintable* interface defined the *Print* method, you could create the *PrintableCollection* class like this:

```
public class PrintableCollection<T> where T : IPrintable
```

When you build this class with a type parameter, the compiler will check to ensure that the type used for *T* actually implements the *IPrintable* interface and will stop with a compilation error if it doesn't.

Creating a Generic Class

The .NET Framework class library contains a number of generic classes readily available for you. You can also define your own generic classes, which is what you will do in this section. Before you do this, I provide a bit of background theory.

The Theory of Binary Trees

In the following exercises, you will define and use a class that represents a binary tree. This is a practical exercise because this class happens to be one that is missing from the *System. Collections.Generic* namespace. A binary tree is a useful data structure used for a variety of operations, including sorting and searching through data very quickly. There are volumes written on the minutiae of binary trees, but it is not the purpose of this book to cover binary trees in detail. Instead, we just look at the pertinent details. If you are interested, you should consult a book such as *The Art of Computer Programming, Volume 3: Sorting and Searching* by Donald E. Knuth (Addison-Wesley Professional, 2nd edition, 1998).

A binary tree is a recursive (self-referencing) data structure that can either be empty or contain three elements: a datum, which is typically referred to as the *node*, and two subtrees, which are themselves binary trees. The two subtrees are conventionally called the *left subtree* and the *right subtree* because they are typically depicted to the left and right of the node, respectively. Each left subtree or right subtree is either empty or contains a node and other subtrees. In theory, the whole structure can continue ad infinitum. Figure 18-1 shows the structure of a small binary tree.

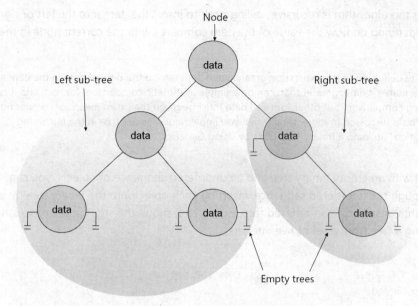

FIGURE 18-1 A binary tree.

The real power of binary trees becomes evident when you use them for sorting data. If you start with an unordered sequence of objects of the same type, you can construct an ordered binary tree and then walk through the tree to visit each node in an ordered sequence. The algorithm for inserting an item *I* into an ordered binary tree *T* is shown here:

```
If the tree, T, is empty
Then
  Construct a new tree T with the new item I as the node, and empty left and
  right subtrees
Else
  Examine the value of the current node, N, of the tree, T
  If the value of N is greater than that of the new item, I
  Then
    If the left subtree of T is empty
    Then
      Construct a new left subtree of T with the item I as the node, and
      empty left and right subtrees
    Else
      Insert I into the left subtree of T
    End If
  Else
    If the right subtree of T is empty
    Then
      Construct a new right subtree of T with the item I as the node, and
      empty left and right subtrees
    Else
      Insert I into the right subtree of T
    End If
  End If
End If
```

Notice that this algorithm is recursive, calling itself to insert the item into the left or right subtree depending on how the value of the item compares with the current node in the tree.

> **Note** The definition of the expression *greater than* depends on the type of data in the item and node. For numeric data, greater than can be a simple arithmetic comparison, for text data it can be a string comparison, but other forms of data must be given their own means of comparing values. This is discussed in more detail when you implement a binary tree in the upcoming section titled "Building a Binary Tree Class by Using Generics."

If you start with an empty binary tree and an unordered sequence of objects, you can iterate through the unordered sequence, inserting each object into the binary tree by using this algorithm, resulting in an ordered tree. Figure 18-2 shows the steps in the process for constructing a tree from a set of five integers.

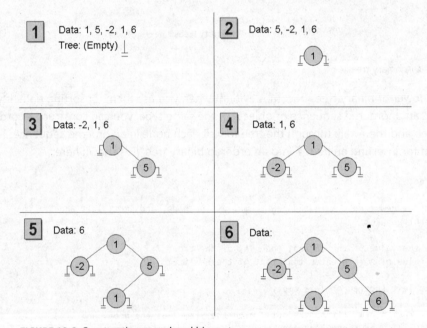

FIGURE 18-2 Constructing an ordered binary tree.

After you have built an ordered binary tree, you can display its contents in sequence by visiting each node in turn and printing the value found. The algorithm for achieving this task is also recursive:

```
If the left subtree is not empty
Then
  Display the contents of the left subtree
End If
Display the value of the node
If the right subtree is not empty
```

```
Then
    Display the contents of the right subtree
End If
```

Figure 18-3 shows the steps in the process for outputting the tree constructed in Figure 18-2. Notice that the integers are now displayed in ascending order.

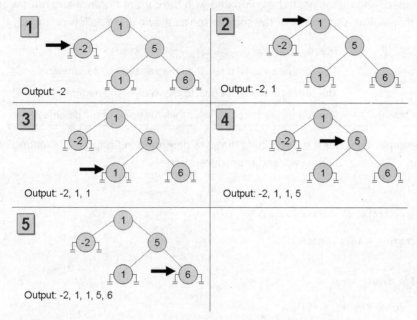

Output: -2

Output: -2, 1

Output: -2, 1, 1

Output: -2, 1, 1, 5

Output: -2, 1, 1, 5, 6

FIGURE 18-3 Printing an ordered binary tree.

Building a Binary Tree Class by Using Generics

In the following exercise, you will use generics to define a binary tree class capable of holding almost any type of data. The only restriction is that the data type must provide a means of comparing values between different instances.

The binary tree class is a class that you might find useful in many different applications. Therefore, you will implement it as a class library rather than as an application in its own right. You can then reuse this class elsewhere without having to copy the source code and recompile it. A class library is a set of compiled classes (and other types such as structures and delegates) stored in an assembly. An assembly is a file that usually has the .dll suffix. Other projects and applications can make use of the items in a class library by adding a reference to its assembly and then bringing its namespaces into scope with *using* statements. You will do this when you test the binary tree class.

The *System.IComparable* and *System.IComparable<T>* Interfaces

If you need to create a class that requires you to be able to compare values according to some natural (or possibly unnatural) ordering, you should implement the *IComparable* interface. This interface contains a method called *CompareTo*, which takes a single parameter specifying the object to be compared with the current instance and returns an integer that indicates the result of the comparison as shown in the following table.

Value	Meaning
Less than 0	The current instance is less than the value of the parameter.
0	The current instance is equal to the value of the parameter.
Greater than 0	The current instance is greater than the value of the parameter.

As an example, consider the *Circle* class that was described in Chapter 7, "Creating and Managing Classes and Objects," and reproduced here:

```
class Circle
{
    public Circle(int initialRadius)
    {
        radius = initialRadius;
    }

    public double Area()
    {
        return Math.PI * radius * radius;
    }

    private double radius;
}
```

You can make the *Circle* class "comparable" by implementing the *System.IComparable* interface and providing the *CompareTo* method. In the example shown, the *CompareTo* method compares *Circle* objects based on their areas. A circle with a larger area is considered to be greater than a circle with a smaller area.

```
class Circle : System.IComparable
{
    ...
    public int CompareTo(object obj)
    {
        Circle circObj = (Circle)obj;  // cast the parameter to its real type
        if (this.Area() == circObj.Area())
            return 0;

        if (this.Area() > circObj.Area())
            return 1;
```

```
            return -1;
        }
}
```

If you examine the *System.IComparable* interface, you will see that its parameter is defined as an *object*. However, this approach is not type-safe. To understand why this is so, consider what happens if you try to pass something that is not a *Circle* to the *CompareTo* method. The *System.IComparable* interface requires the use of a cast to be able to access the *Area* method. If the parameter is not a *Circle* but some other type of object, this cast will fail. However, the *System* namespace also defines the generic *IComparable<T>* interface, which contains the following methods:

```
int CompareTo(T other);
bool Equals(T other);
```

Notice that there is an additional method in this interface called *Equals*, which should return *true* if both instances are equals and *false* if they are not equals.

Also notice that these methods take a type parameter (*T*) rather than an *object* and, therefore, are much safer than is the nongeneric version of the interface. The following code shows how you can implement this interface in the *Circle* class:

```
class Circle : System.IComparable<Circle>
{
    ...
    public int CompareTo(Circle other)
    {
        if (this.Area() == other.Area())
            return 0;

        if (this.Area() > other.Area())
            return 1;

        return -1;
    }

    public bool Equals(Circle other)
    {
        return (this.CompareTo(other) == 0);
    }
}
```

The parameters for the *CompareTo* and *Equals* methods must match the type specified in the interface, *IComparable<Circle>*. In general, it is preferable to implement the *System.IComparable<T>* interface rather than the *System.IComparable* interface. You can also implement both just as many of the types in the .NET Framework do.

Create the *Tree<TItem>* class

1. Start Visual Studio 2008 if it is not already running.

2. If you are using Visual Studio 2008 Standard Edition or Visual Studio 2008 Professional Edition, perform the following tasks to create a new class library project:

    2.1. On the *File* menu, point to *New*, and then click *Project*.

    2.2. In the *New Project* dialog box, select the *Class Library* template.

    2.3. Set the *Name* to *BinaryTree* and set the *Location* to *\Microsoft Press\Visual CSharp Step By Step\Chapter 18* under your Documents folder.

    2.4. Click *OK*.

3. If you are using Microsoft Visual C# 2008 Express Edition, perform the following tasks to create a new class library project:

    3.1. On the *Tools* menu, click *Options*.

    3.2. In the *Options* dialog box, click *Projects and Solutions* in the tree view in the left pane.

    3.3. In the right pane, in the *Visual Studio projects location* text box, specify the location *\Microsoft Press\Visual CSharp Step By Step\Chapter 18* folder under your Documents folder.

    3.4. Click *OK*.

    3.5. On the *File* menu, click *New Project*.

    3.6. In the *New Project* dialog box, click the *Class Library* icon.

    3.7. In the *Name* field, type *BinaryTree*.

    3.8. Click *OK*.

4. In *Solution Explorer*, right-click *Class1.cs* and change the name of the file to Tree.cs. Allow Visual Studio to change the name of the class as well as the name of the file when prompted.

5. In the *Code and Text Editor* window, change the definition of the *Tree* class to *Tree<TItem>*, as shown in bold type in the following code:

```
public class Tree<TItem>
{
}
```

6. In the *Code and Text Editor* window, modify the definition of the *Tree<TItem>* class as follows in bold type to specify that the type parameter *TItem* must denote a type that implements the generic *IComparable<TItem>* interface.

The modified definition of the *Tree<TItem>* class should look like this:

```
public class Tree<TItem> where TItem : IComparable<TItem>
{
}
```

7. Add three public, automatic properties to the *Tree<TItem>* class: a *TItem* property called *NodeData* and two *Tree<TItem>* properties called *LeftTree* and *RightTree*, as follows in bold type:

```
public class Tree<TItem> where TItem : IComparable<TItem>
{
    public TItem NodeData { get; set; }
    public Tree<TItem> LeftTree { get; set; }
    public Tree<TItem> RightTree { get; set; }
}
```

8. Add a constructor to the *Tree<TItem>* class that takes a single *TItem* parameter called *nodeValue*. In the constructor, set the *NodeData* property to *nodeValue*, and initialize the *LeftTree* and *RightTree* properties to *null*, as shown in bold type in the following code:

```
public class Tree<TItem> where TItem : IComparable<TItem>
{
    public Tree(TItem nodeValue)
    {
        this.NodeData = nodeValue;
        this.LeftTree = null;
        this.RightTree = null;
    }
    ...
}
```

Note Notice that the name of the constructor does not include the type parameter; it is called *Tree*, and not *Tree<TItem>*.

9. Add a public method called *Insert* to the *Tree<TItem>* class as shown in bold type in the following code. This method will insert a *TItem* value into the tree.

The method definition should look like this:

```
public class Tree<TItem> where TItem: IComparable<TItem>
{
    ...
    public void Insert(TItem newItem)
    {
    }
    ...
}
```

The *Insert* method will implement the recursive algorithm described earlier for creating an ordered binary tree. The programmer will have used the constructor to create the initial node of the tree (there is no default constructor), so the *Insert* method can assume that the tree is not empty. The part of the algorithm after checking whether the tree is empty is reproduced here to help you understand the code you will write for the *Insert* method in the following steps:

```
...
Examine the value of the node, N, of the tree, T
If the value of N is greater than that of the new item, I
Then
  If the left subtree of T is empty
  Then
    Construct a new left subtree of T with the item I as the node, and empty
    left and right subtrees
  Else
    Insert I into the left subtree of T
End If
...
```

10. In the *Insert* method, add a statement that declares a local variable of type *TItem*, called *currentNodeValue*. Initialize this variable to the value of the *NodeData* property of the tree, as shown here:

```
public void Insert(TItem newItem)
{
    TItem currentNodeValue = this.NodeData;
}
```

11. Add the following *if-else* statement shown in bold type to the *Insert* method after the definition of the *currentNodeValue* variable. This statement uses the *CompareTo* method of the *IComparable<T>* interface to determine whether the value of the current node is greater than the new item is:

```
public void Insert(TItem newItem)
{
    TItem currentNodeValue = this.NodeData;
    if (currentNodeValue.CompareTo(newItem) > 0)
    {
        // Insert the new item into the left subtree
    }
    else
    {
        // Insert the new item into the right subtree
    }
}
```

12. Replace the // Insert the new item into the left subtree comment with the following block of code:

```
if (this.LeftTree == null)
{
```

```
    this.LeftTree = new Tree<TItem>(newItem);
}
else
{
    this.LeftTree.Insert(newItem);
}
```

These statements check whether the left subtree is empty. If so, a new tree is created using the new item and attached as the left subtree of the current node; otherwise, the new item is inserted into the existing left subtree by calling the *Insert* method recursively.

13. Replace the `// Insert the new item into the right subtree` comment with the equivalent code that inserts the new node into the right subtree:

```
if (this.RightTree == null)
{
    this.RightTree = new Tree<TItem>(newItem);
}
else
{
    this.RightTree.Insert(newItem);
}
```

14. Add another public method called *WalkTree* to the *Tree<TItem>* class after the *Insert* method. This method will walk through the tree, visiting each node in sequence and printing out its value.

   The method definition should look like this:

```
public void WalkTree()
{
}
```

15. Add the following statements to the *WalkTree* method. These statements implement the algorithm described earlier for printing the contents of a binary tree:

```
if (this.LeftTree != null)
{
    this.LeftTree.WalkTree();
}

Console.WriteLine(this.NodeData.ToString());

if (this.RightTree != null)
{
    this.RightTree.WalkTree();
}
```

16. On the *Build* menu, click *Build Solution*. The class should compile cleanly, but correct any errors that are reported and rebuild the solution if necessary.

17. If you are using Visual C# 2008 Express Edition, on the *File* menu, click *Save All*. If the *Save Project* dialog box appears, click *Save*.

In the next exercise, you will test the *Tree<TItem>* class by creating binary trees of integers and strings.

Test the *Tree<TItem>* class

1. In *Solution Explorer*, right-click the BinaryTree solution, point to *Add*, and then click *New Project*.

   Note Make sure you right-click the BinaryTree solution rather than the BinaryTree project.

2. Add a new project using the Console Application template. Name the project *BinaryTreeTest*. Set the *Location* to *\Microsoft Press\Visual CSharp Step By Step\Chapter 18* under your Documents folder, and then click *OK*.

   Note Remember that a Visual Studio 2008 solution can contain more than one project. You are using this feature to add a second project to the BinaryTree solution for testing the *Tree<TItem>* class. This is the recommended way of testing class libraries.

3. Ensure that the BinaryTreeTest project is selected in *Solution Explorer*. On the *Project* menu, click *Set as Startup Project*.

   The BinaryTreeTest project is highlighted in *Solution Explorer*. When you run the application, this is the project that will actually execute.

4. Ensure that the BinaryTreeTest project is still selected in *Solution Explorer*. On the *Project* menu, click *Add Reference*. In the *Add Reference* dialog box, click the *Projects* tab. Select the *BinaryTree* project, and then click *OK*.

   The BinaryTree assembly appears in the list of references for the BinaryTreeTest project in *Solution Explorer*. You will now be able to create *Tree<TItem>* objects in the BinaryTreeTest project.

   Note If the class library project is not part of the same solution as the project that uses it, you must add a reference to the assembly (the .dll file) and not to the class library project. You do this by selecting the assembly from the *Browse* tab in the *Add Reference* dialog box. You will use this technique in the final set of exercises in this chapter.

5. In the *Code and Text Editor* window displaying the *Program* class, add the following *using* directive to the list at the top of the class:

```
using BinaryTree;
```

6. Add the statements in bold type in the following code to the *Main* method:

```
static void Main(string[] args)
{
    Tree<int> tree1 = new Tree<int>(10);
    tree1.Insert(5);
    tree1.Insert(11);
    tree1.Insert(5);
    tree1.Insert(-12);
    tree1.Insert(15);
    tree1.Insert(0);
    tree1.Insert(14);
    tree1.Insert(-8);
    tree1.Insert(10);
    tree1.Insert(8);
    tree1.Insert(8);
    tree1.WalkTree();
}
```

These statements create a new binary tree for holding *int*s. The constructor creates an initial node containing the value 10. The *Insert* statements add nodes to the tree, and the *WalkTree* method prints out the contents of the tree, which should appear sorted in ascending order.

> **Note**   Remember that the *int* keyword in C# is actually just an alias for the *System.Int32* type; whenever you declare an *int* variable, you are actually declaring a *struct* variable of type *System.Int32*. The *System.Int32* type implements the *IComparable* and *IComparable<T>* interfaces, which is why you can create *Tree<int>* variables. Similarly, the *string* keyword is an alias for *System.String*, which also implements *IComparable* and *IComparable<T>*.

7. On the *Build* menu, click *Build Solution*. Verify that the solution compiles, and correct any errors if necessary.

8. Save the project, and then on the *Debug* menu, click *Start Without Debugging*.

The program runs and displays the values in the following sequence:

–12, –8, 0, 5, 5, 8, 8, 10, 10, 11, 14, 15

9. Press the Enter key to return to Visual Studio 2008.

10. Add the following statements shown in bold type to the end of the *Main* method in the *Program* class, after the existing code:

```
static void Main(string[] args)
{
    ...
    Tree<string> tree2 = new Tree<string>("Hello");
    tree2.Insert("World");
    tree2.Insert("How");
    tree2.Insert("Are");
    tree2.Insert("You");
```

```
                    tree2.Insert("Today");
                    tree2.Insert("I");
                    tree2.Insert("Hope");
                    tree2.Insert("You");
                    tree2.Insert("Are");
                    tree2.Insert("Feeling");
                    tree2.Insert("Well");
                    tree2.Insert("!");
                    tree2.WalkTree();
                }
```

These statements create another binary tree for holding *strings*, populate it with some test data, and then print the tree. This time, the data is sorted alphabetically.

11. On the *Build* menu, click *Build Solution*. Verify that the solution compiles, and correct any errors if necessary.

12. On the *Debug* menu, click *Start Without Debugging*.

The program runs and displays the integer values as before, followed by the strings in the following sequence:

!, Are, Are, Feeling, Hello, Hope, How, I, Today, Well, World, You, You

13. Press the Enter key to return to Visual Studio 2008.

Creating a Generic Method

As well as defining generic classes, you can also use the .NET Framework to create generic methods.

With a generic method, you can specify parameters and the return type by using a type parameter in a manner similar to that used when defining a generic class. In this way, you can define generalized methods that are type-safe and avoid the overhead of casting (and boxing in some cases). Generic methods are frequently used in conjunction with generic classes—you need them for methods that take a generic class as a parameter or that have a return type that is a generic class.

You define generic methods by using the same type parameter syntax that you use when creating generic classes (you can also specify constraints). For example, you can call the following generic *Swap<T>* method to swap the values in its parameters. Because this functionality is useful regardless of the type of data being swapped, it is helpful to define it as a generic method:

```
static void Swap<T>(ref T first, ref T second)
{
    T temp = first;
    first = second;
    second = temp;
}
```

You invoke the method by specifying the appropriate type for its type parameter. The following examples show how to invoke the *Swap<T>* method to swap over two *int*s and two *string*s:

```
int a = 1, b = 2;
Swap<int>(ref a, ref b);
...
string s1 = "Hello", s2 = "World";
Swap<string>(ref s1, ref s2);
```

> **Note**   Just as instantiating a generic class with different type parameters causes the compiler to generate different types, each distinct use of the *Swap<T>* method causes the compiler to generate a different version of the method. *Swap<int>* is not the same method as *Swap<string>*; both methods just happen to have been generated from the same generic method, so they exhibit the same behavior, albeit over different types.

Defining a Generic Method to Build a Binary Tree

The preceding exercise showed you how to create a generic class for implementing a binary tree. The *Tree<TItem>* class provides the *Insert* method for adding data items to the tree. However, if you want to add a large number of items, repeated calls to the *Insert* method are not very convenient. In the following exercise, you will define a generic method called *InsertIntoTree* that you can use to insert a list of data items into a tree with a single method call. You will test this method by using it to insert a list of characters into a tree of characters.

Write the *InsertIntoTree* method

1. Using Visual Studio 2008, create a new project by using the Console Application template. In the *New Project* dialog box, name the project *BuildTree*. If you are using Visual Studio 2008 Standard Edition or Visual Studio 2008 Professional Edition, set the *Location* to *\Microsoft Press\Visual CSharp Step By Step\Chapter 18* under your Documents folder, and select *Create a new Solution* from the *Solution* drop-down list. Click *OK*.

2. On the *Project* menu, click *Add Reference*. In the *Add Reference* dialog box, click the *Browse* tab. Move to the folder *\Microsoft Press\Visual CSharp Step By Step\Chapter 18 \BinaryTree\BinaryTree\bin\Debug*, click *BinaryTree.dll*, and then click *OK*.

   The *BinaryTree* assembly is added to the list of references shown in *Solution Explorer*.

3. In the *Code and Text Editor* window displaying the Program.cs file, add the following *using* directive to the top of the Program.cs file:

   ```
   using BinaryTree;
   ```

   This namespace contains the *Tree<TItem>* class.

4. Add a method called *InsertIntoTree* to the *Program* class after the *Main* method. This should be a *static* method that takes a *Tree<TItem>* variable and a *params* array of *TItem* elements called *data*.

The method definition should look like this:

```
static void InsertIntoTree<TItem>(Tree<TItem> tree, params TItem[] data)
{
}
```

> **Tip** An alternative way of implementing this method is to create an extension method of the *Tree<TItem>* class by prefixing the *Tree<TItem>* parameter with the *this* keyword and defining the *InsertIntoTree* method in a static class, like this:
>
> ```
> public static class TreeMethods
> {
>     public static void InsertIntoTree<TItem>(this Tree<TItem> tree,
>                                              params TItem[] data)
>     {
>         ...
>     }
>     ...
> }
> ```
>
> The principal advantage of this approach is that you can invoke the *InsertIntoTree* method directly on a *Tree<TItem>* object rather than pass the *Tree<TItem>* in as a parameter. However, for this exercise, we will keep things simple.

5. The *TItem* type used for the elements being inserted into the binary tree must implement the *IComparable<TItem>* interface. Modify the definition of the *InsertIntoTree* method and add the appropriate *where* clause, as shown in bold type in the following code.

```
static void InsertIntoTree<TItem>(Tree<TItem> tree, params TItem[] data) where TItem :
IComparable<TItem>
{
}
```

6. Add the following statements shown in bold type to the *InsertIntoTree* method. These statements check to make sure that the user has actually passed some parameters into the method (the *data* array might be empty), and then they iterate through the *params* list, adding each item to the tree by using the *Insert* method. The tree is passed back as the return value:

```
static void InsertIntoTree<TItem>(Tree<TItem> tree, params TItem[] data) where TItem :
IComparable<TItem>
{
    if (data.Length == 0)
        throw new ArgumentException("Must provide at least one data value");
```

```
        foreach (TItem datum in data)
        {
            tree.Insert(datum);
        }
    }
```

Test the *InsertIntoTree* method

1. In the *Main* method of the *Program* class, add the following statements shown in bold type that create a new *Tree* for holding character data, populate it with some sample data by using the *InsertIntoTree* method, and then display it by using the *WalkTree* method of *Tree*:

```
static void Main(string[] args)
{
    Tree<char> charTree = new Tree<char>('M');
    InsertIntoTree<char>(charTree, 'X', 'A', 'M', 'Z', 'Z', 'N');
    charTree.WalkTree();
}
```

2. On the *Build* menu, click *Build Solution*. Verify that the solution compiles, and correct any errors if necessary.

3. On the *Debug* menu, click *Start Without Debugging*.

   The program runs and displays the character values in the following order:

   A, M, M, N, X, Z, Z

4. Press the Enter key to return to Visual Studio 2008.

- If you want to continue to the next chapter

  Keep Visual Studio 2008 running and turn to Chapter 19.

- If you want to exit Visual Studio 2008 now

  On the *File* menu, click *Exit*. If you see a *Save* dialog box, click *Yes* (if you are using Visual Studio 2008) or *Save* (if you are using Visual C# 2008 Express Edition) and save the project.

Chapter 18 Quick Reference

To	Do this
Instantiate an object by using a generic type	Specify the appropriate generic type parameter. For example: `Queue<int> myQueue = new Queue<int>();`

Create a new generic type	Define the class using a type parameter. For example:
	
	`public class Tree<TItem>`
	`{`
	`    ...`
	`}`

Restrict the type that can be substituted for the generic type parameter	Specify a constraint by using a *where* clause when defining the class. For example:
	
	`public class Tree<TItem>`
	`where TItem : IComparable<TItem>`
	`{`
	`    ...`
	`}`

Define a generic method	Define the method by using type parameters. For example:
	
	`static void InsertIntoTree<TItem>`
	`(Tree<TItem> tree, params TItem[] data)`
	`{`
	`    ...`
	`}`

Invoke a generic method	Provide types for each of the type parameters. For example: `InsertIntoTree<char>(charTree, 'Z', 'X');`

Chapter 19
Enumerating Collections

After completing this chapter, you will be able to:

- Manually define an enumerator that can be used to iterate over the elements in a collection.

- Implement an enumerator automatically by creating an iterator.

- Provide additional iterators that can step through the elements of a collection in different sequences.

In Chapter 10, "Using Arrays and Collections," you learned about arrays and collection classes for holding sequences or sets of data. Chapter 10 also introduced the *foreach* statement that you can use for stepping through, or iterating over, the elements in a collection. At the time, you just used the *foreach* statement as a quick and convenient way of accessing the contents of a collection, but now it is time to learn a little more about how this statement actually works. This topic becomes important when you start defining your own collection classes. Fortunately, C# provides iterators to help you automate much of the process.

Enumerating the Elements in a Collection

In Chapter 10, you saw an example of using the *foreach* statement to list the items in a simple array. The code looked like this:

```
int[] pins = { 9, 3, 7, 2 };
foreach (int pin in pins)
{
    Console.WriteLine(pin);
}
```

The *foreach* construct provides an elegant mechanism that greatly simplifies the code that you need to write, but it can be exercised only under certain circumstances—you can use *foreach* only to step through an *enumerable* collection. So, what exactly is an enumerable collection? The quick answer is that it is a collection that implements the *System.Collections. IEnumerable* interface.

 Note Remember that all arrays in C# are actually instances of the *System.Array* class. The *System.Array* class is a collection class that implements the *IEnumerable* interface.

The *IEnumerable* interface contains a single method called *GetEnumerator*:

```
IEnumerator GetEnumerator();
```

The *GetEnumerator* method should return an enumerator object that implements the *System.Collections.IEnumerator* interface. The enumerator object is used for stepping through (enumerating) the elements of the collection. The *IEnumerator* interface specifies the following property and methods:

```
object Current { get; }
bool MoveNext();
void Reset();
```

Think of an enumerator as a pointer pointing to elements in a list. Initially, the pointer points *before* the first element. You call the *MoveNext* method to move the pointer down to the next (first) item in the list; the *MoveNext* method should return *true* if there actually is another item and *false* if there isn't. You use the *Current* property to access the item currently pointed to, and you use the *Reset* method to return the pointer back to *before* the first item in the list. By creating an enumerator by using the *GetEnumerator* method of a collection and repeatedly calling the *MoveNext* method and retrieving the value of the *Current* property by using the enumerator, you can move forward through the elements of a collection one item at a time. This is exactly what the *foreach* statement does. So if you want to create your own enumerable collection class, you must implement the *IEnumerable* interface in your collection class and also provide an implementation of the *IEnumerator* interface to be returned by the *GetEnumerator* method of the collection class.

Important At first glance, it is easy to confuse the *IEnumerable<T>* and the *IEnumerator<T>* interfaces because of the similarity of their names. Don't get them mixed up.

If you are observant, you will have noticed that the *Current* property of the *IEnumerator* interface exhibits non-type-safe behavior in that it returns an *object* rather than a specific type. However, you should be pleased to know that the Microsoft .NET Framework class library also provides the generic *IEnumerator<T>* interface, which has a *Current* property that returns a *T* instead. Likewise, there is also an *IEnumerable<T>* interface containing a *GetEnumerator* method that returns an *Enumerator<T>* object. If you are building applications for the .NET Framework version 2.0 or later, you should make use of these generic interfaces when defining enumerable collections rather than using the nongeneric definitions.

Note The *IEnumerator<T>* interface has some further differences from the *IEnumerator* interface; it does not contain a *Reset* method but extends the *IDisposable* interface.

Manually Implementing an Enumerator

In the next exercise, you will define a class that implements the generic *IEnumerator<T>* interface and create an enumerator for the binary tree class that you built in Chapter 18, "Introducing Generics." In Chapter 18, you saw how easy it is to traverse a binary tree and display its contents. You would therefore be inclined to think that defining an enumerator that retrieves each element in a binary tree in the same order would be a simple matter. Sadly, you would be mistaken. The main problem is that when defining an enumerator you need to remember where you are in the structure so that subsequent calls to the *MoveNext* method can update the position appropriately. Recursive algorithms, such as that used when walking a binary tree, do not lend themselves to maintaining state information between method calls in an easily accessible manner. For this reason, you will first preprocess the data in the binary tree into a more amenable data structure (a queue) and actually enumerate this data structure instead. Of course, this deviousness is hidden from the user iterating through the elements of the binary tree!

Create the *TreeEnumerator* class

1. Start Microsoft Visual Studio 2008 if it is not already running.

2. Open the BinaryTree solution located in the \Microsoft Press\Visual CSharp Step by Step\Chapter 19\BinaryTree folder in your Documents folder. This solution contains a working copy of the BinaryTree project you created in Chapter 18.

3. Add a new class to the project: On the *Project* menu, click *Add Class*, select the *Class* template, type **TreeEnumerator.cs** in the *Name* text box, and then click *Add*.

4. The *TreeEnumerator* class will generate an enumerator for a *Tree<TItem>* object. To ensure that the class is typesafe, you must provide a type parameter and implement the *IEnumerator<T>* interface. Also, the type parameter must be a valid type for the *Tree<TItem>* object that the class enumerates, so it must be constrained to implement the *IComparable<TItem>* interface.

   In the *Code and Text Editor* window displaying the TreeEnumerator.cs file, modify the definition of the *TreeEnumerator* class to satisfy these requirements, as shown in bold type in the following example.

   ```
   class TreeEnumerator<TItem> : IEnumerator<TItem> where TItem : IComparable<TItem>
   {
   }
   ```

5. Add the following three private variables shown in bold type to the *TreeEnumerator<TItem>* class:

```
class TreeEnumerator<TItem> : IEnumerator<TItem> where TItem : IComparable<TItem>
{
    private Tree<TItem> currentData = null;
    private TItem currentItem = default(TItem);
    private Queue<TItem> enumData = null;
}
```

The *currentData* variable will be used to hold a reference to the tree being enumerated, and the *currentItem* variable will hold the value returned by the *Current* property. You will populate the *enumData* queue with the values extracted from the nodes in the tree, and the *MoveNext* method will return each item from this queue in turn. The *default* keyword is explained in the section titled "Initializing a Variable Defined with a Type Parameter" later in this chapter.

6. Add a *TreeEnumerator* constructor that takes a single *Tree<TItem>* parameter called *data*. In the body of the constructor, add a statement that initializes the *currentData* variable to *data*:

```
class TreeEnumerator<TItem> : IEnumerator<TItem> where TItem : IComparable<TItem>
{
    public TreeEnumerator(Tree<TItem> data)
    {
        this.currentData = data;
    }
    ...
}
```

7. Add the following private method, called *populate*, to the *TreeEnumerator<TItem>* class immediately after the constructor:

```
private void populate(Queue<TItem> enumQueue, Tree<TItem> tree)
{
    if (tree.LeftTree != null)
    {
        populate(enumQueue, tree.LeftTree);
    }

    enumQueue.Enqueue(tree.NodeData);

    if (tree.RightTree != null)
    {
        populate(enumQueue, tree.RightTree);
    }
}
```

This method walks a binary tree, adding the data it contains to the queue. The algorithm used is similar to that used by the *WalkTree* method in the *Tree<TItem>* class, which was described in Chapter 18. The main difference is that rather than the method outputting *NodeData* values to the screen, it stores these values in the queue.

8. Return to the definition of the *TreeEnumerator<TItem>* class. Right-click anywhere in the *IEnumerator<TItem>* interface in the class declaration, point to *Implement Interface*, and then click *Implement Interface Explicitly*.

This action generates stubs for the methods of the *IEnumerator<TItem>* interface and the *IEnumerator* interface and adds them to the end of the class. It also generates the *Dispose* method for the *IDisposable* interface.

> **Note** The *IEnumerator<TItem>* interface inherits from the *IEnumerator* and *IDisposable* interfaces, which is why their methods also appear. In fact, the only item that belongs to the *IEnumerator<TItem>* interface is the generic *Current* property. The *MoveNext* and *Reset* methods belong to the nongeneric *IEnumerator* interface. The *IDisposable* interface was described in Chapter 14, "Using Garbage Collection and Resource Management."

9. Examine the code that has been generated. The bodies of the properties and methods contain a default implementation that simply throws a *NotImplementedException*. You will replace this code with a real implementation in the following steps.

10. Replace the body of the *MoveNext* method with the code shown in bold type here:

```
bool System.Collections.IEnumerator.MoveNext()
{
    if (this.enumData == null)
    {
        this.enumData = new Queue<TItem>();
        populate(this.enumData, this.currentData);
    }

    if (this.enumData.Count > 0)
    {
        this.currentItem = this.enumData.Dequeue();
        return true;
    }

    return false;
}
```

The purpose of the *MoveNext* method of an enumerator is actually twofold. The first time it is called, it should initialize the data used by the enumerator and advance to the first piece of data to be returned. (Prior to *MoveNext* being called for the first time, the value returned by the *Current* property is undefined and should result in an exception.) In this case, the initialization process consists of instantiating the queue and then calling the *populate* method to fill the queue with data extracted from the tree.

Subsequent calls to the *MoveNext* method should just move through data items until there are no more left, dequeuing items from the queue until the queue is empty in this example. It is important to bear in mind that *MoveNext* does not actually return data items—that is the purpose of the *Current* property. All *MoveNext* does is update internal state in the enumerator (the value of the *currentItem* variable is set to the data item extracted from the queue) for use by the *Current* property, returning *true* if there is a next value and *false* otherwise.

11. Modify the definition of the *get* accessor of the generic *Current* property as follows:

```
TItem IEnumerator<TItem>.Current
{
    get
    {
        if (this.enumData == null)
            throw new InvalidOperationException
                ("Use MoveNext before calling Current");

        return this.currentItem;
    }
}
```

> **Important** Be sure to add the code to the correct implementation of the *Current* property. Leave the nongeneric version, *System.Collections.IEnumerator.Current*, with its default implementation.

The *Current* property examines the *enumData* variable to ensure that *MoveNext* has been called. (This variable will be null prior to the first call to *MoveNext*.) If this is not the case, the property throws an *InvalidOperationException*—this is the conventional mechanism used by .NET Framework applications to indicate that an operation cannot be performed in the current state. If *MoveNext* has been called beforehand, it will have updated the *currentItem* variable, so all the *Current* property needs to do is return the value in this variable.

12. Locate the *IDisposable.Dispose* method. Comment out the `throw new NotImplementedException();` statement as follows in bold type. The enumerator does not use any resources that require explicit disposal, so this method does not need to do anything. It must still be present, however. For more information about the *Dispose* method, refer to Chapter 14.

```
void IDisposable.Dispose()
{
    // throw new NotImplementedException();
}
```

13. Build the solution, and fix any errors that are reported.

Initializing a Variable Defined with a Type Parameter

You should have noticed that the statement that defines and initializes the *currentItem* variable uses the *default* keyword. The *currentItem* variable is defined by using the type parameter *TItem*. When the program is written and compiled, the actual type that will be substituted for *TItem* might not be known—this issue is resolved only when the code is executed. This makes it difficult to specify how the variable should be initialized. The temptation is to set it to *null*. However, if the type substituted for *TItem* is a value type, this is an illegal assignment. (You cannot set value types to *null*, only reference types.) Similarly, if you set it to 0 in the expectation that the type will be numeric, this will be illegal if the type used is actually a reference type. There are other possibilities as well—*TItem* could be a *boolean*, for example. The *default* keyword solves this problem. The value used to initialize the variable will be determined when the statement is executed; if *TItem* is a reference type, *default(TItem)* returns *null*; if *TItem* is numeric, *default(TItem)* returns 0; if *TItem* is a *boolean*, *default(TItem)* returns *false*. If *TItem* is a *struct*, the individual fields in the *struct* are initialized in the same way (reference fields are set to *null*, numeric fields are set to 0, and *boolean* fields are set to *false*).

Implementing the *IEnumerable* Interface

In the following exercise, you will modify the binary tree class to implement the *IEnumerable* interface. The *GetEnumerator* method will return a *TreeEnumerator<TItem>* object.

Implement the *IEnumerable<TItem>* interface in the *Tree<TItem>* class

1. In *Solution Explorer*, double-click the file Tree.cs to display the *Tree<TItem>* class in the *Code and Text Editor* window.

2. Modify the definition of the *Tree<TItem>* class so that it implements the *IEnumerable<TItem>* interface, as shown in bold type in the following code:

   ```
   public class Tree<TItem> : IEnumerable<TItem> where TItem : IComparable<TItem>
   ```

   Notice that constraints are always placed at the end of the class definition.

3. Right-click the *IEnumerable<TItem>* interface in the class definition, point to *Implement Interface*, and then click *Implement Interface Explicitly*.

   This action generates implementations of the *IEnumerable<TItem>.GetEnumerator* and *IEnumerable.GetEnumerator* methods and adds them to the class. The non-generic *IEnumerable* interface method is implemented because the generic *IEnumerable<TItem>* interface inherits from *IEnumerable*.

4. Locate the generic *IEnumerable<TItem>.GetEnumerator* method near the end of the class. Modify the body of the *GetEnumerator()* method, replacing the existing *throw* statement as shown in bold type here:

```
IEnumerator<TItem> IEnumerable<TItem>.GetEnumerator()
{
    return new TreeEnumerator<TItem>(this);
}
```

The purpose of the *GetEnumerator* method is to construct an enumerator object for iterating through the collection. In this case, all you need to do is build a new *TreeEnumerator<TItem>* object by using the data in the tree.

5. Build the solution.

The project should compile cleanly, but correct any errors that are reported and rebuild the solution if necessary.

You will now test the modified *Tree<TItem>* class by using a *foreach* statement to iterate through a binary tree and display its contents.

Test the enumerator

1. In *Solution Explorer*, right-click the BinaryTree solution, point to *Add*, and then click *New Project*. Add a new project by using the Console Application template. Name the project *EnumeratorTest*, set the *Location* to *\Microsoft Press\Visual CSharp Step By Step\ Chapter 19* in your Documents folder, and then click *OK*.

2. Right-click the EnumeratorTest project in *Solution Explorer*, and then click *Set as Startup Project*.

3. On the *Project* menu, click *Add Reference*. In the *Add Reference* dialog box, click the *Projects* tab. Select the BinaryTree project, and then click *OK*.

The BinaryTree assembly appears in the list of references for the EnumeratorTest project in *Solution Explorer*.

4. In the *Code and Text Editor* window displaying the *Program* class, add the following *using* directive to the list at the top of the file:

```
using BinaryTree;
```

5. Add to the *Main* method the following statements shown in bold type that create and populate a binary tree of integers:

```
static void Main(string[] args)
{
    Tree<int> tree1 = new Tree<int>(10);
    tree1.Insert(5);
    tree1.Insert(11);
    tree1.Insert(5);
    tree1.Insert(-12);
    tree1.Insert(15);
    tree1.Insert(0);
    tree1.Insert(14);
    tree1.Insert(-8);
    tree1.Insert(10);
}
```

6. Add a *foreach* statement, as follows in bold type, that enumerates the contents of the tree and displays the results:

```
static void Main(string[] args)
{
    ...
    foreach (int item in tree1)
        Console.WriteLine(item);
}
```

7. Build the solution, correcting any errors if necessary.

8. On the *Debug* menu, click *Start Without Debugging*.

    The program runs and displays the values in the following sequence:

    –12, –8, 0, 5, 5, 10, 10, 11, 14, 15

9. Press Enter to return to Visual Studio 2008.

Implementing an Enumerator by Using an Iterator

As you can see, the process of making a collection enumerable can become complex and potentially error-prone. To make life easier, C# includes iterators that can automate much of this process.

An *iterator* is a block of code that yields an ordered sequence of values. Additionally, an iterator is not actually a member of an enumerable class. Rather, it specifies the sequence that an enumerator should use for returning its values. In other words, an iterator is just a description of the enumeration sequence that the C# compiler can use for creating its own enumerator. This concept requires a little thought to understand it properly, so consider a basic example before returning to binary trees and recursion.

A Simple Iterator

The following *BasicCollection<T>* class illustrates the principles of implementing an iterator. The class uses a *List<T>* object for holding data and provides the *FillList* method for populating this list. Notice also that the *BasicCollection<T>* class implements the *IEnumerable<T>* interface. The *GetEnumerator* method is implemented by using an iterator:

```
using System;
using System.Collections.Generic;
using System.Collections;

class BasicCollection<T> : IEnumerable<T>
{
    private List<T> data = new List<T>();

    public void FillList(params T [] items)
    {
        foreach (var datum in items)
            data.Add(datum);
    }

    IEnumerator<T> IEnumerable<T>.GetEnumerator()
    {
        foreach (var datum in data)
            yield return datum;
    }

    IEnumerator IEnumerable.GetEnumerator()
    {
        // Not implemented in this example
    }
}
```

The *GetEnumerator* method appears to be straightforward, but it bears closer examination. The first thing you should notice is that it doesn't appear to return an *IEnumerator<T>* type. Instead, it loops through the items in the *data* array, returning each item in turn. The key point is the use of the *yield* keyword. The *yield* keyword indicates the value that should be returned by each iteration. If it helps, you can think of the *yield* statement as calling a temporary halt to the method, passing back a value to the caller. When the caller needs the next value, the *GetEnumerator* method continues at the point it left off, looping around and then yielding the next value. Eventually, the data is exhausted, the loop finishes, and the *GetEnumerator* method terminates. At this point, the iteration is complete.

Remember that this is not a normal method in the usual sense. The code in the *GetEnumerator* method defines an *iterator*. The compiler uses this code to generate an implementation of the *IEnumerator<T>* class containing a *Current* and a *MoveNext* method. This implementation exactly matches the functionality specified by the *GetEnumerator* method. You don't actually get to see this generated code (unless you decompile the assembly containing the compiled code), but that is a small price to pay for the convenience and reduction in code that you need to write. You can invoke the enumerator generated by the iterator in the usual manner, as shown in this block of code:

```
BasicCollection<string> bc = new BasicCollection<string>();
bc.FillList("Twas", "brillig", "and", "the", "slithy", "toves");
foreach (string word in bc)
    Console.WriteLine(word);
```

This code simply outputs the contents of the *bc* object in this order:

Twas, brillig, and, the, slithy, toves

If you want to provide alternative iteration mechanisms presenting the data in a different sequence, you can implement additional properties that implement the *IEnumerable* interface and that use an iterator for returning data. For example, the *Reverse* property of the *BasicCollection<T>* class, shown here, emits the data in the list in reverse order:

```
public IEnumerable<T> Reverse
{
    get
    {
        for (int i = data.Count - 1; i >= 0; i--)
            yield return data[i];
    }
}
```

You can invoke this property as follows:

```
BasicCollection<string> bc = new BasicCollection<string>();
bc.FillList("Twas", "brillig", "and", "the", "slithy", "toves");
foreach (string word in bc.Reverse)
    Console.WriteLine(word);
```

This code outputs the contents of the *bc* object in reverse order:

toves, slithy, the, and, brillig, Twas

Defining an Enumerator for the *Tree\<TItem>* Class by Using an Iterator

In the next exercise, you will implement the enumerator for the *Tree\<TItem>* class by using an iterator. Unlike the preceding set of exercises, which required the data in the tree to be preprocessed into a queue by the *MoveNext* method, you can define an iterator that traverses the tree by using the more natural recursive mechanism, similar to the *WalkTree* method discussed in Chapter 18.

Add an enumerator to the *Tree\<TItem>* class

1. Using Visual Studio 2008, open the BinaryTree solution located in the \Microsoft Press \Visual CSharp Step by Step\Chapter 19\IterarorBinaryTree folder in your Documents folder. This solution contains another copy of the BinaryTree project you created in Chapter 18.

2. Display the file Tree.cs in the *Code and Text Editor* window. Modify the definition of the *Tree\<TItem>* class so that it implements the *IEnumerable\<TItem>* interface, as shown in bold type here:

```
public class Tree<TItem> : IEnumerable<TItem> where TItem : IComparable<TItem>
{
    ...
}
```

3. Right-click the *IEnumerable\<TItem>* interface in the class definition, point to *Implement Interface*, and then click *Implement Interface Explicitly*.

   The *IEnumerable\<TItem>.GetEnumerator* and *IEnumerable.GetEnumerator* methods are added to the class.

4. Locate the generic *IEnumerable\<TItem>.GetEnumerator* method. Replace the contents of the *GetEnumerator* method as shown in bold type in the following code:

```
IEnumerator<TItem> IEnumerable<TItem>.GetEnumerator()
{
    if (this.LeftTree != null)
    {
        foreach (TItem item in this.LeftTree)
        {
            yield return item;
        }
    }

    yield return this.NodeData;
```

```
if (this.RightTree != null)
{
    foreach (TItem item in this.RightTree)
    {
        yield return item;
    }
}
}
}
```

It might not look like it at first glance, but this code follows the same recursive algorithm that you used in Chapter 18 for printing the contents of a binary tree. If the *LeftTree* is not empty, the first *foreach* statement implicitly calls the *GetEnumerator* method (which you are currently defining) over it. This process continues until a node is found that has no left subtree. At this point, the value in the *NodeData* property is yielded, and the right subtree is examined in the same way. When the right subtree is exhausted, the process unwinds to the parent node, outputting the parent's *NodeData* property and examining the right subtree of the parent. This course of action continues until the entire tree has been enumerated and all the nodes have been output.

Test the new enumerator

1. In *Solution Explorer*, right-click the BinaryTree solution, point to *Add*, and then click *Existing Project*. In the *Add Existing Project* dialog box, move to the folder \Microsoft Press\Visual CSharp Step By Step\Chapter 19\EnumeratorTest, select the EnumeratorTest project file, and then click *Open*.

   This is the project that you created to test the enumerator you developed manually earlier in this chapter.

2. Right-click the EnumeratorTest project in *Solution Explorer*, and then click *Set as Startup Project*.

3. Expand the References node for the EnumeratorTest project in *Solution Explorer*. Right-click the BinaryTree assembly, and then click *Remove*.

   This action removes the reference to the old BinaryTree assembly (from Chapter 18) from the project.

4. On the *Project* menu, click *Add Reference*. In the *Add Reference* dialog box, click the *Projects* tab. Select the BinaryTree project, and then click *OK*.

   The new BinaryTree assembly appears in the list of references for the EnumeratorTest project in *Solution Explorer*.

Note These two steps ensure that the EnumeratorTest project references the version of the BinaryTree assembly that uses the iterator to create its enumerator rather than the earlier version.

5. Display the Program.cs file for the EnumeratorTest project in the *Code and Text Editor* window. Review the *Main* method in the Program.cs file. Recall from testing the earlier enumerator that this method instantiates a *Tree<int>* object, fills it with some data, and then uses a *foreach* statement to display its contents.

6. Build the solution, correcting any errors if necessary.

7. On the *Debug* menu, click *Start Without Debugging*.

    The program runs and displays the values in the same sequence as before:

    –12, –8, 0, 5, 5, 10, 10, 11, 14, 15

8. Press Enter and return to Visual Studio 2008.

■ If you want to continue to the next chapter

    Keep Visual Studio 2008 running, and turn to Chapter 20.

■ If you want to exit Visual Studio 2008 now

    On the *File* menu, click *Exit*. If you see a *Save* dialog box, click *Yes* (if you are using Visual Studio 2008) or *Save* (if you are using Microsoft Visual C# 2008 Express Edition) and save the project.

Chapter 19 Quick Reference

To	Do this
Make a class enumerable, allowing it to support the *foreach* construct	Implement the *IEnumerable* interface, and provide a *GetEnumerator* method that returns an *IEnumerator* object. For example:

```
public class Tree<TItem> : IEnumerable<TItem>
{
    ...
    IEnumerator<TItem> GetEnumerator()
    {
        ...
    }
}
```

Implement an enumerator not by using an iterator	Define an enumerator class that implements the *IEnumerator* interface and that provides the *Current* property and the *MoveNext* method (and optionally the *Reset* method). For example:

```
public class TreeEnumerator<TItem> : IEnumerator<TItem>
{
    ...
    TItem Current
    {
        get
        {
            ...
        }
    }

    bool MoveNext()
    {
        ...
    }
}
```

Define an enumerator by using an iterator	Implement the enumerator to indicate which items should be returned (using the *yield* statement) and in which order. For example:

```
IEnumerator<TItem> GetEnumerator()
{
    for (...)
        yield return ...
}
```

Chapter 20

Querying In-Memory Data by Using Query Expressions

After completing this chapter, you will be able to:

- Define Language Integrated Query (LINQ) queries to examine the contents of enumerable collections.

- Use LINQ extension methods and query operators.

- Explain how LINQ defers evaluation of a query and how you can force immediate execution and cache the results of a LINQ query.

You have now met most of the features of the C# language. However, we have glossed over one important aspect of the language that is likely to be used by many applications—the support that C# provides for querying data. You have seen that you can define structures and classes for modeling data and that you can use collections and arrays for temporarily storing data in memory. However, how do you perform common tasks such as searching for items in a collection that match a specific set of criteria? For example, if you have a collection of *Customer* objects, how do you find all customers that are located in London, or how can you find out which town has the most customers for your services? You can write your own code to iterate through a collection and examine the fields in each object, but these types of tasks occur so often that the designers of C# decided to include features to minimize the amount of code you need to write. In this chapter, you will learn how to use these advanced C# language features to query and manipulate data.

What Is Language Integrated Query (LINQ)?

All but the most trivial of applications need to process data. Historically, most applications provided their own logic for performing these operations. However, this strategy can lead to the code in an application becoming very tightly coupled to the structure of the data that it processes; if the data structures change, you might need to make a significant number of changes to the code that handles the data. The designers of the Microsoft .NET Framework thought long and hard about these issues and decided to make the life of an application developer easier by providing features that abstract the mechanism that an application uses to query data from application code itself. These features are called Language Integrated Query, or LINQ.

The designers of LINQ took an unabashed look at the way in which relational database management systems, such as Microsoft SQL Server, separate the language used to query a database from the internal format of the data in the database. Developers accessing a SQL Server database issue Structured Query Language (SQL) statements to the database management system. SQL provides a high-level description of the data that the developer wants to retrieve but does not indicate exactly how the database management system should retrieve this data. These details are controlled by the database management system itself. Consequently, an application that invokes SQL statements does not care how the database management system physically stores or retrieves data. The format used by the database management system can change (for example, if a new version is released) without the application developer needing to modify the SQL statements used by the application.

LINQ provides syntax and semantics very reminiscent of SQL, and with many of the same advantages. You can change the underlying structure of the data being queried without needing to change the code that actually performs the queries. You should be aware that although LINQ looks similar to SQL, it is far more flexible and can handle a wider variety of logical data structures. For example, LINQ can handle data organized hierarchically, such as that found in an XML document. However, this chapter concentrates on using LINQ in a relational manner.

Using LINQ in a C# Application

Perhaps the easiest way to explain how to use the C# features that support LINQ is to work through some simple examples based on the following sets of customer and address information:

Customer Information

CustomerID	FirstName	LastName	CompanyName
1	Orlando	Gee	A Bike Store
2	Keith	Harris	Bike World
3	Donna	Carreras	A Bike Store
4	Janet	Gates	Fitness Hotel
5	Lucy	Harrington	Grand Industries
6	David	Liu	Bike World
7	Donald	Blanton	Grand Industries
8	Jackie	Blackwell	Fitness Hotel
9	Elsa	Leavitt	Grand Industries
10	Eric	Lang	Distant Inn

Address Information

CompanyName	City	Country
A Bike Store	New York	United States
Bike World	Chicago	United States
Fitness Hotel	Ottawa	Canada
Grand Industries	London	United Kingdom
Distant Inn	Tetbury	United Kingdom

LINQ requires the data to be stored in a data structure that implements the *IEnumerable* interface, as described in Chapter 19, "Enumerating Collections." It does not matter what structure you use (an array, a *HashTable*, a *Queue*, or any of the other collection types, or even one that you define yourself) as long as it is enumerable. However, to keep things straightforward, the examples in this chapter assume that the customer and address information is held in the *customers* and *addresses* arrays shown in the following code example.

Note In a real-world application, you would populate these arrays by reading the data from a file or a database. You will learn more about the features provided by the .NET Framework for retrieving information from a database in Part V of this book, "Managing Data".

```
var customers = new[] {
    new { CustomerID = 1, FirstName = "Orlando", LastName = "Gee",
        CompanyName = "A Bike Store" },
    new { CustomerID = 2, FirstName = "Keith", LastName = "Harris",
        CompanyName = "Bike World" },
    new { CustomerID = 3, FirstName = "Donna", LastName = "Carreras",
        CompanyName = "A Bike Store" },
    new { CustomerID = 4, FirstName = "Janet", LastName = "Gates",
        CompanyName = "Fitness Hotel" },
    new { CustomerID = 5, FirstName = "Lucy", LastName = "Harrington",
        CompanyName = "Grand Industries" },
    new { CustomerID = 6, FirstName = "David", LastName = "Liu",
        CompanyName = "Bike World" },
    new { CustomerID = 7, FirstName = "Donald", LastName = "Blanton",
        CompanyName = "Grand Industries" },
    new { CustomerID = 8, FirstName = "Jackie", LastName = "Blackwell",
        CompanyName = "Fitness Hotel" },
    new { CustomerID = 9, FirstName = "Elsa", LastName = "Leavitt",
        CompanyName = "Grand Industries" },
    new { CustomerID = 10, FirstName = "Eric", LastName = "Lang",
        CompanyName = "Distant Inn" }
};
```

```
var addresses = new[] {
    new { CompanyName = "A Bike Store", City = "New York", Country = "United States"},
    new { CompanyName = "Bike World", City = "Chicago", Country = "United States"},
    new { CompanyName = "Fitness Hotel", City = "Ottawa", Country = "Canada"},
    new { CompanyName = "Grand Industries", City = "London",
          Country = "United Kingdom"},
    new { CompanyName = "Distant Inn", City = "Tetbury", Country = "United Kingdom"}
};
```

> **Note** The following sections, "Selecting Data," "Filtering Data," "Ordering, Grouping, and Aggregating Data," and "Joining Data," show you the basic capabilities and syntax for querying data by using LINQ methods. The syntax can become a little complex at times, and you will see when you reach the section "Using Query Operators" that it is not actually necessary to remember how the syntax all works. However, it is useful for you to at least take a look at the following sections so that you can fully appreciate how the query operators provided with C# perform their tasks.

Selecting Data

Suppose you want to display a list comprising the first name of each customer in the *customers* array. You can achieve this task with the following code:

```
IEnumerable<string> customerFirstNames =
    customers.Select(cust => cust.FirstName);
foreach (string name in customerFirstNames)
{
    Console.WriteLine(name);
}
```

Although this block of code is quite short, it does a lot, and it requires a degree of explanation, starting with the use of the *Select* method of the *customers* array.

The *Select* method enables you to retrieve specific data from the array—in this case, just the value in the *FirstName* field of each item in the array. How does it work? The parameter to the *Select* method is actually another method that takes a row from the *customers* array and returns the selected data from that row. You could define your own custom method to perform this task, but the simplest mechanism is to use a lambda expression to define an anonymous method, as shown in the preceding example. There are three important things that you need to understand at this point:

- The type *cust* is the type of the parameter passed in to the method. You can think of *cust* as an alias for the type of each row in the *customers* array. The compiler deduces this from the fact that you are calling the *Select* method on the *customers* array. You can use any legal C# identifier in place of *cust*.

- The *Select* method does not actually retrieve the data at this time; it simply returns an enumerable object that will fetch the data identified by the *Select* method when you iterate over it later. We will return to this aspect of LINQ in the section "LINQ and Deferred Evaluation" later in this chapter.

- The *Select* method is not actually a method of the *Array* type. It is an extension method of the *Enumerable* class. The *Enumerable* class is located in the *System.Linq* namespace and provides a substantial set of static methods for querying objects that implement the generic *IEnumerable<T>* interface.

The preceding example uses the *Select* method of the *customers* array to generate an *IEnumerable<string>* object named *customerFirstNames*. (It is of type *IEnumerable<string>* because the *Select* method returns an enumerable collection of customer first names, which are strings.) The *foreach* statement iterates through this collection of strings, printing out the first name of each customer in the following sequence:

```
Orlando
Keith
Donna
Janet
Lucy
David
Donald
Jackie
Elsa
Eric
```

You can now display the first name of each customer. How do you fetch the first and last name of each customer? This task is slightly trickier. If you examine the definition of the *Enumerable.Select* method in the *System.Linq* namespace in the documentation supplied with Microsoft Visual Studio 2008, you will see that it looks like this:

```
public static IEnumerable<TResult> Select<TSource, TResult> (
        IEnumerable<TSource> source,
        Func<TSource, TResult> selector
)
```

What this actually says is that *Select* is a generic method that takes two type parameters named *TSource* and *TResult*, as well as two ordinary parameters named *source* and *selector*. *TSource* is the type of the collection that you are generating an enumerable set of results for (*customer* objects in our example), and *TResult* is the type of the data in the enumerable set of results (*string* objects in our example). Remember that *Select* is an extension method, so the *source* parameter is actually a reference to the type being extended (a generic collection of *customer* objects that implements the *IEnumerable* interface in our example). The *selector* parameter specifies a generic method that identifies the fields to be retrieved. (*Func* is the name of a generic delegate type in the .NET Framework that you can use for encapsulating a generic method.) The method referred to by the *selector* parameter takes a *TSource* (in this case, *customer*) parameter and yields a collection of *TResult* (in this case, *string*) objects. The value returned by the *Select* method is an enumerable collection of *TResult* (again *string*) objects.

 Note If you need to review how extension methods work and the role of the first parameter to an extension method, go back and revisit Chapter 12, "Working with Inheritance."

The important point to understand from the preceding paragraph is that the *Select* method returns an enumerable collection based on a single type. If you want the enumerator to return multiple items of data, such as the first and last name of each customer, you have at least two options:

- You can concatenate the first and last names together into a single string in the *Select* method, like this:

```
IEnumerable<string> customerFullName =
    customers.Select(cust => cust.FirstName + " " + cust.LastName);
```

- You can define a new type that wraps the first and last names and use the *Select* method to construct instances of this type, like this:

```
class Names
{
    public string FirstName{ get; set; }
    public string LastName{ get; set; }
}
...
IEnumerable<Names> customerName =
    customers.Select(cust => new Names { FirstName = cust.FirstName,
                                         LastName = cust.LastName } );
```

The second option is arguably preferable, but if this is the only use that your application makes of the *Names* type, you might prefer to use an anonymous type instead of defining a new type specifically for a single operation, like this:

```
var customerName =
    customers.Select(cust => new { FirstName = cust.FirstName, LastName = cust.LastName } );
```

Notice the use of the *var* keyword here to define the type of the enumerable collection. The type of objects in the collection is anonymous, so you cannot specify a specific type for the objects in the collection.

Filtering Data

The *Select* method enables you to specify, or *project*, the fields that you want to include in the enumerable collection. However, you might also want to restrict the rows that the enumerable collection contains. For example, suppose you want to list the names of all companies in the *addresses* array that are located in the United States only. To do this, you can use the *Where* method, as follows:

```
IEnumerable<string> usCompanies =
    addresses.Where(addr => String.Equals(addr.Country, "United States"))
            .Select(usComp => usComp.CompanyName);

foreach (string name in usCompanies)
{
    Console.WriteLine(name);
}
```

Syntactically, the *Where* method is similar to *Select*. It expects a parameter that defines a method that filters the data according to whatever criteria you specify. This example makes use of another lambda expression. The type *addr* is an alias for a row in the *addresses* array, and the lambda expression returns all rows where the *Country* field matches the string *"United States"*. The *Where* method returns an enumerable collection of rows containing every field from the original collection. The *Select* method is then applied to these rows to project only the *CompanyName* field from this enumerable collection to return another enumerable collection of *string* objects. (The type *usComp* is an alias for the type of each row in the enumerable collection returned by the *Where* method.) The type of the result of this complete expression is therefore *IEnumerable<string>*. It is important to understand this sequence of operations—the *Where* method is applied first to filter the rows, followed by the *Select* method to specify the fields. The *foreach* statement that iterates through this collection displays the following companies:

```
A Bike Store
Bike World
```

Ordering, Grouping, and Aggregating Data

If you are familiar with SQL, you are aware that SQL enables you to perform a wide variety of relational operations besides simple projection and filtering. For example, you can specify that you want data to be returned in a specific order, you can group the rows returned according to one or more key fields, and you can calculate summary values based on the rows in each group. LINQ provides the same functionality.

To retrieve data in a particular order, you can use the *OrderBy* method. Like the *Select* and *Where* methods, *OrderBy* expects a method as its argument. This method identifies the expressions that you want to use to sort the data. For example, you can display the names of each company in the *addresses* array in ascending order, like this:

```
IEnumerable<string> companyNames =
    addresses.OrderBy(addr => addr.CompanyName).Select(comp => comp.CompanyName);

foreach (string name in companyNames)
{
    Console.WriteLine(name);
}
```

This block of code displays the companies in the addresses table in alphabetical order:

```
A Bike Store
Bike World
Distant Inn
Fitness Hotel
Grand Industries
```

If you want to enumerate the data in descending order, you can use the *OrderByDescending* method instead. If you want to order by more than one key value, you can use the *ThenBy* or *ThenByDescending* method after *OrderBy* or *OrderByDescending*.

To group data according to common values in one or more fields, you can use the *GroupBy* method. The next example shows how to group the companies in the *addresses* array by country:

```
var companiesGroupedByCountry =
    addresses.GroupBy(addrs => addrs.Country);

foreach (var companiesPerCountry in companiesGroupedByCountry)
{
    Console.WriteLine("Country: {0}\t{1} companies",
            companiesPerCountry.Key, companiesPerCountry.Count());
    foreach (var companies in companiesPerCountry)
    {
        Console.WriteLine("\t{0}", companies.CompanyName);
    }
}
```

By now you should recognize the pattern! The *GroupBy* method expects a method that specifies the fields to group the data by. There are some subtle differences between the *GroupBy* method and the other methods that you have seen so far, though. The main point of interest is that you don't need to use the *Select* method to project the fields to the result. The enumerable set returned by *GroupBy* contains all the fields in the original source collection, but the rows are ordered into a set of enumerable collections based on the field identified by the method specified by *GroupBy*. In other words, the result of the *GroupBy* method is an enumerable set of groups, each of which is an enumerable set of rows. In the example just shown, the enumerable set *companiesGroupedByCountry* is a set of countries. The items in this set are themselves enumerable collections containing the companies for each country in turn. The code that displays the companies in each country uses a *foreach* loop to iterate through the *companiesGroupedByCountry* set to yield and display each country in turn and then uses a nested *foreach* loop to iterate through the set of companies in each country. Notice in the outer *foreach* loop that you can access the value that you are grouping by using the *Key* field of each item, and you can also calculate summary data for each group by using methods such as *Count, Max, Min,* and many others. The output generated by the example code looks like this:

```
Country: United States   2 companies
        A Bike Store
        Bike World
Country: Canada 1 companies
        Fitness Hotel
Country: United Kingdom  2 companies
        Grand Industries
        Distant Inn
```

You can use many of the summary methods such as *Count, Max,* and *Min* directly over the results of the *Select* method. If you want to know how many companies there are in the *addresses* array, you can use a block of code such as this:

```
int numberOfCompanies = addresses.Select(addr => addr.CompanyName).Count();
Console.WriteLine("Number of companies: {0}", numberOfCompanies);
```

Notice that the result of these methods is a single scalar value rather than an enumerable collection. The output from this block of code looks like this:

```
Number of companies: 5
```

I should utter a word of caution at this point. These summary methods do not distinguish between rows in the underlying set that contain duplicate values in the fields you are projecting. What this means is that, strictly speaking, the preceding example shows you only how many rows in the *addresses* array contain a value in the *CompanyName* field. If you wanted to find out how many different countries are mentioned in this table, you might be tempted to try this:

```
int numberOfCountries = addresses.Select(addr => addr.Country).Count();
Console.WriteLine("Number of countries: {0}", numberOfCountries);
```

The output looks like this:

```
Number of countries: 5
```

In fact, there are only three different countries in the *addresses* array; it just so happens that United States and United Kingdom both occur twice. You can eliminate duplicates from the calculation by using the *Distinct* method, like this:

```
int numberOfCountries =
    addresses.Select(addr => addr.Country).Distinct().Count();
```

The *Console.WriteLine* statement will now output the expected result:

```
Number of countries: 3
```

Joining Data

Just like SQL, LINQ enables you to join multiple sets of data together over one or more common key fields. The following example shows how to display the first and last name of each customer, together with the names of the countries where they are located:

```
var citiesAndCustomers = customers
  .Select(c => new { c.FirstName, c.LastName, c.CompanyName })
  .Join(addresses, custs => custs.CompanyName, addrs => addrs.CompanyName,
  (custs, addrs) => new {custs.FirstName, custs.LastName, addrs.Country });

foreach (var row in citiesAndCustomers)
{
    Console.WriteLine(row);
}
```

The customers' first and last names are available in the *customers* array, but the country for each company that customers work for is stored in the *addresses* array. The common key between the *customers* array and the *addresses* array is the company name. The *Select* method specifies the fields of interest in the *customers* array (*FirstName* and *LastName*), together with the field containing the common key (*CompanyName*). You use the *Join* method to join the data identified by the *Select* method with another enumerable collection. The parameters to the *Join* method are:

- The enumerable collection with which to join.

- A method that identifies the common key fields from the data identified by the *Select* method.

- A method that identifies the common key fields on which to join the selected data.

- A method that specifies the columns you require in the enumerable result set returned by the *Join* method.

In this example, the *Join* method joins the enumerable collection containing the *FirstName*, *LastName*, and *CompanyName* fields from the *customers* array with the rows in the *addresses* array. The two sets of data are joined where the value in the *CompanyName* field in the *customers* array matches the value in the *CompanyName* field in the *addresses* array. The result set comprises rows containing the *FirstName* and *LastName* fields from the *customers* array with the *Country* field from the *addresses* array. The code that outputs the data from the *citiesAndCustomers* collection displays the following information:

```
{ FirstName = Orlando, LastName = Gee, Country = United States }
{ FirstName = Keith, LastName = Harris, Country = United States }
```

```
{ FirstName = Donna, LastName = Carreras, Country = United States }
{ FirstName = Janet, LastName = Gates, Country = Canada }
{ FirstName = Lucy, LastName = Harrington, Country = United Kingdom }
{ FirstName = David, LastName = Liu, Country = United States }
{ FirstName = Donald, LastName = Blanton, Country = United Kingdom }
{ FirstName = Jackie, LastName = Blackwell, Country = Canada }
{ FirstName = Elsa, LastName = Leavitt, Country = United Kingdom }
{ FirstName = Eric, LastName = Lang, Country = United Kingdom }
```

Note It is important to remember that collections in memory are not the same as tables in a relational database and that the data that they contain is not subject to the same data integrity constraints. In a relational database, it could be acceptable to assume that every customer had a corresponding company and that each company had its own unique address. Collections do not enforce the same level of data integrity, meaning that you could quite easily have a customer referencing a company that does not exist in the *addresses* array, and you might even have the same company occurring more than once in the *addresses* array. In these situations, the results that you obtain might be accurate but unexpected. Join operations work best when you fully understand the relationships between the data you are joining.

Using Query Operators

The preceding sections have shown you many of the features available for querying in-memory data by using the extension methods for the *Enumerable* class defined in the *System.Linq* namespace. The syntax makes use of several advanced C# language features, and the resultant code can sometimes be quite hard to understand and maintain. To relieve you of some of this burden, the designers of C# added query operators to the language to enable you to employ LINQ features by using a syntax more akin to SQL.

As you saw in the examples shown earlier in this chapter, you can retrieve the first name for each customer like this:

```
IEnumerable<string> customerFirstNames =
    customers.Select(cust => cust.FirstName);
```

You can rephrase this statement by using the *from* and *select* query operators, like this:

```
var customerFirstNames = from cust in customers
                         select cust.FirstName;
```

At compile time, the C# compiler resolves this expression into the corresponding *Select* method. The *from* operator defines an alias for the source collection, and the *select* operator specifies the fields to retrieve by using this alias. The result is an enumerable collection of customer first names. If you are familiar with SQL, notice that the *from* operator occurs before the *select* operator.

Continuing in the same vein, to retrieve the first and last name for each customer, you can use the following statement. (You might want to refer to the earlier example of the same statement based on the *Select* extension method.)

```
var customerNames = from c in customers
                    select new { c.FirstName, c.LastName };
```

You use the *where* operator to filter data. The following example shows how to return the names of the companies based in the United States from the *addresses* array:

```
var usCompanies = from a in addresses
                  where String.Equals(a.Country, "United States")
                  select a.CompanyName;
```

To order data, use the *orderby* operator, like this:

```
var companyNames = from a in addresses
                   orderby a.CompanyName
                   select a.CompanyName;
```

You can group data by using the *group* operator:

```
var companiesGroupedByCountry = from a in addresses
                                group a by a.Country;
```

Notice that, as with the earlier example showing how to group data, you do not provide the *select* operator, and you can iterate through the results by using exactly the same code as the earlier example, like this:

```
foreach (var companiesPerCountry in companiesGroupedByCountry)
{
    Console.WriteLine("Country: {0}\t{1} companies",
            companiesPerCountry.Key, companiesPerCountry.Count());
    foreach (var companies in companiesPerCountry)
    {
        Console.WriteLine("\t{0}", companies.CompanyName);
    }
}
```

You can invoke the summary functions, such as *Count*, over the collection returned by an enumerable collection, like this:

```
int numberOfCompanies = (from a in addresses
                         select a.CompanyName).Count();
```

Notice that you wrap the expression in parentheses. If you want to ignore duplicate values, use the *Distinct* method, like this:

```
int numberOfCountries = (from a in addresses
                         select a.Country).Distinct().Count();
```

> **Tip** In many cases, you probably want to count just the number of rows in a collection rather than the number of values in a field across all the rows in the collection. In this case, you can invoke the *Count* method directly over the original collection, like this:

```
int numberOfCompanies = addresses.Count();
```

You can use the *join* operator to combine two collections across a common key. The following example shows the query returning customers and addresses over the *CompanyName* column in each collection, this time rephrased using the *join* operator. You use the *on* clause with the *equals* operator to specify how the two collections are related. (LINQ currently supports equi-joins only.)

```
var citiesAndCustomers = from a in addresses
                         join c in customers
                         on a.CompanyName equals c.CompanyName
                         select new { c.FirstName, c.LastName, a.Country };
```

> **Note** In contrast with SQL, the order of the expressions in the *on* clause of a LINQ expression is important. You must place the item you are joining from (referencing the data in the collection in the *from* clause) to the left of the *equals* operator and the item you are joining with (referencing the data in the collection in the *join* clause) to the right.

LINQ provides a large number of other methods for summarizing information, joining, grouping, and searching through data; this section has covered just the most common features. For example, LINQ provides the *Intersect* and *Union* methods, which you can use to perform setwide operations. It also provides methods such as *Any* and *All* that you can use to determine whether at least one item in a collection or every item in a collection matches a specified predicate. You can partition the values in an enumerable collection by using the *Take* and *Skip* methods. For more information, see the documentation provided with Visual Studio 2008.

Querying Data in *Tree<TItem>* Objects

The examples you've seen so far in this chapter have shown how to query the data in an array. You can use exactly the same techniques for any collection class that implements the *IEnumerable* interface. In the following exercise, you will define a new class for modeling employees for a company. You will create a *BinaryTree* object containing a collection of *Employee* objects, and then you will use LINQ to query this information. You will initially call the LINQ extension methods directly, but then you will modify your code to use query operators.

Retrieve data from a *BinaryTree* by using the extension methods

1. Start Visual Studio 2008 if it is not already running.

2. Open the *QueryBinaryTree* solution, located in the \Microsoft Press\Visual CSharp Step by Step\Chapter 20\QueryBinaryTree folder in your Documents folder. The project contains the Program.cs file, which defines the *Program* class with the *Main* and *Entrance* methods that you have seen in previous exercises.

3. In *Solution Explorer*, right-click the *QueryBinaryTree* project, point to *Add*, and then click *Class*. In the *Add New Item—Query BinaryTree* dialog box, type **Employee.cs** in the *Name* box, and then click *Add*.

4. Add the automatic properties shown here in bold to the *Employee* class:

```
class Employee
{
    public string FirstName { get; set; }
    public string LastName { get; set; }
    public string Department { get; set; }
    public int Id { get; set; }
}
```

5. Add the *ToString* method shown here in bold to the *Employee* class. Classes in the .NET Framework use this method when converting the object to a string representation, such as when displaying it by using the *Console.WriteLine* statement.

```
class  Employee
{
    ...
    public override string ToString()
    {
        return String.Format("Id: {0}, Name: {1} {2}, Dept: {3}",
                        this.Id, this.FirstName, this.LastName,
                        this.Department);
    }
}
```

6. Modify the definition of the *Employee* class in the Employee.cs file to implement the *IComparable<Employee>* interface, as shown here:

```
class Employee : IComparable<Employee>
{
}
```

This step is necessary because the *BinaryTree* class specifies that its elements must be "comparable."

7. Right-click the *IComparable<Employee>* interface in the class definition, point to *Implement Interface*, and then click *Implement Interface Explicitly*.

This action generates a default implementation of the *CompareTo* method. Remember that the *BinaryTree* class calls this method when it needs to compare elements when inserting them into the tree.

8. Replace the body of the *CompareTo* method with the code shown here in bold. This implementation of the *CompareTo* method compares *Employee* objects based on the value of the *Id* field.

```
int IComparable<Employee>.CompareTo(Employee other)
{
    if (other == null)
        return 1;

    if (this.Id > other.Id)
        return 1;

    if (this.Id < other.Id)
        return -1;

    return 0;
}
```

 Note For a description of the *IComparable* interface, refer to Chapter 18, "Introducing Generics."

9. In *Solution Explorer*, right-click the *QueryBinaryTree* solution, point to *Add*, and then click *Existing Project*. In the *Add Existing Project* dialog box, move to the folder Microsoft Press\Visual CSharp Step By Step\Chapter 20\BinaryTree in your Documents folder, click the *BinaryTree* project, and then click *Open*.

The *BinaryTree* project contains a copy of the enumerable *BinaryTree* class that you implemented in Chapter 19.

10. In *Solution Explorer*, right-click the *QueryBinaryTree* project, and then click *Add Reference*. In the *Add Reference* dialog box, click the *Projects* tab, select the *BinaryTree* project, and then click *OK*.

11. In *Solution Explorer*, open the Program.cs file, and verify that the list of *using* statements at the top of the file includes the following line of code:

```
using System.Linq;
```

12. Add the following *using* statement to the list at the top of the Program.cs file to bring the *BinaryTree* namespace into scope:

```
using BinaryTree;
```

13. In the *Entrance* method in the *Program* class, add the following statements shown in bold type to construct and populate an instance of the *BinaryTree* class:

```
static void Entrance()
{
    Tree<Employee> empTree = new Tree<Employee>(new Employee
        { Id = 1, FirstName = "Janet", LastName = "Gates", Department = "IT"});
    empTree.Insert(new Employee
        { Id = 2, FirstName = "Orlando", LastName = "Gee", Department = "Marketing"});
    empTree.Insert(new Employee
        { Id = 4, FirstName = "Keith", LastName = "Harris", Department = "IT" });
    empTree.Insert(new Employee
        { Id = 6, FirstName = "Lucy", LastName = "Harrington", Department = "Sales" });
    empTree.Insert(new Employee
        { Id = 3, FirstName = "Eric", LastName = "Lang", Department = "Sales" });
    empTree.Insert(new Employee
        { Id = 5, FirstName = "David", LastName = "Liu", Department = "Marketing" });
}
```

14. Add the following statements shown in bold to the end of the *Entrance* method. This code uses the *Select* method to list the departments found in the binary tree.

```
static void Entrance()
{
    ...
    Console.WriteLine("List of departments");
    var depts = empTree.Select(d => d.Department);

    foreach (var dept in depts)
        Console.WriteLine("Department: {0}", dept);
}
```

15. On the *Debug* menu, click *Start Without Debugging*.

The application should output the following list of departments:

```
List of departments
Department: IT
Department: Marketing
Department: Sales
Department: IT
Department: Marketing
Department: Sales
```

Each department occurs twice because there are two employees in each department. The order of the departments is determined by the *CompareTo* method of the *Employee* class, which uses the *Id* property of each employee to sort the data. The first department is for the employee with the *Id* value 1, the second department is for the employee with the *Id* value 2, and so on.

16. Press Enter to return to Visual Studio 2008.

17. Modify the statement that creates the enumerable collection of departments as shown here in bold:

```
var depts = empTree.Select(d => d.Department).Distinct();
```

The *Distinct* method removes duplicate rows from the enumerable collection.

18. On the *Debug* menu, click *Start Without Debugging*.

Verify that the application now displays each department only once, like this:

```
List of departments
Department: IT
Department: Marketing
Department: Sales
```

19. Press Enter to return to Visual Studio 2008.

20. Add the following statements to the end of the *Entrance* method. This block of code uses the *Where* method to filter the employees and return only those in the IT department. The *Select* method returns the entire row rather than projecting specific columns.

```
Console.WriteLine("\nEmployees in the IT department");
var ITEmployees =
    empTree.Where(e => String.Equals(e.Department, "IT")).Select(emp => emp);

foreach (var emp in ITEmployees)
    Console.WriteLine(emp);
```

21. Add the code shown here to the end of the *Entrance* method, after the code from the preceding step. This code uses the *GroupBy* method to group the employees found in the binary tree by department. The outer *foreach* statement iterates through each group, displaying the name of the department. The inner *foreach* statement displays the names of the employees in each department.

```
Console.WriteLine("\nAll employees grouped by department");
var employeesByDept = empTree.GroupBy(e => e.Department);

foreach (var dept in employeesByDept)
{
    Console.WriteLine("Department: {0}", dept.Key);
    foreach (var emp in dept)
    {
        Console.WriteLine("\t{0} {1}", emp.FirstName, emp.LastName);
    }
}
```

22. On the *Debug* menu, click *Start Without Debugging*. Verify that the output of the application looks like this:

```
List of departments
Department: IT
Department: Marketing
Department: Sales

Employees in the IT department
Id: 1, Name: Janet Gates, Dept: IT
Id: 4, Name: Keith Harris, Dept: IT

All employees grouped by department
Department: IT
        Janet Gates
        Keith Harris
Department: Marketing
        Orlando Gee
        David Liu
Department: Sales
        Eric Lang
        Lucy Harrington
```

23. Press Enter to return to Visual Studio 2008.

Retrieve data from a *BinaryTree* by using query operators

1. In the *Entrance* method, comment out the statement that generates the enumerable collection of departments, and replace it with the following statement shown in bold, based on the *from* and *select* query operators:

```
//var depts = empTree.Select(d => d.Department).Distinct();
var depts = (from d in empTree
            select d.Department).Distinct();
```

2. Comment out the statement that generates the enumerable collection of employees in the IT department, and replace it with the following code shown in bold:

```
//var ITEmployees =
//   empTree.Where(e => String.Equals(e.Department, "IT")).Select(emp => emp);
var ITEmployees = from e in empTree
                  where String.Equals(e.Department, "IT")
                  select e;
```

3. Comment out the statement that generates the enumerable collection grouping employees by department, and replace it with the statement shown here in bold:

```
//var employeesByDept = empTree.GroupBy(e => e.Department);
var employeesByDept = from e in empTree
                      group e by e.Department;
```

4. On the *Debug* menu, click *Start Without Debugging*. Verify that the output of the application is the same as before.

5. Press Enter to return to Visual Studio 2008.

LINQ and Deferred Evaluation

When you use LINQ to define an enumerable collection, either by using the LINQ extension methods or by using query operators, you should remember that the application does not actually build the collection at the time that the LINQ extension method is executed; the collection is enumerated only when you iterate over the collection. This means that the data in the original collection can change between executing a LINQ query and retrieving the data that the query identifies; you will always fetch the most up-to-date data. For example, the following query (which you saw earlier) defines an enumerable collection of U.S. companies:

```
var usCompanies = from a in addresses
            where String.Equals(a.Country, "United States")
            select a.CompanyName;
```

The data in the *addresses* array is not retrieved and any conditions specified in the *Where* filter are not evaluated until you iterate through the *usCompanies* collection:

```
foreach (string name in usCompanies)
{
    Console.WriteLine(name);
}
```

If you modify the data in the *addresses* array between defining the *usCompanies* collection and iterating through the collection (for example, if you add a new company based in the United States), you will see this new data. This strategy is referred to as *deferred evaluation*.

You can force evaluation of a LINQ query and generate a static, cached collection. This collection is a copy of the original data and will not change if the data in the collection changes. LINQ provides the *ToList* method to build a static *List* object containing a cached copy of the data. You use it like this:

```
var usCompanies = from a in addresses.ToList()
            where String.Equals(a.Country, "United States")
            select a.CompanyName;
```

This time, the list of companies is fixed when you define the query. If you add more U.S. companies to the *addresses* array, you will not see them when you iterate through the *usCompanies* collection. LINQ also provides the *ToArray* method that stores the cached collection as an array.

In the final exercise in this chapter, you will compare the effects of using deferred evaluation of a LINQ query to generating a cached collection.

Examine the effects of deferred and cached evaluation of a LINQ query

1. Return to Visual Studio 2008, displaying the *QueryBinaryTree* project, and edit the Program.cs file.

2. Comment out the contents of the *Entrance* method apart from the statements that construct the *empTree* binary tree, as shown here:

```
static void Entrance()
{
  Tree<Employee> empTree = new Tree<Employee>(new Employee
    { Id = 1, FirstName = "Janet", LastName = "Gates", Department = "IT" });
  empTree.Insert(new Employee
    { Id = 2, FirstName = "Orlando", LastName = "Gee", Department = "Marketing" });
  empTree.Insert(new Employee
    { Id = 4, FirstName = "Keith", LastName = "Harris", Department = "IT" });
  empTree.Insert(new Employee
    { Id = 6, FirstName = "Lucy", LastName = "Harrington", Department = "Sales" });
  empTree.Insert(new Employee
    { Id = 3, FirstName = "Eric", LastName = "Lang", Department = "Sales" });
  empTree.Insert(new Employee
    { Id = 5, FirstName = "David", LastName = "Liu", Department = "Marketing" });

  // comment out the rest of the method
  ...
}
```

> **Tip** You can comment out a block of code by selecting the entire block in the *Code and Text Editor* window and then clicking the *Comment Out The Selected Lines* button on the toolbar or by pressing Ctrl+E and then pressing C.

3. Add the following statements to the *Entrance* method, after building the *empTree* binary tree:

```
Console.WriteLine("All employees");
var allEmployees = from e in empTree
                   select e;

foreach (var emp in allEmployees)
    Console.WriteLine(emp);
```

This code generates an enumerable collection of employees named *allEmployees* and then iterates through this collection, displaying the details of each employee.

4. Add the following code immediately after the statements you typed in the preceding step:

```
empTree.Insert(new Employee { Id = 7, FirstName = "Donald", LastName = "Blanton",
Department = "IT" });
Console.WriteLine("\nEmployee added");

Console.WriteLine("All employees");
foreach (var emp in allEmployees)
    Console.WriteLine(emp);
```

These statements add a new employee to the *empTree* tree and then iterate through the *allEmployees* collection again.

5. On the *Debug* menu, click *Start Without Debugging*. Verify that the output of the application looks like this:

```
All employees
Id: 1, Name: Janet Gates, Dept: IT
Id: 2, Name: Orlando Gee, Dept: Marketing
Id: 3, Name: Eric Lang, Dept: Sales
Id: 4, Name: Keith Harris, Dept: IT
Id: 5, Name: David Liu, Dept: Marketing
Id: 6, Name: Lucy Harrington, Dept: Sales

Employee added
All employees
Id: 1, Name: Janet Gates, Dept: IT
Id: 2, Name: Orlando Gee, Dept: Marketing
Id: 3, Name: Eric Lang, Dept: Sales
Id: 4, Name: Keith Harris, Dept: IT
Id: 5, Name: David Liu, Dept: Marketing
Id: 6, Name: Lucy Harrington, Dept: Sales
Id: 7, Name: Donald Blanton, Dept: IT
```

Notice that the second time the application iterates through the *allEmployees* collection, the list displayed includes Donald Blanton, even though this employee was added only after the *allEmployees* collection was defined.

6. Press Enter to return to Visual Studio 2008.

7. In the *Entrance* method, change the statement that generates the *allEmployees* collection to identify and cache the data immediately, as shown here in bold:

```
var allEmployees = from e in empTree.ToList<Employee>()
                   select e;
```

LINQ provides generic and nongeneric versions of the *ToList* and *ToArray* methods. If possible, it is better to use the generic versions of these methods to ensure the type safety of the result. The data returned by the *select* operator is an *Employee* object, and the code shown in this step generates *allEmployees* as a generic *List<Employee>* collection. If you specify the nongeneric *ToList* method, the *allEmployees* collection will be a *List* of *object* types.

8. On the *Debug* menu, click *Start Without Debugging*. Verify that the output of the application looks like this:

```
All employees
Id: 1, Name: Janet Gates, Dept: IT
Id: 2, Name: Orlando Gee, Dept: Marketing
Id: 3, Name: Eric Lang, Dept: Sales
Id: 4, Name: Keith Harris, Dept: IT
Id: 5, Name: David Liu, Dept: Marketing
Id: 6, Name: Lucy Harrington, Dept: Sales

Employee added
All employees
```

```
Id: 1, Name: Janet Gates, Dept: IT
Id: 2, Name: Orlando Gee, Dept: Marketing
Id: 3, Name: Eric Lang, Dept: Sales
Id: 4, Name: Keith Harris, Dept: IT
Id: 5, Name: David Liu, Dept: Marketing
Id: 6, Name: Lucy Harrington, Dept: Sales
```

Notice that this time, the second time the application iterates through the *allEmployees* collection, the list displayed does not include Donald Blanton. This is because the query is evaluated and the results cached before Donald Blanton is added to the *empTree* binary tree.

9. Press Enter to return to Visual Studio 2008.

- If you want to continue to the next chapter:

  Keep Visual Studio 2008 running, and turn to Chapter 21.

- If you want to exit Visual Studio 2008 now:

  On the *File* menu, click *Exit*. If you see a *Save* dialog box, click *Yes* (if you are using Visual Studio 2008) or *Save* (if you are using Visual C# 2008 Express Edition) and save the project.

Chapter 20 Quick Reference

To	Do this

Project specified fields from an enumerable collection	Use the *Select* method, and specify a lambda expression that identifies the fields to project. For example:
	
	`var customerFirstNames = customers.Select(cust => cust.FirstName);`
	
	Or use the *from* and *select* query operators. For example:
	
	`var customerFirstNames =`
	`    from cust in customers`
	`    select cust.FirstName;`

Filter rows from an enumerable collection	Use the *Where* method, and specify a lambda expression containing the criteria that rows should match. For example:
	
	`var usCompanies =`
	`    addresses.Where(addr =>`
	`        String.Equals(addr.Country, "United States")).`
	`    Select(usComp => usComp.CompanyName);`
	
	Or use the *where* query operator. For example:
	
	`var usCompanies =`
	`    from a in addresses`
	`    where String.Equals(a.Country, "United States")`
	`    select a.CompanyName;`

Enumerate data in a specific order	Use the *OrderBy* method, and specify a lambda expression identifying the field to use to order rows. For example:

```
var companyNames =
    addresses.OrderBy(addr -> addr.CompanyName).
    Select(comp => comp.CompanyName);
```

Or use the *orderby* query operator. For example:

```
var companyNames =
    from a in addresses
    orderby a.CompanyName
    select a.CompanyName;
```

Group data by the values in a field	Use the *GroupBy* method, and specify a lambda expression identifying the field to use to group rows. For example:

```
var companiesGroupedByCountry =
    addresses.GroupBy(addrs => addrs.Country);
```

Or use the *group by* query operator. For example:

```
var companiesGroupedByCountry =
    from a in addresses
    group a by a.Country;
```

Join data held in two different collections	Use the *Join* method specifying the collection to join with, the join riteria, and the fields for the result. For example:

```
var citiesAndCustomers =
  customers.
    Select(c => new { c.FirstName, c.LastName, c.CompanyName }).
  Join(addresses, custs => custs.CompanyName,
      addrs => addrs.CompanyName,
      (custs, addrs) => new {custs.FirstName, custs.LastName,
                             addrs.Country });
```

Or use the *join* query operator. For example:

```
var citiesAndCustomers =
    from a in addresses
    join c in customers
    on a.CompanyName equals c.CompanyName
    select new { c.FirstName, c.LastName, a.Country };
```

Force immediate generation of the results for a LINQ query	Use the *ToList* or *ToArray* method to generate a list or an array containing the results. For example:

```
var allEmployees =
    from e in empTree.ToList<Employee>()
    select e;
```

Chapter 21
Operator Overloading

After completing this chapter, you will be able to:

- Implement binary operators for your own types.

- Implement unary operators for your own types.

- Write increment and decrement operators for your own types.

- Understand the need to implement some operators as pairs.

- Implement implicit conversion operators for your own types.

- Implement explicit conversion operators for your own types.

You have made a great deal of use of the standard operator symbols (such as + and –) to perform standard operations (such as addition and subtraction) on types (such as *int* and *double*). Many of the built-in types come with their own predefined behaviors for each operator. You can also define how operators should behave for your own structures and classes, which is the subject of this chapter.

Understanding Operators

You use operators to combine operands together into expressions. Each operator has its own semantics, dependent on the type it works with. For example, the + operator means "add" when used with numeric types or "concatenate" when used with strings.

Each operator symbol has a *precedence*. For example, the * operator has a higher precedence than the + operator. This means that the expression *a + b * c* is the same as *a + (b * c)*.

Each operator symbol also has an *associativity* to define whether the operator evaluates from left to right or from right to left. For example, the = operator is right-associative (it evaluates from right to left), so *a = b = c* is the same as *a = (b = c)*.

A *unary operator* is an operator that has just one operand. For example, the increment operator (++) is a unary operator.

A *binary operator* is an operator that has two operands. For example, the multiplication operator (*) is a binary operator.

Operator Constraints

You have seen throughout this book that C# enables you to overload methods when defining your own types. C# also allows you to overload many of the existing operator symbols for your own types, although the syntax is slightly different. When you do this, the operators you implement automatically fall into a well-defined framework with the following rules:

- You cannot change the precedence and associativity of an operator. The precedence and associativity are based on the operator symbol (for example, +) and not on the type (for example, *int*) on which the operator symbol is being used. Hence, the expression *a + b * c* is *always* the same as *a + (b * c)*, regardless of the types of *a*, *b*, and *c*.

- You cannot change the multiplicity (the number of operands) of an operator. For example, * (the symbol for multiplication), is a binary operator. If you declare a * operator for your own type, it must be a binary operator.

- You cannot invent new operator symbols. For example, you can't create a new operator symbol, such as ** for raising one number to the power of another number. You'd have to create a method for that.

- You can't change the meaning of operators when applied to built-in types. For example, the expression *1 + 2* has a predefined meaning, and you're not allowed to override this meaning. If you could do this, things would be too complicated!

- There are some operator symbols that you can't overload. For example, you can't overload the dot (.) operator, which indicates access to a class member. Again, if you could do this, it would lead to unnecessary complexity.

Tip You can use indexers to simulate *[ ]* as an operator. Similarly, you can use properties to simulate assignment (=) as an operator, and you can use delegates to simulate a function call as an operator.

Overloaded Operators

To define your own operator behavior, you must overload a selected operator. You use methodlike syntax with a return type and parameters, but the name of the method is the keyword *operator* together with the operator symbol you are declaring. For example, here's a user-defined structure named *Hour* that defines a binary + operator to add together two instances of *Hour*:

```
struct Hour
{
    public Hour(int initialValue)
    {
        this.value = initialValue;
    }
}
```

```
public static Hour operator+ (Hour lhs, Hour rhs)
{
    return new Hour(lhs.value + rhs.value);
}
...
private int value;
}
```

Notice the following:

- The operator is *public*. All operators *must* be public.

- The operator is *static*. All operators *must* be static. Operators are never polymorphic and cannot use the *virtual*, *abstract*, *override*, or *sealed* modifier.

- A binary operator (such as the + operator, shown earlier) has two explicit arguments, and a unary operator has one explicit argument. (C++ programmers should note that operators never have a hidden *this* parameter.)

> **Tip** When declaring highly stylized functionality (such as operators), it is useful to adopt a naming convention for the parameters. For example, developers often use *lhs* and *rhs* (acronyms for left-hand side and right-hand side, respectively) for binary operators.

When you use the + operator on two expressions of type *Hour*, the C# compiler automatically converts your code to a call to the user-defined operator. The C# compiler converts this:

```
Hour Example(Hour a, Hour b)
{
    return a + b;
}
```

to this:

```
Hour Example(Hour a, Hour b)
{
    return Hour.operator+(a,b); // pseudocode
}
```

Note, however, that this syntax is pseudocode and not valid C#. You can use a binary operator only in its standard infix notation (with the symbol between the operands).

There is one final rule that you must follow when declaring an operator (otherwise, your code will not compile): at least one of the parameters must always be of the containing type. In the preceding *operator+* example for the *Hour* class, one of the parameters, *a* or *b*, must be an *Hour* object. In this example, both parameters are *Hour* objects. However, there could be times when you want to define additional implementations of *operator+* that add, for example, an integer (a number of hours) to an *Hour* object—the first parameter could be *Hour*, and the second parameter could be the integer. This rule makes it easier for the compiler to know where to look when trying to resolve an operator invocation, and it also ensures that you can't change the meaning of the built-in operators.

Creating Symmetric Operators

In the preceding section, you saw how to declare a binary + operator to add together two instances of type *Hour*. The *Hour* structure also has a constructor that creates an *Hour* from an *int*. This means that you can add together an *Hour* and an *int*—you just have to first use the *Hour* constructor to convert the *int* to an *Hour*. For example:

```
Hour a = ...;
int b = ...;
Hour sum = a + new Hour(b);
```

This is certainly valid code, but it is not as clear or as concise as adding together an *Hour* and an *int* directly, like this:

```
Hour a = ...;
int b = ...;
Hour sum = a + b;
```

To make the expression *(a + b)* valid, you must specify what it means to add together an *Hour* (*a*, on the left) and an *int* (*b*, on the right). In other words, you must declare a binary + operator whose first parameter is an *Hour* and whose second parameter is an *int*. The following code shows the recommended approach:

```
struct Hour
{
    public Hour(int initialValue)
    {
        this.value = initialValue;
    }
    ...
    public static Hour operator+ (Hour lhs, Hour rhs)
    {
        return new Hour(lhs.value + rhs.value);
    }

    public static Hour operator+ (Hour lhs, int rhs)
    {
        return lhs + new Hour(rhs);
    }
    ...
    private int value;
}
```

Notice that all the second version of the operator does is construct an *Hour* from its *int* argument and then call the first version. In this way, the real logic behind the operator is held in a single place. The point is that the extra *operator+* simply makes existing functionality easier to use. Also, notice that you should not provide many different versions of this operator, each with a different second parameter type—cater to the common and meaningful cases only, and let the user of the class take any additional steps if an unusual case is required.

This *operator+* declares how to add together an *Hour* as the left-hand operand and an *int* as the right-hand operator. It does not declare how to add together an *int* as the left-hand operand and an *Hour* as the right-hand operand:

```
int a = ...;
Hour b = ...;
Hour sum = a + b; // compile-time error
```

This is counterintuitive. If you can write the expression *a + b*, you expect to also be able to write *b + a*. Therefore, you should provide another overload of *operator+*:

```
struct Hour
{
    public Hour(int initialValue)
    {
        this.value = initialValue;
    }
    ...
    public static Hour operator+ (Hour lhs, int rhs)
    {
        return lhs + new Hour(rhs);
    }

    public static Hour operator+ (int lhs, Hour rhs)
    {
        return new Hour(lhs) + rhs;
    }
    ...
    private int value;
}
```

> **Note** C++ programmers should notice that you must provide the overload yourself. The compiler won't write the overload for you or silently swap the sequence of the two operands to find a matching operator.

Operators and Language Interoperability

Not all languages that execute using the common language runtime (CLR) support or understand operator overloading. Microsoft Visual Basic is a common example. If you are creating classes that you want to be able to use from other languages, if you overload an operator, you should provide an alternative mechanism that supports the same functionality. For example, suppose you implement *operator+* for the *Hour* structure:

```
public static Hour operator+ (Hour lhs, int rhs)
{
    ...
}
```

If you need to be able to use your class from a Visual Basic application, you should also provide an *Add* method that achieves the same thing:

```
public static Hour Add(Hour lhs, int rhs)
{
    ...
}
```

Understanding Compound Assignment

A compound assignment operator (such as +=) is always evaluated in terms of its associated operator (such as +). In other words, this:

```
a += b;
```

is automatically evaluated as this:

```
a = a + b;
```

In general, the expression *a @= b* (where @ represents any valid operator) is always evaluated as *a = a @ b*. If you have overloaded the appropriate simple operator, the overloaded version is automatically called when you use its associated compound assignment operator. For example:

```
Hour a = ...;
int b = ...;
a += a; // same as a = a + a
a += b; // same as a = a + b
```

The first compound assignment expression (*a += a*) is valid because *a* is of type *Hour*, and the *Hour* type declares a binary *operator+* whose parameters are both *Hour*. Similarly, the second compound assignment expression (*a += b*) is also valid because *a* is of type *Hour* and *b* is of type *int*. The *Hour* type also declares a binary *operator+* whose first parameter is an *Hour* and whose second parameter is an *int*. Note, however, that you cannot write the expression *b += a* because that's the same as *b = b + a*. Although the addition is valid, the assignment is not, because there is no way to assign an *Hour* to the built-in *int* type.

Declaring Increment and Decrement Operators

C# allows you to declare your own version of the increment (++) and decrement (−−) operators. The usual rules apply when declaring these operators: they must be public, they must be static, and they must be unary. Here is the increment operator for the *Hour* structure:

```
struct Hour
{
    ...
    public static Hour operator++ (Hour arg)
    {
        arg.value++;
        return arg;
    }
    ...
    private int value;
}
```

The increment and decrement operators are unique in that they can be used in prefix and postfix forms. C# cleverly uses the same single operator for both the prefix and postfix versions. The result of a postfix expression is the value of the operand *before* the expression takes place. In other words, the compiler effectively converts this:

```
Hour now = new Hour(9);
Hour postfix = now++;
```

to this:

```
Hour now = new Hour(9);
Hour postfix = now;
now = Hour.operator++(now); // pseudocode, not valid C#
```

The result of a prefix expression is the return value of the operator. The C# compiler effectively converts this:

```
Hour now = new Hour(9);
Hour prefix = ++now;
```

to this:

```
Hour now = new Hour(9);
now = Hour.operator++(now); // pseudocode, not valid C#
Hour prefix = now;
```

This equivalence means that the return type of the increment and decrement operators must be the same as the parameter type.

Operators in Structures and Classes

It is important to realize that the implementation of the increment operator in the *Hour* structure works only because *Hour* is a structure. If you change *Hour* into a class but leave the implementation of its increment operator unchanged, you will find that the postfix translation won't give the correct answer. If you remember that a class is a reference type and revisit the compiler translations explained earlier, you can see why this occurs:

```
Hour now = new Hour(9);
Hour postfix = now;
now = Hour.operator++(now); // pseudocode, not valid C#
```

If *Hour* is a class, the assignment statement *postfix = now* makes the variable *postfix* refer to the same object as *now*. Updating *now* automatically updates *postfix*! If *Hour* is a structure, the assignment statement makes a copy of *now* in *postfix*, and any changes to *now* leave *postfix* unchanged, which is what we want.

The correct implementation of the increment operator when *Hour* is a class is as follows:

```
class Hour
{
    public Hour(int initialValue)
    {
        this.value = initialValue;
    }
    ...
    public static Hour operator++(Hour arg)
    {
        return new Hour(arg.value + 1);
    }
    ...
    private int value;
}
```

Notice that *operator++* now creates a new object based on the data in the original. The data in the new object is incremented, but the data in the original is left unchanged. Although this works, the compiler translation of the increment operator results in a new object being created each time it is used. This can be expensive in terms of memory use and garbage collection overhead. Therefore, it is recommended that you limit operator overloads when you define types. This recommendation applies to all operators, and not just to the increment operator.

Defining Operator Pairs

Some operators naturally come in pairs. For example, if you can compare two *Hour* values by using the *!=* operator, you would expect to be able to also compare two *Hour* values by using the *==* operator. The C# compiler enforces this very reasonable expectation by insisting that if you define either *operator==* or *operator!=*, you must define them both. This neither-or-both rule also applies to the *<* and *>* operators and the *<=* and *>=* operators. The C# compiler does not write any of these operator partners for you. You must write them all explicitly yourself, regardless of how obvious they might seem. Here are the *==* and *!=* operators for the *Hour* structure:

```
struct Hour
{
    public Hour(int initialValue)
    {
        this.value = initialValue;
    }
    ...
    public static bool operator==(Hour lhs, Hour rhs)
    {
        return lhs.value == rhs.value;
    }

    public static bool operator!=(Hour lhs, Hour rhs)
    {
        return lhs.value != rhs.value;
    }
    ...
    private int value;
}
```

The return type from these operators does not actually have to be Boolean. However, you would have to have a very good reason for using some other type, or these operators could become very confusing!

Note If you define *operator==* and *operator!=*, you should also override the *Equals* and *GetHashCode* methods inherited from *System.Object*. The *Equals* method should exhibit *exactly* the same behavior as *operator==*. (You should define one in terms of the other.) The *GetHashCode* method is used by other classes in the Microsoft .NET Framework. (When you use an object as a key in a hash table, for example, the *GetHashCode* method is called on the object to help calculate a hash value. For more information, see the .NET Framework Reference documentation supplied with Visual Studio 2008.) All this method needs to do is return a distinguishing integer value. (Don't return the same integer from the *GetHashCode* method of all your objects, however, as this will nullify the effectiveness of the hashing algorithms.)

Implementing an Operator

In the following exercise, you will complete another digital clock application. This version of the code is similar to the exercise in Chapter 17, "Interrupting Program Flow and Handling Events." However, in this version, the *delegate* method (which is called every second) does not receive the current *hour*, *minute*, and *second* values when the event is raised. Instead, the *delegate* method keeps track of the time itself by updating three fields, one each for the *hour*, *minute*, and *second* values. The type of these three fields is *Hour*, *Minute*, and *Second*, respectively, and they are all structures. However, the application will not yet compile, because the *Minute* structure is not finished. In the first exercise, you will finish the *Minute* structure by implementing its missing addition operators.

Write the *operator+* overloads

1. Start Microsoft Visual Studio 2008 if it is not already running.

2. Open the *Operators* project, located in the \Microsoft Press\Visual CSharp Step by Step \Chapter 21\Operators folder in your Documents folder.

3. In the *Code and Text Editor* window, open the Clock.cs file and locate the declarations of the *hour*, *minute*, and *second* fields at the end of the class.

   These fields hold the clock's current time:

   ```
   class Clock
   {
       ...
       private Hour hour;
       private Minute minute;
       private Second second;
   }
   ```

4. Locate the *tock* method of the *Clock* class. This method is called every second to update the *hour*, *minute*, and *second* fields.

   The *tock* method looks like this:

   ```
   private void tock()
   {
       this.second++;
       if (this.second == 0)
       {
           this.minute++;
           if (this.minute == 0)
           {
               this.hour++;
           }
       }
   }
   ```

The constructors for the *Clock* class contain the following statement that subscribes to the *tick* event of the *pulsed* field so that this method is called whenever the event is raised. (The *pulsed* field is a *Ticker* that uses a *DispatcherTimer* object to generate an event every second, as described in the exercises in Chapter 17.)

```
this.pulsed.tick += tock;
```

5. On the *Build* menu, click *Build Solution*.

The build fails and displays the following error message:

```
Operator '==' cannot be applied to operands of type 'Operators.Minute' and 'int'.
```

The problem is that the *tock* method contains the following *if* statement, but the appropriate *operator==* is not declared in the *Minute* structure:

```
if (minute == 0)
{
    hour++;
}
```

Your first task is to implement this operator for the *Minute* structure.

6. In the *Code and Text Editor* window, open the Minute.cs file.

7. In the *Minute* structure, implement a version of *operator==* that accepts a *Minute* as its left-hand operand and an *int* as its right-hand operand. Don't forget that the return type of this operator should be a *bool*.

The completed operator should look exactly as shown in bold here:

```
struct Minute
{
    ...
    public static bool operator==(Minute lhs, int rhs)
    {
        return lhs.value == rhs;
    }
    ...
    private int value;
}
```

8. On the *Build* menu, click *Build Solution*.

The build fails again and displays a different error message:

```
The operator 'Operators.Minute.operator ==(Operators.Minute, int)' requires a matching
operator "!=" to also be defined.
```

The problem now is that you have implemented a version of *operator==* but have not implemented its required *operator!=* partner.

9. Implement a version of *operator!=* that accepts a *Minute* as its left-hand operand and an *int* as its right-hand operand.

The completed operator should look exactly as shown in bold here:

```
struct Minute
{
    ...
    public static bool operator!=(Minute lhs, int rhs)
    {
        return lhs.value != rhs;
    }
    ...
    private int value;
}
```

10. On the *Build* menu, click *Build Solution*.

This time, the project builds without errors.

11. On the *Debug* menu, click *Start Without Debugging*.

The application runs and displays a digital clock that updates itself every second.

12. Close the application, and return to the Visual Studio 2008 programming environment.

Understanding Conversion Operators

Sometimes it is necessary to convert an expression of one type to another. For example, the following method is declared with a single *double* parameter:

```
class Example
{
    public static void MyDoubleMethod(double parameter)
    {
        ...
    }
}
```

You might reasonably expect that only values of type *double* could be used as arguments when calling *MyDoubleMethod*, but this is not so. The C# compiler also allows *MyDoubleMethod* to be called with an argument whose type is not *double*, but only as long as that value can be converted to a *double*. The compiler will generate code that performs this conversion when the method is called.

Providing Built-In Conversions

The built-in types have some built-in conversions. For example, an *int* can be implicitly converted to a *double*. An implicit conversion requires no special syntax and never throws an exception:

```
Example.MyDoubleMethod(42); // implicit int-to-double conversion
```

An implicit conversion is sometimes called a *widening conversion*, as the result is *wider* than the original value—it contains at least as much information as the original value, and nothing is lost.

On the other hand, a *double* cannot be implicitly converted to an *int*:

```
class Example
{
    public static void MyIntMethod(int parameter)
    {
        ...
    }
}
...
Example.MyIntMethod(42.0); // compile-time error
```

Converting from a *double* to an *int* runs the risk of losing information, so it will not be done automatically. (Consider what would happen if the argument to *MyIntMethod* were 42.5—how should this be converted?) A *double* can be converted to an *int*, but the conversion requires an explicit notation (a cast):

```
Example.MyIntMethod((int)42.0);
```

An explicit conversion is sometimes called a *narrowing conversion*, as the result is *narrower* than the original value (it can contain less information) and can throw an *OverflowException*. C# allows you to provide conversion operators for your own user-defined types to control whether it is sensible to convert values to other types and whether these conversions are implicit or explicit.

Implementing User-Defined Conversion Operators

The syntax for declaring a user-defined conversion operator is similar to that for declaring an overloaded operator. A conversion operator must be *public* and must also be *static*. Here's a conversion operator that allows an *Hour* object to be implicitly converted to an *int*:

```
struct Hour
{
    ...
    public static implicit operator int (Hour from)
    {
        return this.value;
    }

    private int value;
}
```

The type you are converting from is declared as the single parameter (in this case, *Hour*), and the type you are converting to is declared as the type name after the keyword *operator* (in this case, *int*). There is no return type specified before the keyword *operator*.

When declaring your own conversion operators, you must specify whether they are implicit conversion operators or explicit conversion operators. You do this by using the *implicit* and *explicit* keywords. For example, the *Hour* to *int* conversion operator mentioned earlier is implicit, meaning that the C# compiler can use it implicitly (without requiring a cast):

```
class Example
{
    public static void MyOtherMethod(int parameter) { ... }
    public static void Main()
    {
        Hour lunch = new Hour(12);
        Example.MyOtherMethod(lunch); // implicit Hour to int conversion
    }
}
```

If the conversion operator had been declared *explicit*, the preceding example would not have compiled, because an explicit conversion operator requires an explicit cast:

```
Example.MyOtherMethod((int)lunch); // explicit Hour to int conversion
```

When should you declare a conversion operator as explicit or implicit? If a conversion is always safe, does not run the risk of losing information, and cannot throw an exception, it can be defined as an *implicit* conversion. Otherwise, it should be declared as an *explicit* conversion. Converting from an *Hour* to an *int* is always safe—every *Hour* has a corresponding *int* value—so it makes sense for it to be implicit. An operator that converts a *string* to an *Hour* should be explicit, as not all strings represent valid *Hours*. (The string *"7"* is fine, but how would you convert the string *"Hello, World"* to an *Hour*?)

Creating Symmetric Operators, Revisited

Conversion operators provide you with an alternative way to resolve the problem of providing symmetric operators. For example, instead of providing three versions of *operator+* (*Hour + Hour*, *Hour + int*, and *int + Hour*) for the *Hour* structure, as shown earlier, you can provide a single version of *operator+* (that takes two *Hour* parameters) and an implicit *int* to *Hour* conversion, like this:

```
struct Hour
{
    public Hour(int initialValue)
    {
        this.value = initialValue;
    }

    public static Hour operator+(Hour lhs, Hour rhs)
    {
        return new Hour(lhs.value + rhs.value);
    }
```

```
public static implicit operator Hour (int from)
{
    return new Hour (from);
}
...
private int value;
}
```

If you add an *Hour* to an *int* (in either order), the C# compiler automatically converts the *int* to an *Hour* and then calls *operator+* with two *Hour* arguments:

```
void Example(Hour a, int b)
{
    Hour eg1 = a + b; // b converted to an Hour
    Hour eg2 = b + a; // b converted to an Hour
}
```

Adding an Implicit Conversion Operator

In the following exercise, you will modify the digital clock application from the preceding exercise. You will add an *implicit* conversion operator to the *Second* structure and remove the operators that it replaces.

Write the conversion operator

1. Return to Visual Studio 2008, displaying the *Operators* project. Display the Clock.cs file in the *Code and Text Editor* window, and examine the *tock* method again:

```
private void tock()
{
    this.second++;
    if (this.second == 0)
    {
        this.minute++;
        if (this.minute == 0)
        {
            this.hour++;
        }
    }
}
```

Notice the statement *if (this.second == 0)* shown in bold in the preceding code example. This fragment of code compares a *Second* to an *int* using the == operator.

2. Display the Second.cs file in the *Code and Text Editor* window.

The *Second* structure currently contains three overloaded implementations of *operator==* and three overloaded implementations of *operator!=*. Each operator is overloaded for the parameter type pairs (*Second*, *Second*), (*Second*, *int*), and (*int*, *Second*).

3. In the *Second* structure, comment out the four versions of *operator*== and operator!= that take one *Second* and one *int* parameter. (Do not comment out the operators that take two *Second* parameters.) The following two operators should be the only versions of *operator*== and *operator!*= remaining in the *Second* structure:

```
struct Second
{
    ...
    public static bool operator==(Second lhs, Second rhs)
    {
        return lhs.value == rhs.value;
    }

    public static bool operator!=(Second lhs, Second rhs)
    {
        return lhs.value != rhs.value;
    }

    ...
}
```

4. On the *Build* menu, click *Build Solution*.

The build fails with the following error message:

```
Operator '==' cannot be applied to the operands of type 'Operators.Second' and 'int'
```

Removing the operators that compare a *Second* and an *int* cause the statement *if (this. second == 0)* highlighted in step 1 to fail to compile.

5. In the *Code and Text Editor* window, add an *implicit* conversion operator to the *Second* structure that converts from an *int* to a *Second*.

The conversion operator should appear as shown in bold here:

```
struct Second
{
    ...
    public static implicit operator Second (int arg)
    {
        return new Second(arg);
    }
    ...
}
```

6. On the *Build* menu, click *Build Solution*.

The program successfully builds this time because the conversion operator and the remaining two operators together provide the same functionality as the four deleted operator overloads. The only difference is that using an *implicit* conversion operator is potentially a little slower than not using an *implicit* conversion operator.

7. On the *Debug* menu, click *Start Without Debugging*.

Verify that the application still works correctly.

8. Close the application, and return to the Visual Studio 2008 programming environment.

- If you want to continue to the next chapter:

  Keep Visual Studio 2008 running, and turn to Chapter 22.

- If you want to exit Visual Studio 2008 now:

  On the *File* menu, click *Exit*. If you see a *Save* dialog box, click *Yes* (if you are using Visual Studio 2008) or *Save* (if you are using Visual C# 2008 Express Edition) and save the project.

Chapter 21 Quick Reference

To	Do this
Implement an operator	Write the keywords *public* and *static*, followed by the return type, followed by the *operator* keyword, followed by the operator symbol being declared, followed by the appropriate parameters between parentheses. For example: ```csharp\nstruct Hour\n{\n ...\n public static bool operator==(Hour lhs, Hour rhs)\n {\n ...\n }\n ...\n}\n```
Declare a conversion operator	Write the keywords *public* and *static*, followed by the keyword *implicit* or *explicit*, followed by the operator keyword, followed by the type being converted to, followed by the type being converted from as a single parameter between parentheses. For example: ```csharp\nstruct Hour\n{\n ...\n public static implicit operator Hour(int arg)\n {\n ...\n }\n ...\n}\n```

Part IV
Working with Windows Applications

Chapter 22
Introducing Windows Presentation Foundation

After completing this chapter, you will be able to:

- Create Microsoft Windows Presentation Foundation (WPF) applications.

- Use common WPF controls such as labels, text boxes, and buttons.

- Define styles for WPF controls.

- Change the properties of WPF forms and controls at design time and through code at run time.

- Handle events exposed by WPF forms and controls.

Now that you have completed the exercises and examined the examples in the first three parts of this book, you should be well versed in the C# language. You have learned how to write programs and create components by using C#, and you should understand many of the finer points of the language, such as extension methods, lambda expressions, and the distinction between value and reference types. You now have the essential language skills, and in Part IV you will expand upon them and use C# to take advantage of the graphical user interface (GUI) libraries provided as part of the Microsoft .NET Framework. In particular, you will see how to use the objects in the *System.Windows* namespace to create WPF applications.

In this chapter, you learn how to build a basic WPF application by using the common components that are a feature of most GUI applications. You see how to set the properties of WPF forms and controls by using the *Design View* and *Properties* windows, and also by using Extensible Application Markup Language, or XAML. You also learn how to use WPF styles to build user interfaces that can be easily adapted to conform to your organization's presentation standards. Finally, you learn how to intercept and handle some of the events that WPF forms and controls expose.

Creating a WPF Application

As an example, you are going to create an application that a user can use to input and display details for members of the Middleshire Bell Ringers Association, an esteemed group of the finest campanologists. Initially, you will keep the application very simple, concentrating on laying out the form and making sure that it all works. On the way, you learn about some of the features that WPF provides for building highly adaptable user interfaces. In later chapters, you will provide menus and learn how to implement validation to ensure that the data that is entered makes sense. The following graphic shows what the application will look like after you have completed it. (You can see the completed version by building and running the BellRingers project in the \Microsoft Press\Visual CSharp Step by Step\Chapter 22 \Completed BellRingers\ folder in your Documents folder.)

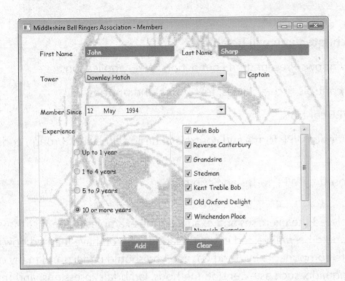

Creating a Windows Presentation Foundation Application

In this exercise, you'll start building the Middleshire Bell Ringers Association application by creating a new project, laying out the form, and adding controls to the form. You have been using existing WPF applications in Microsoft Visual Studio 2008 in previous chapters, so much of the first couple of exercises will be a review for you.

Create the Middleshire Bell Ringers Association project

1. Start Visual Studio 2008 if it is not already running.

2. If you are using Visual Studio 2008 Standard Edition or Visual Studio 2008 Professional Edition, perform the following operations to create a new WPF application:

   2.1. On the *File* menu, point to *New*, and then click *Project*.

   The *New Project* dialog box opens.

   2.2. In the *Project Types* pane, click *Visual C#*.

   2.3. In the *Templates* pane, click the *WPF Application* icon.

   2.4. In the *Location* field, type **\Microsoft Press\Visual CSharp Step By Step\ Chapter 22** under your Documents folder.

2.5. In the *Name* field, type **BellRingers**.

2.6. Click *OK*.

3. If you are using Microsoft Visual C# 2008 Express Edition, perform the following tasks to create a new graphical application.

3.1. On the *Tools* menu, click *Options*.

3.2. In the *Options* dialog box, click *Projects and Solutions* in the tree view in the left pane.

3.3. In the right pane, in the *Visual Studio projects location* text box, specify the location **Microsoft Press\Visual CSharp Step By Step\Chapter 22** under your Documents folder.

3.4. Click *OK*.

3.5. On the *File* menu, click *New Project*.

3.6. In the *New Project* dialog box, click the *WPF Application* icon.

3.7. In the *Name* field, type *BellRingers*.

3.8. Click *OK*.

The new project is created and contains a blank form called Window1.

Examine the form and the Grid layout

1. Examine the form in the *XAML* pane underneath the *Design View* window. Notice that the XAML definition of the form looks like this:

```
<Window x:Class="BellRingers.Window1"
    xmlns="http://schemas.microsoft.com/winfx/2006/xaml/presentation"
    xmlns:x="http://schemas.microsoft.com/winfx/2006/xaml"
    Title="Window1" Height="300" Width="300">
    <Grid>

    </Grid>
</Window>
```

The *Class* attribute specifies the fully qualified name of the class that implements the form. In this case, it is called Window1 in the *BellRingers* namespace. The WPF Application template uses the name of the application as the default namespace for forms. The *xmlns* attributes specify the XML namespaces that define the schemas used by WPF; all the controls and other items that you can incorporate into a WPF application have definitions that live in these namespaces. (If you are not familiar with XML namespaces, you can ignore these *xmlns* attributes for now.) The *Title* attribute specifies the text that appears in the title bar of the form, and the *Height* and *Width* attributes specify the default height and width of the form. You can modify these values either by changing them in the *XAML* pane or by using the *Properties* window. You can also change the value of these and many other properties dynamically by writing C# code that executes when the form runs.

2. Click the Window1 form in the *Design View* window. In the *Properties* window, locate and click the *Title* property, and then type **Middleshire Bell Ringers Association – Members** to change the text in the title bar of the form.

   Notice that the value in the *Title* attribute of the form changes in the *XAML* pane, and the new title is displayed in the title bar of the form in the *Design View* window.

> **Note** The Window1 form contains a child control that you will examine in the next step. If the *Properties* window displays the properties for a *System.Windows.Controls.Grid* control, click the *Window1* text on the Window1 form. This action selects the form rather than the grid, and the *Properties* window then displays the properties for the *System.Windows. Window* control.

3. In the *XAML* pane, notice that the *Window* element contains a child element called *Grid*.

   In a WPF application, you place controls such as buttons, text boxes, and labels in a panel on a form. The panel manages the layout of the controls it contains. The default panel added by the WPF Application template is the *Grid*, with which you can specify exactly the location of your controls at design time. Other panels are available that provide different styles of layout. For example, *StackPanel* automatically places controls in a vertical arrangement, with each control arranged directly beneath its immediate predecessor. Another example is *WrapPanel*, which arranges controls in a row from left to right and then wraps the content to the next line when the current row is full. A primary purpose of a layout panel is to govern how the controls are positioned if the user resizes the window at run time; the controls are automatically resized and repositioned according to the type of the panel.

> **Note** The *Grid* panel is flexible but complex. By default, you can think of the *Grid* panel as defining a single cell into which you can drop controls. However, you can set the properties of a *Grid* panel to define multiple rows and columns (hence its name), and you can drop controls into each of the cells defined by these rows and columns. In this chapter, we keep things simple and use only a single cell.

4. In the *Design View* window, click the Window1 form, and then click the *Toolbox* tab.

5. In the *Common* section, click *Button*, and then click in the upper-right part of the form.

A button control that displays two connectors anchoring it to the top and right edges of the form is added to the form, like this:

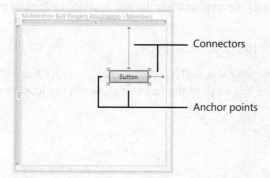

Although you clicked the form, the *Button* control is added to the *Grid* control contained in the form. The grid occupies the entire form apart from the title bar at the top. The connectors show that the button is anchored to the top and right edges of the grid.

6. Examine the code in the *XAML* pane. The *Grid* element and its contents should now look something like this (your values for the *Margin* property might vary):

```
<Grid>
    <Button HorizontalAlignment="Right" Margin="0,84,34,0"
        Name="button1" Width="75" Height="23" VerticalAlignment="Top">Button</Button>
</Grid>
```

> **Note** Throughout this chapter, lines from the *XAML* pane are shown split and indented so they will fit on the printed page.

When you place a control on a grid, you can connect any or all of the anchor points to the corresponding edge of the grid. By default, the *Design View* window connects the control to the nearest edges. If you place the control toward the lower left of the grid, it will be connected to the bottom and left edges of the grid.

The *HorizontalAlignment* and *VerticalAlignment* properties of the button indicate the edges to which the button is currently connected, and the *Margin* property indicates the distance to those edges. Recall from Chapter 1, "Welcome to C#," that the *Margin* property contains four values specifying the distance from the left, top, right, and bottom edges of the grid, respectively. In the XAML fragment just shown, the button is 84 units from the top edge of the grid and 34 units from the right edge. (Each unit is 1/96th of an inch.) Margin values of 0 indicate that the button is not connected to the corresponding edge. When you run the application, the WPF runtime will endeavor to maintain these distances even if you resize the form.

7. On the *Debug* menu, click *Start Without Debugging* to build and run the application.

8. When the form appears, resize the window. Notice that as you drag the edges of the form around, the distance of the button from the top and right edges of the form remains fixed.

9. Close the form, and return to Visual Studio 2008.

10. In the *Design View* window, click the button control, and then click the left anchor point to attach the control to the left edge of the form, as shown in the following image:

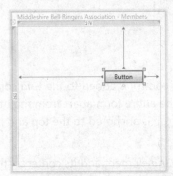

In the *XAML* pane, notice that the *HorizontalAlignment* property is no longer specified. The default value for the *HorizontalAlignment* and *VerticalAlignment* properties is a value called *Stretch*, which indicates that the control is anchored to both opposite edges. Also notice that the *Margin* property now specifies a nonzero value for the left margin.

> **Note** You can click the anchor point that is connected to the edge of the grid to remove the connection.

11. On the *Debug* menu, click *Start Without Debugging* to build and run the application again.

12. When the form appears, experiment by making the form narrower and wider. Notice that the button no longer moves because it is anchored to the left and right edges of the form. Instead, the button gets wider or narrower as the edges move.

13. Close the form, and return to Visual Studio 2008.

14. In the *Design View* window, add a second *Button* control to the form from the *Toolbox*, and position it near the middle of the form.

15. In the *XAML* pane, set the *Margin* property values to 0, remove the *VerticalAlignment* and *HorizontalAlignment* properties if they appear, and set the *Width* and *Height* properties, as shown here:

```
<Button Margin="0,0,0,0" Name="button2" Width="75" Height="23">Button</Button>
```

> **Tip** You can also set many of the properties of a control, such as *Margin*, by using the *Properties* window. However, you cannot set all properties by using the *Properties* window, and sometimes it is simply easier to type values directly into the *XAML* pane as long as you enter the values carefully.

> **Note** If you don't set the *Width* and *Height* properties of the button control, the button fills the entire form.

16. On the *Debug* menu, click *Start Without Debugging* to build and run the application once more.

17. When the form appears, resize the form. Notice that as the form shrinks or grows the new button relocates itself to try to maintain its relative position on the form with respect to all four sides (it tries to stay in the center of the form). The new button control even travels over the top of the first button control if you shrink the height of the form.

18. Close the form, and return to Visual Studio 2008.

As long as you are consistent in your approach, by using layout panes, such as the *Grid*, you can build forms that look right regardless of the user's screen resolution without having to write complex code to determine when the user has resized a window. Additionally, with WPF, you can modify the look and feel of the controls an application uses, again without having to write lots of complex code. With these features together, you can build applications that can easily be customized to conform to any house style required by your organization. You will examine some of these features in the following exercises.

Add a background image to the form

1. In the *Design View* window, click the Window1 form.

2. In the *Toolbox*, in the *Common* section, click *Image*, and then click anywhere on the form. You will use this image control to display an image on the background of the form.

> **Note**   You can use many other techniques to display an image in the background of a *Grid*. The method shown in this exercise is probably the simplest, although other strategies can provide more flexibility.

3. In the *XAML* pane, set the *Margin* property of the image control, and remove any other property values apart from the *Name*, as shown here:

```
<Image Margin="0,0,0,0" Name="image1"/>
```

The image control expands to occupy the grid fully, although the two button controls remain visible.

4. In *Solution Explorer*, right-click the BellRingers project, point to *Add*, and then click *Existing Item*. In the *Add Existing Item – BellRingers* dialog box, move to the folder Microsoft Press\Visual CSharp Step By Step\Chapter 22 under your Documents folder. In the *File name* box, type **Bell.gif**, and then click *Add*.

This action adds the image file Bell.gif as a resource to your application. The Bell.gif file contains a sketch of a ringing bell.

5. In the *XAML* pane, modify the definition of the image control as shown here. Notice that you must replace the closing tag delimiter (/>) of the image control with an ordinary tag delimiter character (>) and add a closing </Image> tag:

```
<Image Margin="0,0,0,0" Name="image1" >
    <Image.Source>
        <BitmapImage UriSource="Bell.gif" />
    </Image.Source>
</Image>
```

The purpose of an image control is to display an image. You can specify the source of the image in a variety of ways. The example shown here loads the image from the file Bell.gif that you just added as a resource to the project.

The image should now appear on the form, like this:

There is a problem, however. The image is not in the background, and it totally obscures the two button controls. The issue is that, unless you specify otherwise, all controls placed on a layout panel have an implied z-order that renders controls added lower down in the XAML description over the top of controls added previously.

 Note The term *z-order* refers to the relative depth positions of items on the z-axis of a three-dimensional space (the y-axis being vertical and the x-axis being horizontal). Items with a higher value for the z-order appear in front of those items with a lower value.

There are at least two ways you can move the image control behind the buttons. The first is to move the XAML definitions of the buttons so that they appear after the image control, and the second is to explicitly specify a value for the *ZIndex* property for the control. Controls with a higher *ZIndex* value appear in front of those on the same panel with a lower *ZIndex*. If two controls have the same *ZIndex* value, their relative precedence is determined by the order in which they occur in the XAML description, as before.

 Note A *panel* is a control that acts as a container for other controls and determines how they are laid out with respect to one another. The *Grid* control is an example of a panel control. You will see other examples of panel controls later in this section of the book. You can place more than one panel on a form.

6. In the *XAML* pane, set the *ZIndex* properties of the button and image controls as shown in bold type in the following code:

```
<Button Panel.ZIndex="1" Margin="169,84,34,0"
    Name="button1" Height="23" VerticalAlignment="Top">Button</Button>
<Button Panel.ZIndex="1" Height="23" Margin="0,0,0,0"
    Name="button2" Width="76">Button</Button>
<Image Panel.ZIndex="0" Margin="0,0,0,0" Name="image1" >
    <Image.Source>
        <BitmapImage UriSource="Bell.gif" />
    </Image.Source>
</Image>
```

The two buttons should now reappear in front of the image.

With WPF, you can modify the way in which controls such as buttons, text boxes, and labels present themselves on a form. You will investigate this feature in the next exercise.

Create a style to manage the look and feel of controls on the form

1. In the *XAML* pane, modify the definition of the first button on the form, as shown in bold type in the following code. Notice that it is good practice to split the XAML description of a control that contains composite child property values such as *Button. Resource* over multiple lines to make the code easier to read and maintain:

```
<Button Style="{DynamicResource buttonStyle}" Panel.ZIndex="1" Margin ="169,84,34,0"
Name="button1" Height="23" VerticalAlignment="Top">
    <Button.Resources>
        <Style x:Key="buttonStyle">
            <Setter Property="Button.Background" Value="Gray"/>
            <Setter Property="Button.Foreground" Value="White"/>
            <Setter Property="Button.FontFamily" Value="Comic Sans MS"/>
        </Style>
    </Button.Resources>
    Button
</Button>
```

You can use the *<Style>* element of a control to set default values for the properties of that control. (Styles can do other things as well, as you will see later in this chapter.) This example specifies the values for the background and foreground colors of the button as well as the font used for the text on the button. You should notice that the button displayed in the *Design View* window changes its appearance to match the property values specified for the style.

Styles are resources, and you add them to a *Resources* element for the control. You can give each style a unique name by using the *Key* property. You can then reference the new style from the *Style* property of the control. The syntax {DynamicResource buttonStyle} creates a new style object based on the named style and then applies this style to the button.

> **Note** When you compile a WPF window, Visual Studio adds any resources included with the window to a collection associated with the window. Strictly speaking, the *Key* property doesn't specify the name of the style but rather an identifier for the resource in this collection. You can specify the *Name* property as well if you want to manipulate the resource in your C# code, but controls reference resources by specifying the *Key* value for that resource. Controls and other items that you add to a form should have their *Name* property set because, as with resources, this is how you reference these items in code.

Styles have scope. If you attempt to reference the *buttonStyle* style from the second button on the form, it will have no effect. Instead, you can create a copy of this style and add it to the *Resources* element of the second button, and then reference it, like this:

```
<Grid>
    <Button Style="{DynamicResource buttonStyle}"  Panel.ZIndex="1"
      Margin ="169,84,34,0" Name="button1" Height="23"
      VerticalAlignment="Top">
```

```
                <Button.Resources>
                    <Style x:Key="buttonStyle">
                        <Setter Property="Button.Background" Value="Gray"/>
                        <Setter Property="Button.Foreground" Value="White"/>
                        <Setter Property="Button.FontFamily" Value="Comic Sans MS"/>
                    </Style>
                </Button.Resources>
                Button
            </Button>
            <Button Style="{DynamicResource buttonStyle}" Panel.ZIndex="1" Height="23"
                Margin="0,0,0,0" Name="button2" Width="76">
                <Button.Resources>
                    <Style x:Key="buttonStyle">
                        <Setter Property="Button.Background" Value="Gray"/>
                        <Setter Property="Button.Foreground" Value="White"/>
                        <Setter Property="Button.FontFamily" Value="Comic Sans MS"/>
                    </Style>
                </Button.Resources>
                Button
            </Button>
            ...
    </Grid>
```

However, this approach can get very repetitive and becomes a maintenance nightmare if you need to change the style of buttons. A much better strategy is to define the style as a resource for the window, and then you can reference it from all controls in that window.

2. In the *XAML* pane, add a *<Window.Resources>* element above the grid, move the definition of the *buttonStyle* style to this new element, and then delete the *<Button. Resources>* element from both buttons. Reference the new style from both buttons, and split the definition of the *button2* control over multiple lines to make it more readable. The updated code for the entire XAML description of the form is as follows, with the resource definition and references to the resource shown in bold type:

```
<Window x:Class="BellRingers.Window1"
    xmlns="http://schemas.microsoft.com/winfx/2006/xaml/presentation"
    xmlns:x="http://schemas.microsoft.com/winfx/2006/xaml"
    Title="Middleshire Bell Ringers Association - Members"
    Height="300" Width="300">
    <Window.Resources>
        <Style x:Key="buttonStyle">
            <Setter Property="Button.Background" Value="Gray"/>
            <Setter Property="Button.Foreground" Value="White"/>
            <Setter Property="Button.FontFamily" Value="Comic Sans MS"/>
        </Style>
    </Window.Resources>
    <Grid>
        <Button Style="{StaticResource buttonStyle}"  Panel.ZIndex="1"
            Margin ="169,84,34,0" Name="button1" Height="23" VerticalAlignment="Top">
            Button
        </Button>
```

```
    <Button Style="{StaticResource buttonStyle}" Panel.ZIndex="1" Height="23"
        Margin="0,0,0,0" Name="button2" Width="76">
        Button
    </Button>
    <Image Panel.ZIndex="0" Margin="0,0,0,0" Name ="image1">
        <Image.Source>
            <BitmapImage UriSource="Bell.gif" />
        </Image.Source>
    </Image>
  </Grid>
</Window>
```

Notice that both buttons now appear in the *Design View* window using the same style.

> **Note**   The code you have just entered references the button style by using the
> *StaticResource* rather than the *DynamicResource* keyword. The scoping rules of static re-
> sources are like those of C# in that they require you to define a resource before you can
> reference it. In step 1 of this exercise, you referenced the *buttonStyle* style above the XAML
> code that defined it, so the style name was not actually in scope. This out-of-scope refer-
> ence works because using *DynamicResource* defers until run time the time at which the
> resource reference is resolved, at which point the resource should have been created.
>
> Generally speaking, static resources are more efficient than dynamic ones are because
> they are resolved when the application is built, but dynamic resources give you more flex-
> ibility. For example, if the resource itself changes as the application executes (you can write
> code to change styles at run time), any controls referencing the style using *StaticResource*
> will not be updated, but any controls referencing the style using *DynamicResource* will be.
>
> There are many other differences between the behavior of static and dynamic resources
> and restrictions on when you can reference a resource dynamically. For more information,
> consult the .NET Framework documentation provided with Visual Studio 2008.

There is still a little bit of repetition involved in the definition of the style; each of the
properties (background, foreground, and font family) explicitly state that they are button properties. You can remove this repetition by specifying the *TargetType* attribute in
the *Style* tag.

3. Modify the definition of the style to specify the *TargetType* attribute, like this:

```
<Style x:Key="buttonStyle" TargetType="Button">
    <Setter Property="Background" Value="Gray"/>
    <Setter Property="Foreground" Value="White"/>
    <Setter Property="FontFamily" Value="Comic Sans MS"/>
</Style>
```

You can add as many buttons as you like to the form, and you can style them all using
the *buttonStyle* style. But what about other controls, such as labels and text boxes?

4. In the *Design View* window, click the Window1 form, and then click the *Toolbox* tab.
In the *Common* section, click *TextBox*, and then click anywhere in the lower half of
the form.

5. In the *XAML* pane, change the definition of the text box control and specify the *Style* attribute shown in bold type in the following example, attempting to apply the *buttonStyle* style:

```
<TextBox Style="{StaticResource buttonStyle}" Height="21" Margin="114,0,44,58"
Name="textBox1" VerticalAlignment="Bottom" />
```

Not surprisingly, attempting to set the style of a text box to a style intended for a button fails. The *Design View* window displays the error message "The document root element has been altered or an unexpected error has been encountered in updating the designer. Click here to reload." If you click the message as indicated, the form disappears from the *Design View* window and is replaced with the following message:

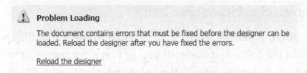

> ⚠ **Problem Loading**
>
> The document contains errors that must be fixed before the designer can be loaded. Reload the designer after you have fixed the errors.
>
> Reload the designer

Don't panic; you will now fix your mistake!

6. In the *XAML* pane, modify the *Key* property and change the *TargetType* to *Control* in the definition of the style, and then modify the references to the style in the button and text box controls as shown in bold type here:

```
<Window x:Class="BellRingers.Window1"
    ...>
    <Window.Resources>
        <Style x:Key="bellRingersStyle" TargetType="Control">
            <Setter Property="Background" Value="Gray"/>
            <Setter Property="Foreground" Value="White"/>
            <Setter Property="FontFamily" Value="Comic Sans MS"/>
        </Style>
    </Window.Resources>
    <Grid>
        <Button Style="{StaticResource bellRingersStyle}" ...>
            Button
        </Button>
        <Button Style="{StaticResource bellRingersStyle}" ...>
            Button
        </Button>
        ...
        <TextBox ... Style="{StaticResource bellRingersStyle}" ... />
    </Grid>
</Window>
```

Because the style no longer applies exclusively to buttons, it makes sense to rename it. Setting the *TargetType* attribute of a style to *Control* specifies that the style can be applied to any control that inherits from the *Control* class. In the WPF model, many different types of controls, including text boxes and buttons, inherit from the *Control* class. However, you can provide *Setter* elements only for properties that explicitly belong to the *Control* class. (Buttons have some additional properties that are not part of the *Control* class; if you specify any of these button-only properties, you cannot set the *TargetType* to *Control*.)

7. In the *Design View* window, click the *Reload the designer* link. The form should now appear. Notice that the text box is rendered with a gray background.

8. On the *Debug* menu, click *Start Without Debugging* to build and run the application. Type some text in the text box, and verify that it appears in white using the Comic Sans MS font.

   Unfortunately, the choice of colors makes it a little difficult to see the text caret when you click the text box and type text. You will fix this in a following step.

9. Close the form, and return to Visual Studio 2008.

10. In the *XAML* pane, edit the *bellRingersStyle* style and add the *<Style.Triggers>* element shown in bold type in the following code. (If you get an error message that the *TriggerCollection* is sealed, simply rebuild the solution.)

```
<Style x:Key="bellRingersStyle" TargetType="Control">
    <Setter Property="Background" Value="Gray"/>
    <Setter Property="Foreground" Value="White"/>
    <Setter Property="FontFamily" Value="Comic Sans MS"/>
    <Style.Triggers>
        <Trigger Property="IsMouseOver" Value="True">
            <Setter Property="Background" Value="Blue" />
        </Trigger>
    </Style.Triggers>
</Style>
```

A trigger specifies an action to perform when a property value changes. The *bellRingersStyle* style detects a change in the *IsMouseOver* property to temporarily modify the background color of the control the mouse is over.

> **Note** Don't confuse triggers with events. Triggers respond to transient changes in property values. If the value in the triggering property reverts, the triggered action is undone. In the example shown previously, when the *IsMouseOver* property is no longer *true* for a control, the *Background* property is set back to its original value. Events specify an action to perform when a significant incident (such as the user clicking a button) occurs in an application; the actions performed by an event are not undone when the incident is finished.

11. On the *Debug* menu, click *Start Without Debugging* to build and run the application again. This time, when you click the text box, it turns blue so that you can see the text caret more easily. The text box reverts to its original gray color when you move the mouse away. Notice that the buttons do not behave in quite the same way. Button controls already implement this functionality and turn a paler shade of blue when you place the mouse over them. This default behavior overrides the trigger specified in the style.

12. Close the form, and return to Visual Studio 2008.

> **Note** An alternative approach that you can use to apply a font globally to all controls on a form is to set the text properties of the window holding the controls. These properties include *FontFamily*, *FontSize*, and *FontWeight*. However, styles provide additional facilities, such as triggers, and you are not restricted to setting font-related properties. If you specify the text properties for a window and apply a style to controls in the window, the controls' style takes precedence over the window's text properties.

How a WPF Application Runs

A WPF application can contain any number of forms—you can add forms to an application by using the *Add Window* command on the *Project* menu in Visual Studio 2008. How does an application know which form to display when an application starts? If you recall from Chapter 1, this is the purpose of the App.xaml file. If you open the App.xaml file for the BellRingers project, you will see that it looks like this:

```
<Application x:Class="BellRingers.App"
    xmlns="http://schemas.microsoft.com/winfx/2006/xaml/presentation"
    xmlns:x="http://schemas.microsoft.com/winfx/2006/xaml"
    StartupUri="Window1.xaml">
    <Application.Resources>

    </Application.Resources>
</Application>
```

When you build a WPF application, the compiler converts this XAML definition to an *Application* object. The *Application* object controls the lifetime of the application and is responsible for creating the initial form that the application displays. You can think of the *Application* object as providing the *Main* method for the application. The key property is *StartupUri*, which specifies the XAML file for the window that the *Application* object should create. When you build the application, this property is converted to code that creates and opens the specified WPF form. If you want to display a different form, you simply need to change the value of the *StartupUri* property.

It is important to realize that the *StartupUri* property refers to the name of the XAML file and not the class implementing the window in this XAML file. If you rename the class from the default (*Window1*), the file name does not change (it is still Window1.xaml). Similarly, if you change the name of the file, the name of the window class defined in this file does not change. It can become confusing if the window class and XAML file have different names, so if you do want to rename things, be consistent and change both the file name and the window class name.

Adding Controls to the Form

So far, you have created a form, set some properties, added a few controls, and defined a style. To make the form useful, you need to add some more controls and write some code of your own. The WPF library contains a varied collection of controls. The purposes of some are fairly obvious—for example, *TextBox*, *ListBox*, *CheckBox*, and *ComboBox*—whereas other, more powerful, controls might not be so familiar.

Using WPF Controls

In the next exercise, you will add controls to the form that a user can use to input details about members of the bell ringers association. You will use a variety of controls, each suited to a particular type of data entry.

You will use *TextBox* controls for entering the first name and last name of the member. Each member belongs to a "tower" (where bells hang). The Middleshire district has several towers, but the list is static—new towers are not built very often, and hopefully, old towers do not to fall down with any great frequency either. The ideal control for handling this type of data is a *ComboBox*. The form also records whether the member is the tower "captain" (the person in charge of the tower who conducts the other ringers). A *CheckBox* is the best sort of control for this; it can be either selected (*True*) or cleared (*False*).

> **Tip** *CheckBox* controls can actually have three states if the *IsThreeState* property is set to *True*. The three states are *true*, *false*, and *null*. These states are useful if you are displaying information that has been retrieved from a relational database. Some columns in a table in a database allow *null* values, indicating that the value held is not defined or is unknown.

The application also gathers statistical information about when members joined the association and how much bell-ringing experience they have (up to 1 year, between 1 and 4 years, between 5 and 9 years, and 10 or more years). You can use a group of options, or radio buttons, to indicate the member's experience—radio buttons provide a mutually exclusive set of values. The older Microsoft Windows Forms library provides the *DateTimePicker* control for selecting and displaying dates, and this control is ideal for indicating the date that the member joined the association. There is one small snag, however: The WPF library does not provide an equivalent control. You can either implement your own custom control to provide this functionality or use Windows Forms interoperability and the *WindowsFormsHost* control to add the *DateTimePicker* control to a WPF form. You will adopt the latter approach in this application.

Finally, the application records the tunes the member can ring—rather confusingly, these tunes are referred to as "methods" by the bell-ringing fraternity. Although a bell ringer rings only one bell at a time, a group of bell ringers under the direction of the tower captain can ring their bells in different sequences and play simple music. There are a variety of bell-ringing methods, and they have rather quaint-sounding names such as Plain Bob, Reverse Canterbury, Grandsire, Stedman, Kent Treble Bob, and Old Oxford Delight. New methods are being written with alarming regularity, so the list of methods can vary over time. In a real-world application, you would store this list in a database. In this application, you will use a small selection of methods that you will hard-wire into the form. (You will see how to access and retrieve data from a database in Part V of this book, "Managing Data.") A good control for displaying this information and indicating whether a member can ring a method is a *ListBox* containing a list of *CheckBox* controls.

When the user has entered the member's details, the *Add* button will validate and store the data. The user can click *Clear* to reset the controls on the form and cancel any data entered.

Add controls to the form

1. Ensure that Window1.xaml is displayed in the *Design View* window. Remove the two button controls and the text box control from the form.

2. In the *XAML* pane, change the *Height* property of the form to **470** and the *Width* property to **600**, as shown in bold type here:

   ```
   <Window x:Class="BellRingers.Window1"
       ...
       Title="..." Height="470" Width="600">
       ...
   </Window>
   ```

3. In the *Design View* window, click the Window1 form. From the *Toolbox*, drag a *Label* control onto the form, and place it near the upper-left corner. Do not worry about positioning and sizing the label precisely because you will do this task for several controls later.

4. In the *XAML* pane, change the text for the label to **First Name**, as shown in bold type here:

   ```
   <Label ...>First Name</Label>
   ```

 Tip You can also change the text displayed by a label and many other controls by setting the *Content* property in the *Properties* window.

5. In the *Design View* window, click the Window1 form. From the *Toolbox*, drag a *TextBox* control onto the form to the right of the label.

 Tip You can use the guide lines displayed by the *Design View* window to help align controls. (The guide lines are displayed after you drop the control on the form.)

6. In the *XAML* pane, change the *Name* property of the text box to **firstName**, as shown here in bold type:

```
<TextBox ... Name="firstName" .../>
```

7. Add a second *Label* control to the form. Place it to the right of the *firstName* text box. In the *XAML* pane, change the text for the label to **Last Name**.

8. Add another *TextBox* control to the form, and position it to the right of the *Last Name* label. In the *XAML* pane, change the *Name* property of this text box to **lastName**.

9. Add a third *Label* control to the form, and place it directly under the *First Name* label. In the *XAML* pane, change the text for the label to **Tower**.

10. Add a *ComboBox* control to the form. Place it under the *firstName* text box and to the right of the *Tower* label. In the *XAML* pane, change the *Name* property of this combo box to **towerNames**.

11. Add a *CheckBox* control to the form. Place it under the *lastName* text box and to the right of the *towerNames* combo box. In the *XAML* pane, change the *Name* property of the check box to **isCaptain**, and change the text displayed by this check box to **Captain**.

12. Add a fourth *Label* to the form, and place it under the *Tower* label. In the *XAML* pane, change the text for this label to **Member Since**.

13. In *Solution Explorer*, right-click the References folder under the BellRingers project, and then click *Add Reference*. In the *Add Reference* dialog box, click the *.NET* tab, hold down the CTRL key while you select the *System.Windows.Forms* and *WindowsFormsIntegration* assemblies, and then click *OK*.

    In the following steps, you will add a *WindowsFormsHost* control to the form to hold a *DateTimePicker* control. These controls require the application to reference the *System. Windows.Forms* and *WindowsFormsIntegration* assemblies.

14. In the *XAML* pane, add the following XML namespace declaration shown in bold type to the *Window1* form. This declaration brings the types in the Windows Forms library into scope and establishes *wf* as an alias for this namespace:

```
<Window x:Class="BellRingers.Window1"
    xmlns="http://schemas.microsoft.com/winfx/2006/xaml/presentation"
    xmlns:x="http://schemas.microsoft.com/winfx/2006/xaml"
    xmlns:wf="clr-namespace:System.Windows.Forms;assembly=System.Windows.Forms"
    Title=...>
    ...
</Window>
```

15. On the *Build* menu, click *Build Solution*.

This step is necessary to enable the Visual Studio 2008 window to resolve the references to the *System.Windows.Forms* namespace correctly before you add Windows Forms controls to the application.

16. From the *Toolbox*, in the *Controls* section, add a *WindowsFormsHost* control to the form, and place it under the *towerNames* combo box.

> **Note** After you position the *WindowsFormsHost* control, it becomes invisible. Don't worry too much about placing this control because you will modify the control's properties in the XAML description in the next exercise. Additionally, the *XAML* pane might display a warning stating that the type *WindowsFormsHost* was not found. As long as you have added the references to the *System.Windows.Forms* and *WindowsFormsIntegration* assemblies to the project, you can ignore this warning.

17. In the *XAML* pane, change the *Name* property of the *WindowsFormsHost* control to *hostMemberSince*. Still in the *XAML* pane, add a Windows Forms *DateTimePicker* control as a child property to the *WindowsFormsHost* control, and name it **memberSince**. (As with other child properties, you must change the closing tag delimiter (/>) of the *WindowsFormsHost* control to an ordinary delimiter character (>) and add a closing </WindowsFormsHost> tag for the *WindowsFormsHost* control.) The completed XAML code for the *WindowsFormsHost* control should look like this:

```
<WindowsFormsHost ... Name="hostMemberSince" ...>
    <wf:DateTimePicker Name="memberSince"/>
</WindowsFormsHost>
```

The *DateTimePicker* control should appear on the form and display the current date. Depending on how your Visual Studio windows are arranged, you might need to scroll to see the new control.

18. Add a *GroupBox* control from the *Containers* section of the *Toolbox* to the form, and place it under the *Member Since* label. In the *XAML* pane, change the *Name* property of the group box to **yearsExperience**, and change the *Header* property to **Experience**. The *Header* property changes the label that appears on the form for the group box.

19. Add a *StackPanel* control to the form, and place it inside the *yearsExperience* group box. In the *XAML* pane, verify that the *StackPanel* control occurs inside the XAML code for the *GroupBox* control, like this:

```
<GroupBox Header="Experience" ... Name="yearsExperience" ...>
    <StackPanel ... Name="stackPanel1" ... />
</GroupBox>
```

20. Add a *RadioButton* control to the form, and place it inside the *StackPanel* control you just added. Add three more *RadioButton* controls to the *StackPanel* control. They should automatically be arranged vertically.

21. In the *XAML* pane, change the *Name* property of each radio button and the text it displays, as shown here in bold type:

```
<GroupBox...>
    <StackPanel ...>
        <RadioButton ... Name="novice" ...>Up to 1 year</RadioButton>
        <RadioButton ... Name="intermediate" ...>1 to 4 years</RadioButton>
        <RadioButton ... Name="experienced" ...>5 to 9 years</RadioButton>
        <RadioButton ... Name="accomplished" ...>10 or more years</RadioButton>
    </StackPanel>
</GroupBox>
```

22. Add a *ListBox* control to the form, and place it to the right of the *GroupBox* control. In the *XAML* pane, change the *Name* property of the list box to *methods*.

23. Add a *Button* control to the form, and place it near the bottom on the lower-left side of the form, underneath the *GroupBox* control. In the *XAML* pane, change the *Name* property of this button to *add*, and change the text displayed by this button to *Add*.

24. Add another *Button* control to the form, and place it near the bottom to the right of the *Add* button. In the *XAML* pane, change the *Name* property of this button to *clear*, and change the text displayed by this button to *Clear*.

You have now added all the required controls to the form. The next step is to tidy up the layout. The following table lists the layout properties and values you should assign to each of the controls. Using the *XAML* pane or the *Properties* window, make these changes. The margins and alignment of the controls are designed to keep the controls in place if the user resizes the form. Also notice that the margin values specified for the radio buttons are relative to each preceding item in the *StackPanel* control containing them; the first radio button is 10 units from the top of the *StackPanel* control, and the remaining radio buttons have a gap between them of 20 units vertically.

Control	Property	Value
label1	*Height*	23
	Margin	29, 25, 0, 0
	VerticalAlignment	Top
	HorizontalAlignment	Left
	Width	75
firstName	*Height*	21
	Margin	121, 25, 0, 0

Control	Property	Value
	VerticalAlignment	Top
	HorizontalAlignment	Left
	Width	175
label2	Height	23
	Margin	305, 25, 0, 0
	VerticalAlignment	Top
	HorizontalAlignment	Left
	Width	75
lastName	Height	21
	Margin	380, 25, 0, 0
	VerticalAlignment	Top
	HorizontalAlignment	Left
	Width	175
label3	Height	23
	Margin	29, 72, 0, 0
	VerticalAlignment	Top
	HorizontalAlignment	Left
	Width	75
towerNames	Height	21
	Margin	121, 72, 0, 0
	VerticalAlignment	Top
	HorizontalAlignment	Left
	Width	275
isCaptain	Height	21
	Margin	420, 72, 0, 0
	VerticalAlignment	Top
	HorizontalAlignment	Left
	Width	75
Label4	Height	23
	Margin	29, 134, 0, 0
	VerticalAlignment	Top

Control	Property	Value
	HorizontalAlignment	Left
	Width	90
hostMemberSince	Height	23
	Margin	121, 134, 0, 0
	VerticalAlignment	Top
	HorizontalAlignment	Left
	Width	275
yearsExperience	Height	200
	Margin	29, 174, 0, 0
	VerticalAlignment	Top
	HorizontalAlignment	Left
	Width	258
stackPanel1	Height	151
	Width	224
Novice	Height	16
	Margin	0, 10, 0, 0
	Width	120
Intermediate	Height	16
	Margin	0, 20, 0, 0
	Width	120
Experienced	Height	16
	Margin	0, 20, 0, 0
	Width	120
Accomplished	Height	16
	Margin	0, 20, 0, 0
	Width	120
Methods	Height	200
	Margin	310, 174, 0, 0
	VerticalAlignment	Top
	HorizontalAlignment	Left
	Width	245

Control	Property	Value
Add	*Height*	23
	Margin	188, 388, 0, 0
	VerticalAlignment	Top
	HorizontalAlignment	Left
	Width	75
Clear	*Height*	23
	Margin	313, 388, 0, 0
	VerticalAlignment	Top
	HorizontalAlignment	Left
	Width	75

As a finishing touch, you will next apply a style to the controls. You can use the *bellRingers-Style* style for controls such as the buttons and text boxes, but the labels, combo box, group box, and radio buttons should probably not be displayed on a gray background.

Apply styles to the controls, and test the form

1. In the *XAML* pane, add the *bellRingersFontStyle* shown in bold type in the following code to the *<Windows.Resources>* element. Leave the existing *bellRingersStyle* style in place. Notice that this style changes the font only of controls that reference this style.

```
<Window.Resources>
    <Style x:Key="bellRingersFontStyle" TargetType="Control">
        <Setter Property="FontFamily" Value="Comic Sans MS"/>
    </Style>
    <Style x:Key="bellRingersStyle" TargetType="Control">
        ...
    </Style>
</Window.Resources>
```

2. In the *XAML* pane, apply the *bellRingersFontStyle* style to the *label1* control, as shown in bold type here:

```
<Label Style="{StaticResource bellRingersFontStyle}" ...>First Name</Label>
```

Apply the same style to the following controls:

- label2
- label3
- isCaptain
- towerNames

- label4

- yearsExperience

- methods

> **Note** Applying the style to the *yearsExperience* group box and the *methods* list box automatically causes the style to be used by the items displayed in these controls.

3. Apply the *bellRingersStyle* style to the following controls:

- firstName

- lastName

- add

- clear

4. On the *Debug* menu, click *Start Without Debugging*.

The form when it runs should look like the following image:

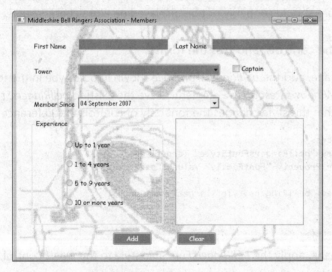

Notice that the *Methods* list box is currently empty. You will add code to populate it in a later exercise.

5. Click the drop-down arrow in the *Tower* combo box. The list of towers is currently empty. Again, you will write code to fill this combo box in a later exercise.

6. Close the form, and return to Visual Studio 2008.

Changing Properties Dynamically

You have been using the *Design View* window, the *Properties* window, and the *XAML* pane to set properties statically. When the form runs, it would be useful to reset the value of each control to an initial default value. To do this, you will need to write some code (at last). In the following exercises, you will create a *private* method called *Reset*. Later, you will invoke the *Reset* method when the form first starts as well as when the user clicks the *Clear* button.

Create the *Reset* method

1. In the *Design View* window, right-click the form, and then click *View Code*. The *Code and Text Editor* window opens and displays the Window1.xaml.cs file so that you can add C# code to the form.

2. Add the following *Reset* method, shown in bold type, to the *Window1* class:

```
public partial class Window1 : Window
{
    ...
    public void Reset()
    {
        firstName.Text = String.Empty;
        lastName.Text = String.Empty;
    }
}
```

The two statements in this method ensure that the *firstName* and *lastName* text boxes are blank by assigning an empty string to their *Text* property.

You also need to initialize the properties for the remaining controls on the form and populate the *towerNames* combo box and the *methods* list box.

If you recall, the *towerName* combo box will contain a list of all the bell towers in the Middleshire district. This information would usually be held in a database, and you would write code to retrieve the list of towers and populate the *ComboBox*. For this example, the application will use a hard-coded collection. A *ComboBox* has a property called *Items* that contains a list of the data to be displayed.

3. Add the following string array called *towers*, shown in bold type, which contains a hard-coded list of tower names, to the *Window1* class:

```
public partial class Window1 : Window
{
    private string[] towers = { "Great Shevington", "Little Mudford",
                                "Upper Gumtree", "Downley Hatch" };
    ...
}
```

4. In the *Reset* method, after the code you have already written, add the following statements shown in bold type to clear the *towerNames* combo box (this is important because otherwise you could end up with many duplicate values in the list) and add the towers found in the *towers* array. The statement after the *foreach* loop causes the first tower to be displayed as the default value:

```
public void Reset()
{
    ...
    towerNames.Items.Clear();
    foreach (string towerName in towers)
    {
        towerNames.Items.Add(towerName);
    }
    towerNames.Text = towerNames.Items[0] as string;
}
```

 Note You can also specify hard-coded values at design time in the XAML description of a combo box, like this:

```
<ComboBox Text="towerNames">
    <ComboBox.Items>
        <ComboBoxItem>
            Great Shevington
        </ComboBoxItem>
        <ComboBoxItem>
            Little Mudford
        </ComboBoxItem>
        <ComboBoxItem>
            Upper Gumtree
        </ComboBoxItem>
        <ComboBoxItem>
            Downley Hatch
        </ComboBoxItem>
    </ComboBox.Items>
</ComboBox>
```

5. You must populate the *methods* list box with a list of bell-ringing methods. Like a combo box, a list box has a property called *Items* that contains a collection of values to be displayed. Also, like the *ComboBox*, it could be populated from a database. However, as before, you will simply supply some hard-coded values for this example. Add the

following string array shown in bold type, which contains the list of methods, to the *Window1* class:

```
public partial class Window1 : Window
{
    ...
    private string[] ringingMethods = { "Plain Bob", "Reverse Canterbury",
        "Grandsire", "Stedman", "Kent Treble Bob", "Old Oxford Delight",
        "Winchendon Place", "Norwich Surprise", "Crayford Little Court" };
    ...
}
```

6. The *methods* list box should display a list of check boxes rather than ordinary text strings. With the flexibility of the WPF model, you can specify a variety of different types of content for controls such as list boxes and combo boxes. Add the following code shown in bold type to the *Reset* method to fill the *methods* list box with the methods in the *ringingMethods* array. Notice that this time each item is a check box. You can specify the text displayed by the check box by setting its *Content* property, and you can specify the spacing between items in the list by setting the *Margin* property; this code inserts a spacing of 10 units after each item:

```
public void Reset()
{
    ...
    methods.Items.Clear();
    CheckBox method;
    foreach (string methodName in ringingMethods)
    {
        method = new CheckBox();
        method.Margin = new Thickness(0, 0, 0, 10);
        method.Content = methodName;
        methods.Items.Add(method);
    }
}
```

 Note Most WPF controls have a *Content* property that you can use to set and read the value displayed by that control. This property is actually an *object*, so you can set it to almost any type, as long as it makes sense to display it!

7. The *isCaptain* check box should default to *false*. To do this, you need to set the *IsChecked* property. Add the following statement shown in bold type to the *Reset* method:

```
public void Reset()
{
    ...
    isCaptain.IsChecked = false;
}
```

8. The form contains four radio buttons that indicate the number of years of bell-ringing experience the member has. A radio button is similar to a *CheckBox* in that it can contain a *true* or *false* value. However, the power of radio buttons increases when you put them together in a *GroupBox*. In this case, the radio buttons form a mutually exclusive collection—at most, only one radio button in a group can be selected (set to *true*), and all the others will automatically be cleared (set to *false*). By default, none of the buttons will be selected. You should rectify this by setting the *IsChecked* property of the *novice* radio button. Add the following statement shown in bold type to the *Reset* method:

```
public void Reset()
{
    ...
    novice.IsChecked = true;
}
```

9. You should ensure that the *Member Since DateTimePicker* control defaults to the current date. You can do this by setting the *Value* property of the control. You can obtain the current date from the static *Today* method of the *DateTime* class.

Add the following code shown in bold type to the *Reset* method to initialize the *DateTimePicker* control.

```
public void Reset()
{
    ...
    System.Windows.Forms.DateTimePicker memberDate =
        hostMemberSince.Child as System.Windows.Forms.DateTimePicker;
    memberDate.Value = DateTime.Today;
}
```

Notice that to access an object in a *WindowsFormsHost* container, you reference the *Child* property of the container and then cast it to the appropriate type. Additionally, notice that the *DateTimePicker* class is defined in the *System.Windows.Forms* namespace. Typically, you add a *using* statement for the file defining a class to bring the namespace of the class into scope, but you should not do this when integrating Windows Forms controls into a WPF application. The reason is that the *System.Windows.Forms* namespace contains many controls that use the same names as those in the WPF library, so adding a *using* statement would make all references to these controls ambiguous!

10. Finally, you need to arrange for the *Reset* method to be called when the form is first displayed. A good place to do this is in the *Window1* constructor. Insert a call to the *Reset* method after the statement that calls the *InitializeComponent* method, as shown in bold type here:

```
public Window1()
{
    InitializeComponent();
    this.Reset();
}
```

11. On the *Debug* menu, click *Start Without Debugging* to verify that the project builds and runs.

12. When the form opens, click the *Tower* combo box.

You will see the list of bell towers, and you can select one of them.

13. Click the drop-down arrow on the right side of the *Member Since* date/time picker.

You will be presented with a calendar of dates. The default value will be the current date. You can click a date and use the arrows to select a month. You can also click the month name to display the months as a drop-down list, and click the year so that you can select a year by using a numeric up-down control.

14. Click each of the radio buttons in the *Experience* group box.

Notice that you cannot select more than one radio button at a time.

15. In the *Methods* list box, click some of the methods to select the corresponding check box. If you click a method a second time, it clears the corresponding check box, just as you would expect.

16. Click the *Add* and *Clear* buttons.

Currently these buttons don't do anything. You will add this functionality in the final set of exercises in this chapter.

17. Close the form, and return to Visual Studio 2008.

Handling Events in a WPF Form

If you are familiar with Microsoft Visual Basic, Microsoft Foundation Classes (MFC), or any of the other tools available for building GUI applications for Windows, you are aware that Windows uses an event-driven model to determine when to execute code. In Chapter 17, "Interrupting Program Flow and Handling Events," you saw how to publish your own events and subscribe to them. WPF forms and controls have their own predefined events that you can subscribe to, and these events should be sufficient to handle the requirements of most user interfaces.

Processing Events in Windows Forms

The developer's task is to capture the events that are relevant to the application and write the code that responds to these events. A familiar example is the *Button* control, which raises a "Somebody clicked me" event when a user clicks it with the mouse or presses Enter when the button has the focus. If you want the button to do something, you write code that responds to this event. This is what you will do in the next exercise.

Handle the *Click* event for the *Clear* button

1. Display the Window1.xaml file in the *Design View* window. Double-click the *Clear* button on the form.

   The *Code and Text Editor* window appears and creates a method called *clear_Click*. This is an event method that will be invoked when the user clicks the *Clear* button. Notice that the event method takes two parameters: the *sender* parameter (an *object*) and an additional arguments parameter (a *RoutedEventArgs* object). The WPF runtime will populate these parameters with information about the source of the event and with any additional information that might be useful when handling the event. You will not use these parameters in this exercise.

   WPF controls can raise a variety of events. When you double-click a control or a form in the *Design View* window, Visual Studio generates the stub of an event method for the default event for the control; for a button, the default event is the *Click* event. (If you double-click a text box control, Visual Studio generates the stub of an event method for handling the *TextChanged* event.)

2. When the user clicks the *Clear* button, you want the form to be reset to its default values. In the body of the *clear_Click* method, call the *Reset* method, as shown here in bold type:

```
private void clear_Click(object sender, RoutedEventArgs e)
{
    this.Reset();
}
```

   Users will click the *Add* button when they have filled in all the data for a member and want to store the information. The *Click* event for the *Add* button should validate the information entered to ensure that it makes sense (for example, should you allow a tower captain to have less than one year of experience?) and, if it is okay, arrange for the data to be sent to a database or other persistent store. You will learn more about validation and storing data in later chapters. For now, the code for the *Click* event of the *Add* button will simply display a message box echoing the data input.

3. Return to the *Design View* window displaying the Window1.xaml form. In the *XAML* pane, locate the element that defines the *Add* button, and begin entering the following code shown in bold type:

```
<Button ... Click="">Add</Button>
```

   Notice that as you type the opening quotation mark after the text Click=, a shortcut menu appears, displaying two items: *<New Event Handler>* and *clear_Click*. If two buttons perform a common action, you can share the same event handler method between them, such as *clear_Click*. If you want to generate an entirely new event handling method, you can select the *<New Event Handler>* command instead.

4. On the shortcut menu, double-click the *<New Event Handler>* command.

The text *add_Click* appears in the XAML code for the button.

> **Note** You are not restricted to handling the *Click* event for a button. When you edit the XAML code for a control, the IntelliSense list displays the properties and events for the control. To handle an event other than the *Click* event, simply type the name of the event, and then select or type the name of the method that you want to handle this event. For a complete list of events supported by each control, see the Visual Studio 2008 documentation.

5. Switch to the *Code and Text Editor* window displaying the Window1.xaml.cs file.

Notice that the *add_Click* method has been added to the *Window1* class.

> **Tip** You don't have to use the default names generated by Visual Studio 2008 for the event handler methods. Rather than clicking the *<New Event Handler>* command on the shortcut menu, you can just type the name of a method. However, you must then manually add the method to the window class. This method must have the correct signature; it should return a *void* and take two arguments—an *object* parameter and a *RoutedEventArgs* parameter.

> **Important** If you later decide to remove an event method such as *add_Click* from the Window1.xaml.cs file, you must also edit the XAML definition of the corresponding control and remove the `Click="add_Click"` reference to the event; otherwise, your application will not compile.

6. Add the following code shown in bold type to the *add_Click* method:

```
private void add_Click(object sender, RoutedEventArgs e)
{
    string nameAndTower = String.Format(
        "Member name: {0} {1} from the tower at {2} rings the following methods:",
        firstName.Text, lastName.Text, towerNames.Text);
    StringBuilder details = new StringBuilder();
    details.AppendLine(nameAndTower);
    foreach (CheckBox cb in methods.Items)
    {
        if (cb.IsChecked.Value)
        {
            details.AppendLine(cb.Content.ToString());
        }
    }
    MessageBox.Show(details.ToString(), "Member Information");
}
```

This block of code creates a *string* variable called *nameAndTower* that it fills with the name of the member and the tower to which the member belongs.

Notice how the code accesses the *Text* property of the text box and combo box controls to read the current values of those controls. Additionally, the code uses the static *String.Format* method to format the result. The *String.Format* method operates in a similar manner to the *Console.WriteLine* method, except that it returns the formatted string as its result rather than displaying it on the screen.

The code then creates a *StringBuilder* object called *details*. The method uses this *StringBuilder* object to build a string representation of the information it will display. The text in the *nameAndTower* string is used to initially populate the *details* object. The code then iterates through the *Items* collection in the *methods* list box. If you recall, this list box contains check box controls. Each check box is examined in turn, and if the user has selected it, the text in the *Content* property of the check box is appended to the *details StringBuilder* object.

> **Note** You could use ordinary string concatenation instead of a *StringBuilder* object, but the *StringBuilder* class is far more efficient and is the recommended approach for performing the kind of tasks required in this code. In the .NET Framework and C#, the *string* data type is immutable; when you modify the value in a string, the runtime actually creates a new string containing the modified value and then discards the old string. Repeatedly modifying a string can cause your code to become inefficient because a new string must be created in memory at each change (the old strings will eventually be garbage collected). The *StringBuilder* class, in the *System.Text* namespace, is designed to avoid this inefficiency. You can add and remove characters from a *StringBuilder* object using the *Append*, *Insert*, and *Remove* methods without creating a new object each time.

Finally, the *MessageBox* class provides static methods for displaying dialog boxes on the screen. The *Show* method used here displays the contents of the *details* string in the body of the message box and will put the text "Member Information" in the title bar. *Show* is an overloaded method, and there are other variants that you can use to specify icons and buttons to display in the message box.

7. On the *Debug* menu, click *Start Without Debugging* to build and run the application.

8. Type some sample data for the member's first name and last name, select a tower, and pick a few methods. Click the *Add* button, and verify that the *Member Information* message box appears, displaying the details of the new member and the methods he or she can ring.

9. Click the *Clear* button, and verify that the controls on the form are reset to the correct default values.

10. Close the form, and return to Visual Studio 2008.

In the final exercise in this chapter, you will add an event handler to handle the *Closing* event for the window so that users can confirm that they really want to quit the application. The *Closing* event is raised when the user attempts to close the form but before the form actually closes. You can use this event to prompt the user to save any unsaved data or even ask the user whether he or she really wants to close the form—if not, you can cancel the event in the event handler and prevent the form from closing.

Handle the *Closing* event for the form

1. In the *Design View* window, in the *XAML* pane, begin entering the code shown in bold type to the XAML description of the *Window1* window:

   ```
   <Window x:Class="BellRingers.Window1"
       ...
       Title="..." ... Closing="">
   ```

2. When the shortcut menu appears after you type the opening quotation mark, double-click the *<New Event Handler>* command.

   Visual Studio generates an event method called *Window_Closing* and associates it with the *Closing* event for the form, like this:

   ```
   <Window x:Class="BellRingers.Window1"
       ...
       Title="..." ... Closing="Window_Closing">
   ```

3. Switch to the *Code and Text Editor* window displaying the Window1.xaml.cs file.

   A stub for the *Window_Closing* event method has been added to the *Window1* class:

   ```
   private void Window_Closing(object sender, System.ComponentModel.CancelEventArgs e)
   {

   }
   ```

   Observe that the second parameter for this method has the type *CancelEventArgs*. The *CancelEventArgs* class has a Boolean property called *Cancel*. If you set *Cancel* to *true* in the event handler, the form will not close. If you set *Cancel* to *false* (the default value), the form will close when the event handler finishes.

4. Add the following statements shown in bold type to the *memberFormClosing* method:

   ```
   private void Window_Closing(object sender, System.ComponentModel.CancelEventArgs e)
   {
       MessageBoxResult key = MessageBox.Show(
           "Are you sure you want to quit",
           "Confirm",
           MessageBoxButton.YesNo,
           MessageBoxImage.Question,
           MessageBoxResult.No);
       e.Cancel = (key == MessageBoxResult.No);
   }
   ```

These statements display a message box asking the user to confirm whether to quit the application. The message box will contain *Yes* and *No* buttons and a question mark icon. The final parameter, *MessageBoxResult.No*, indicates the default button if the user simply presses the Enter key—it is safer to assume that the user does not want to exit the application than to risk accidentally losing the details that the user has just typed. When the user clicks either button, the message box will close and the button clicked will be returned as the value of the method (as a *MessageBoxResult*—an enumeration identifying which button was clicked). If the user clicks *No*, the second statement will set the *Cancel* property of the *CancelEventArgs* parameter (*e*) to *true*, preventing the form from closing.

5. On the *Debug* menu, click *Start Without Debugging* to run the application.

6. Try to close the form. In the message box that appears, click *No*.

   The form should continue running.

7. Try to close the form again. This time, in the message box, click *Yes*.

   The form closes, and the application finishes.

You have now seen how to use the essential features of WPF to build a functional user interface. WPF contains many more features than we have space to go into here, especially concerning some of its really cool capabilities for handling two-dimensional and three-dimensional graphics and animation. If you want to learn more about WPF, you can consult a book such as *Applications = Code + Markup: A Guide to the Microsoft Windows Presentation Foundation*, by Charles Petzold (Microsoft Press, 2006).

- If you want to continue to the next chapter

  Keep Visual Studio 2008 running, and turn to Chapter 23.

- If you want to exit Visual Studio 2008 now

  On the *File* menu, click *Exit*. If you see a *Save* dialog box, click *Yes* (if you are using Visual Studio 2008) or *Save* (if you are using Visual C# 2008 Express Edition) and save the project.

Chapter 22 Quick Reference

To	Do this
Create a WPF application	Use the WPF Application template.
Add controls to a form	Drag the control from the *Toolbox* onto the form.
Change the properties of a form or control	Click the form or control in the *Design View* window. Then do one of the following: ■ In the *Properties* window, select the property you want to change and enter the new value. ■ In the *XAML* pane, specify the property and value in the *<Window>* element or the element defining the control.
View the code behind a form	Do one of the following: ■ On the *View* menu, click *Code*. ■ Right-click in the *Design View* window, and then click *View Code*. ■ In *Solution Explorer*, expand the folder corresponding to the .xaml file for the form, and then double-click the .xaml.cs file that appears.
Define a set of mutually exclusive radio buttons.	Add a panel control, such as *StackPanel*, to the form. Add the radio buttons to the panel. All radio buttons in the same panel are mutually exclusive.

Populate a combo box or a list box by using C# code	Use the *Add* method of the *Items* property. For example:
	
	`towerNames.Items.Add("Upper Gumtree");`
	
	You might need to clear the *Items* property first, depending on whether you want to retain the existing contents of the list. For example:
	
	`towerNames.Items.Clear();`

Initialize a check box or radio button control	Set the *IsChecked* property to *true* or *false*. For example: `novice.IsChecked = true;`
Handle an event for a control or form	In the *XAML* pane, add code to specify the event, and then either select an existing method that has the appropriate signature or click the *<Add New Event>* command on the shortcut menu that appears, and then write the code that handles the event in the event method that is created.

Chapter 23
Working with Menus and Dialog Boxes

After completing this chapter, you will be able to:

- Create menus for Microsoft Windows Presentation Foundation (WPF) applications by using the Menu and MenuItem classes.

- Perform processing in response to menu events when a user clicks a menu command.

- Create context-sensitive pop-up menus by using the ContextMenu class.

- Manipulate menus through code and create dynamic menus.

- Use Windows common dialog boxes in an application to prompt the user for the name of a file.

In Chapter 22, "Introducing Windows Presentation Foundation," you saw how to create a simple WPF application made up of a selection of controls and events. Many professional Microsoft Windows–based applications also provide menus containing commands and options, giving the user the ability to perform various tasks related to the application. In this chapter, you will learn how to create menus and add them to forms by using the *Menu* control. You will see how to respond when the user clicks a command on a menu. You'll learn how to create pop-up menus whose contents vary according to the current context. Finally, you will find out about the common dialog classes supplied as part of the WPF library. With these dialog classes, you can prompt the user for frequently used items, such as files and printers, in a quick, easy, and familiar manner.

Menu Guidelines and Style

If you look at most Windows-based applications, you'll notice that some items on the menu bar tend to appear repeatedly in the same place, and the contents of these items are often predictable. For example, the *File* menu is typically the first item on the menu strip, and on this menu you typically find commands for creating a new document, opening an existing document, saving the document, printing the document, and exiting the application.

 Note The term *document* means the data that the application manipulates. In Microsoft Office Excel, it would be a worksheet; in the Bell Ringers application that you created in Chapter 22, it could be the details of a new member.

451

The order in which these commands appear tends to be the same across applications; for example, the *Exit* command is invariably the last command on the *File* menu. There might be other application-specific commands on the *File* menu as well.

An application often has an *Edit* menu containing commands such as *Cut*, *Paste*, *Clear*, and *Find*. There are usually some additional application-specific menus on the menu bar, but again, convention dictates that the final menu is the *Help* menu, which contains access to Help as well as "about" information, which contains copyright and licensing details for the application. In a well-designed application, most menus are predictable and help ensure that the application is easy to learn and use.

> **Tip** Microsoft publishes a full set of guidelines for building intuitive user interfaces, including menu design, on the Microsoft Web site at *http://msdn2.microsoft.com/en-us/library /Aa286531.aspx.*

Menus and Menu Events

WPF provides the *Menu* control as a container for menu items. The *Menu* control provides a basic shell for defining a menu. Like most aspects of WPF, the *Menu* control is very flexible so that you can define a menu structure consisting of almost any type of WPF control. You are probably familiar with menus that contain text items that you can click to perform a command. WPF menus can also contain buttons, text boxes, combo boxes, and so on. You can define menus by using the *XAML* pane in the *Design View* window, and you can also construct menus at run time by using Microsoft Visual C# code. Laying out a menu is only half of the story. When a user clicks a command on a menu, the user expects something to happen! Your application acts on the commands by trapping menu events and executing code in much the same way as handling control events.

Creating a Menu

In the following exercise, you will use the *XAML* pane to create menus for the Middleshire Bell Ringers Association application. You will learn how to manipulate and create menus through code later in this chapter.

Create the application menu

1. Start Microsoft Visual Studio 2008 if it is not already running.

2. Open the BellRingers solution located in the \Microsoft Press\Visual CSharp Step by Step\Chapter 23\BellRingers folder in your Documents folder. This is a copy of the application that you built in Chapter 22.

3. Display Window1.xaml in the *Design View* window. (Double-click *Window1.xaml* in *Solution Explorer*.)

4. From the *Toolbox*, drag a *DockPanel* control from the *Controls* section anywhere onto the form. In the *Properties* window, set the *Width* property of the *DockPanel* to **Auto**, set the *HorizontalAlignment* property to *Stretch*, set the *VerticalAlignment* property to **Top**, and set the *Margin* property to **0**.

> **Note** Setting the *Margin* property to 0 is the same as setting it to 0, 0, 0, 0.

The *DockPanel* control should appear at the top of the form, occupying the full width of the form. (It will cover the *First Name*, *Last Name*, *Tower*, and *Captain* user interface elements.)

The *DockPanel* control is a panel control that you can use for controlling the arrangement of other controls that you place on it, like the *Grid* and *StackPanel* controls that you met in Chapter 22. You can add a menu directly to a form, but it is better practice to place it on a *DockPanel* because you can then more easily manipulate the menu and its positioning on the form. For example, if you want to place the menu at the bottom or on one side, you can relocate the entire menu elsewhere on the form simply by moving the panel either at design time or at run time by executing code.

5. From the *Toolbox*, drag a *Menu* control from the *Controls* section onto the *DockPanel* control. In the *Properties* window, set the *DockPanel.Dock* property to **Top**, set the *Width* property to **Auto**, set the *HorizontalAlignment* property to **Stretch**, and set the *VerticalAlignment* property to **Top**.

The *Menu* control appears as a gray bar across the top of the *DockPanel*. If you examine the code for the *DockPanel* and *Menu* controls in the *XAML* pane, they should look like this:

```
<DockPanel Height="100" HorizontalAlignment="Stretch" Margin="0"
    Name="dockPanel1" VerticalAlignment="Top" Width="Auto">
    <Menu Height="22" Name="menu1" Width="Auto" DockPanel.Dock="Top"
        VerticalAlignment="Top">
</DockPanel>
```

The *HorizontalAlignment* property does not appear in the XAML code because the value "Stretch" is the default value for this property.

> **Note** Throughout this chapter, lines from the *XAML* pane are shown split and indented so that they fit on the printed page.

6. In the *XAML* pane, modify the definition of the *Menu* control and add the *MenuItem* elements as shown in bold type in the following code. Notice that *MenuItem* elements appear as children of the *Menu* control, so replace the closing tag delimiter (/>) of the *Menu* element with a regular tag delimiter (>), and place a separate closing </Menu> element at the end.

```
<Menu Height="22" Name="menu1" Width="Auto" DockPanel.Dock="Top"
    VerticalAlignment="Top" HorizontalAlignment="Stretch" >
    <MenuItem Header="_File" />
    <MenuItem Header="_Help" />
</Menu>
```

The *Header* attribute of the *MenuItem* element specifies the text that appears for the menu item. The underscore (_) in front of a letter provides fast access to that menu item when the user presses the Alt key and the letter following the underscore (in this case, Alt+F for File or Alt+H for Help). This is another common convention. At run time, when the user presses the Alt key, the F at the start of File appears underscored. Do not use the same access key more than once on any menu because you will confuse the user (and probably the application).

> **Note** The *Properties* window for the *Menu* control displays a property called *Items*. If you click this property and then click the ellipsis button that appears in this property, the *Collection Editor* appears. At the time of writing, the current release of Visual Studio 2008 (Beta 2) allows you to use this window to remove items from a menu, change the order of items on a menu, and set the properties of these items, but it does not allow you to add new items to a menu. Consequently, in this chapter you will use the *XAML* pane to define the structure of your menus.

7. On the *Debug* menu, click *Start Without Debugging* to build and run the application.

When the form appears, you should see the menu at the top of the window underneath the title bar. Press the Alt key; the menu should get the focus, and the "F" in "File" and the "H" in "Help" should both be underscored, like this:

If you click either menu item, nothing currently happens because you have not defined the child menus that each of these items will contain.

8. Close the form and return to Visual Studio 2008.

9. In the *XAML* pane, modify the definition of the *_File* menu item, and add the child menu items together with a closing </MenuItem> element as shown here in bold type:

```
<MenuItem Header="_File" >
    <MenuItem Header="_New Member" Name="newMember" />
    <MenuItem Header="_Save Member Details" Name="saveMember" />
    <Separator/>
    <MenuItem Header="E_xit" Name="exit" />
</MenuItem>
```

This XAML code adds *New Member, Save Member Details,* and *Exit* as commands to the *File* menu. The <Separator/> element appears as a bar when the menu is displayed and is conventionally used to group related menu items.

10. Modify the definition of the *_Help* menu item, and add the child menu item as shown in bold type here:

```
<MenuItem Header="_Help" >
    <MenuItem Header="_About Middleshire Bell Ringers" Name="about" />
</MenuItem>
```

11. On the *Debug* menu, click *Start Without Debugging* to build and run the application.

When the form appears, click the *File* menu. You should see the child menu items, like this:

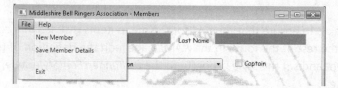

You can also click the *Help* menu to display the *About Middleshire Bell Ringers* child menu item.

12. Close the form, and return to Visual Studio 2008.

As a further touch, you can add icons to menu items. Many applications, including Visual Studio 2008, make use of icons in menus to provide an additional visual cue.

13. In *Solution Explorer*, right-click the BellRingers project, point to *Add*, and then click *Existing Item*. In the *Add Existing Item – BellRingers* dialog box, move to the folder Microsoft Press\Visual CSharp Step By Step\Chapter 23 under your Documents folder, in the *File name* box type **"Ring.bmp" "Face.bmp" "Note.bmp"** (including the quotation marks), and then click *Add*.

This action adds the three image files as resources to your application.

14. In the *XAML* pane, modify the definitions of the *newMember*, *saveMember*, and *about* menu items and add *MenuItem.Icon* child elements that refer to each of the three icon files you added to the project in the preceding step, as shown in bold type here:

```
<Menu Height="22" Name="menu1" ... >
    <MenuItem Header="_File" >
        <MenuItem Header="_New Member" Name="newMember" >
            <MenuItem.Icon>
                <Image Source="face.bmp"/>
            </MenuItem.Icon>
        </MenuItem>
        <MenuItem Header="_Save Member Details" Name="saveMember" >
            <MenuItem.Icon>
                <Image Source="note.bmp"/>
            </MenuItem.Icon>
        </MenuItem>
        <Separator/>
        <MenuItem Header="E_xit" Name="exit"/>
    </MenuItem>
    <MenuItem Header="_Help">
        <MenuItem Header="_About Middleshire Bell Ringers" Name="about" >
            <MenuItem.Icon>
                <Image Source="ring.bmp"/>
            </MenuItem.Icon>
        </MenuItem>
    </MenuItem>
</Menu>
```

15. The final tweak is to ensure that the text for the menu items is styled in a consistent manner with the rest of the form. In the *XAML* pane, edit the definition of the top-level *menu1* element and set the *Style* property to the *BellRingersFontStyle* style, as shown in bold type here:

```
<Menu Style="{StaticResource bellRingersFontStyle}" ... Name="menu1" ... >
```

Note that the child menu items automatically inherit the style from the top-level menu item that contains them.

16. On the *Debug* menu, click *Start Without Debugging* to build and run the application again.

When the form appears, click the *File* menu. You should now see that the text of the menu items is displayed in the correct font and that the icons appear with the child menu items, like this:

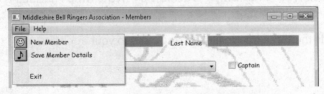

17. Close the form, and return to Visual Studio 2008.

Types of Menu Items

You have been using the *MenuItem* element to add child menu items to a *Menu* control. You have seen that you can specify the items in the top level menu as *MenuItem* elements and then add nested *MenuItem* elements to define your menu structure. The nested *MenuItem* elements can themselves contain further nested *MenuItem* elements if you want to create cascading menus. In theory, you can continue this process to a very deep level, but in practice you should probably not go beyond two levels of nesting.

However, you are not restricted to using the *MenuItem* element. You can also add combo boxes, text boxes, and most other types of controls to WPF menus. For example, the following menu structure contains a button and a combo box:

```
<Menu ...>
    <MenuItem Header="Miscellaneous">
        <Button>Add new member</Button>
        <ComboBox Text="Towers">
            <ComboBox.Items>
                <ComboBoxItem>
                    Great Shevington
                </ComboBoxItem>
                <ComboBoxItem>
                    Little Mudford
                </ComboBoxItem>
                <ComboBoxItem>
                    Upper Gumtree
                </ComboBoxItem>
                <ComboBoxItem>
                    Downley Hatch
                </ComboBoxItem>
            </ComboBox.Items>
        </ComboBox>
    </MenuItem>
</Menu>
```

At run time, the menu structure looks like this:

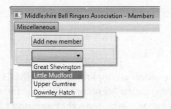

Although you have great freedom when designing your menus, you should endeavor to keep things simple and not be too elaborate. A menu such as this is not very intuitive!

Handling Menu Events

The menu that you have built so far looks very pretty, but none of the items do anything when you click them. To make them functional, you have to write code to handle the various menu events. Several different events can occur when a user selects a menu item. Some are more useful than others are. The most frequently used event is the *Click* event, which occurs when the user clicks the menu item. You typically trap this event to perform the tasks associated with the menu item.

In the following exercise, you will learn more about menu events and how to process them. You will create *Click* events for the *newMember* and *exit* menu items.

The purpose of the *New Member* command is so that the user can enter the details of a new member. Therefore, until the user clicks *New Member*, all fields on the form should be disabled, as should the *Save Member Details* command. When the user clicks the *New Member* command, you want to enable all the fields, reset the contents of the form so that the user can start adding information about a new member, and enable the *Save Member Details* command.

Handle the *New Member* and *Exit* menu item events

1. In the *XAML* pane, click the definition of the *firstName* text box. In the *Properties* window, clear the *IsEnabled* property. (This action sets *IsEnabled* to *False* in the XAML definition.)

   Repeat this process for the *lastName*, *towerNames*, *isCaptain*, *hostMemberSince*, *yearsExperience*, *methods*, *add*, and *clear* controls and for the *saveMember* menu item.

2. In the *Design View* window, in the *XAML* pane, begin entering the code shown here in bold type in the XAML description of the *_New Member* menu item:

   ```
   <MenuItem Header="_New Member" Click="">
   ```

3. When the shortcut menu appears after you type the opening quotation mark, double-click the *<New Event Handler>* command.

   Visual Studio generates an event method called *newMember_Click* and associates it with the *Click* event for the menu item.

> **Tip** Always give a menu item a meaningful name when you define event methods for it. If you don't, Visual Studio generates an event method called *MenuItem_Click* for the *Click* event. If you then create *Click* event methods for other menu items that also don't have names, they are called *MenuItem_Click_1*, *MenuItem_Click_2*, and so on. If you have several of these event methods, it can be difficult to work out which event method belongs to which menu item.

4. Switch to the *Code and Text Editor* window displaying the Window1.xaml.cs file. (On the *View* menu, click *Code*.)

The *newMember_Click* event method will have been added to the bottom of the *Window1* class definition:

```
private void newMember_Click(object sender, RoutedEventArgs e)
{

}
```

5. Add the following statements shown in bold type to the *memberFormClosing* method:

```
private void newMember_Click(object sender, RoutedEventArgs e)
{
    this.Reset();
    saveMember.IsEnabled = true;
    firstName.IsEnabled = true;
    lastName.IsEnabled = true;
    towerNames.IsEnabled = true;
    isCaptain.IsEnabled = true;
    hostMemberSince.IsEnabled = true;
    yearsExperience.IsEnabled = true;
    methods.IsEnabled = true;
    add.IsEnabled = true;
    clear.IsEnabled = true;
}
```

This code calls the *Reset* method and then enables all the controls. If you remember from Chapter 22, the *Reset* method resets the controls on the form to their default values. (If you don't recall how the *Reset* method works, scroll the *Code and Text Editor* window to display the method and refresh your memory.)

Next, you need to create a *Click* event method for the *Exit* command. This method should cause the form to close.

6. Return to the *Design View* window displaying the Window1.xaml file. Use the technique you followed in step 2 to create a *Click* event method for the *exit* menu item called *exit_Click*.

7. Switch to the *Code and Text Editor* window. In the body of the *exit_Click* method, type the statement shown in bold type in the following code:

```
private void exit_Click(object sender, RoutedEventArgs e)
{
    this.Close();
}
```

The *Close* method of a form *attempts* to close the form. Remember that if the form intercepts the *Closing* event, it can prevent the form from closing. The Middleshire Bell Ringers Association application does precisely this, and it asks the user if he or she wants to quit. If the user says no, the form does not close and the application continues to run.

The next step is to handle the *saveMember* menu item. When the user clicks this menu item, the data on the form should be saved to a file. For the time being, you will save the information to an ordinary text file called Members.txt in the current folder. Later, you will modify the code so that the user can select an alternative file name and location.

Handle the *Save Member Information* menu item event

1. Return to the *Design View* window displaying the Window1.xaml file. In the *XAML* pane, locate the definition of the *saveMember* menu item and use the *<New Event Handler>* command to specify a *Click* event method called *saveMember_Click*. (This is the default name generated by the *<New Event Handler>* command.)

2. In the *Code and Text Editor* window displaying the Window1.xaml.cs file, scroll to the top of the file and add the following *using* statement to the list:

   ```
   using System.IO;
   ```

3. Locate the *saveMember_Click* event method at the end of the file. Add the following statements shown in bold type to the body of the method:

   ```
   private void saveMember_Click(object sender, RoutedEventArgs e)
   {
       using (StreamWriter writer = new StreamWriter("Members.txt"))
       {
           writer.WriteLine("First Name: {0}", firstName.Text);
           writer.WriteLine("Last Name: {0}", lastName.Text);
           writer.WriteLine("Tower: {0}", towerNames.Text);
           writer.WriteLine("Captain: {0}", isCaptain.IsChecked.ToString());
           System.Windows.Forms.DateTimePicker memberDate =
               hostMemberSince.Child as System.Windows.Forms.DateTimePicker;
           writer.WriteLine("Member Since: {0}", memberDate.Value.ToString());
           writer.WriteLine("Methods: ");
           foreach (CheckBox cb in methods.Items)
           {
               if (cb.IsChecked.Value)
               {
                   writer.WriteLine(cb.Content.ToString());
               }
           }

           MessageBox.Show("Member details saved", "Saved");
       }
   }
   ```

This block of code creates a *StreamWriter* object that the method uses for writing text to the Member.txt file. Using the *StreamWriter* class is very similar to displaying text in a console application by using the *Console* object—you can simply use the *WriteLine* method.

When the details have all been written out, a message box is displayed giving the user some feedback (always a good idea).

4. The *Add* button and its associated event method are now obsolete, so in the *Design View* window delete the *Add* button. In the *Code and Text Editor* window, comment out the *add_Click* method.

5. In the *newMember_Click* method, comment out the following statement:

```
// add.IsEnabled = true;
```

The remaining menu item is the *about* menu item, which should display a dialog box providing information about the version of the application, the publisher, and any other useful information. You will add an event method to handle this event in the next exercise.

Handle the *About Middleshire Bell Ringers* menu item event

1. On the *Project* menu, click *Add Window*.

2. In the *Add New Item – BellRingers* dialog box, in the *Templates* pane, click **Window (WPF)**. In the *Name* text box, type **About.xaml**, and then click *Add*.

   When you have added the appropriate controls, you will display this window when the user clicks the *About Middleshire Bell Ringers* command on the *Help* menu.

> **Note** Visual Studio provides the *About Box* windows template. However, this template generates a Windows Forms window rather than a WPF window.

3. In the *Design View* window, click the *About.xaml* form. In the *Properties* window, change the *Title* property to **About Middleshire Bell Ringers**, set the *Width* property to **300**, and set the *Height* property to **156**. Set the *ResizeMode* property to **NoResize** to prevent the user from changing the size of the window.

4. In the *Name* box at the top of the *Properties* window, type **AboutBellRingers**.

5. Add two label controls and a button control to the form. In the *XAML* pane, modify the properties of these three controls as shown here in bold type (feel free to change the text displayed by the *buildDate* label if you prefer):

```
<Window x:Class="BellRingers.About"
    xmlns="http://schemas.microsoft.com/winfx/2006/xaml/presentation"
    xmlns:x="http://schemas.microsoft.com/winfx/2006/xaml"
    Title="About Middleshire Bell Ringers" Height="156" Width="300"
        Name="AboutBellRingers" ResizeMode="NoResize">
    <Grid>
        <Label Margin="80,20,0,0" Name="version" Height="30"
            VerticalAlignment="Top" HorizontalAlignment="Left"
            Width="75">Version 1.0</Label>
```

```
        <Label Margin="80,50,0,0" Name="buildDate" Height="30"
            VerticalAlignment="Top" HorizontalAlignment="Left"
            Width="160">Build date: September 2007</Label>
        <Button Margin="100,85,0,0" Name="ok" HorizontalAlignment="Left" Width="78"
            Height="23" VerticalAlignment="Top">OK</Button>
    </Grid>
</Window>
```

The completed form should look like this:

6. In the *Design View* window, double-click the *OK* button.

Visual Studio generates an event method for the *Click* event of the button and adds the *ok_Click* method to the About.xaml.cs file.

7. In the *Code and Text Editor* window displaying the About.xaml.cs file, add the statement shown in bold type to the *ok_Click* method:

```
private void ok_Click(object sender, RoutedEventArgs e)
{
    this.Close();
}
```

When the user clicks the *OK* button, the window will close.

8. Return to the *Design View* window displaying the Window1.xaml file. In the *XAML* pane, locate the definition of the *about* menu item and use the *<New Event Handler>* command to specify a *Click* event method called *about_Click*. (This is the default name generated by the *<New Event Handler>* command.)

9. In the *Code and Text Editor* window displaying the Window1.xaml.cs file, add the following statements shown in bold to the *about_Click* method:

```
private void about_Click(object sender, RoutedEventArgs e)
{
    About aboutWindow = new About();
    aboutWindow.ShowDialog();
}
```

This code creates a new instance of the *About* window and then calls the *ShowDialog* method to display it. The *ShowDialog* method does not return until the *About* window closes (when the user clicks the *OK* button).

Test the menu events

1. On the *Debug* menu, click *Start Without Debugging* to build and run the application.

   Notice that all the fields on the form are disabled.

2. Click the *File* menu.

   The *Save Member Details* command is disabled.

3. On the *File* menu, click *New Member*.

   The fields on the form are now available.

4. Input some details for a new member.

5. Click the *File* menu again.

   The *Save Member Details* command is now available.

6. On the *File* menu, click *Save Member Details*.

   After a short delay, the message "Member details saved" appears. Click *OK* in this message box.

7. Using Windows Explorer, move to the \Microsoft Press\Visual CSharp Step by Step\ Chapter 23\BellRingers\BellRingers\bin\Debug folder under your Documents folder.

   You should see a file called Members.txt in this folder.

8. Double-click *Members.txt* to display its contents using Notepad.

   This file should contain the details of the new member.

9. Close Notepad, and return to the Middleshire Bell Ringers application.

10. On the *Help* menu, click *About Middleshire Bell Ringers*.

    The *About* window appears. Notice that you cannot resize this window, and you cannot click any items on the *Members* form while the *About* window is still visible.

11. Click *OK* to return to the *Members* form.

12. On the *File* menu, click *Exit*.

    The form tries to close. You are asked if you are sure you want to close the form. If you click *No*, the form remains open; if you click *Yes*, the form closes and the application finishes.

13. Click *Yes* to close the form.

Shortcut Menus

Many Windows-based applications make use of pop-up menus that appear when you right-click a form or control. These menus are usually context-sensitive and display commands that are applicable only to the control or form that currently has the focus. They are usually referred to as *context* or *shortcut* menus. You can easily add shortcut menus to a WPF application by using the *ContextMenu* class.

Creating Shortcut Menus

In the following exercises, you will create two shortcut menus. The first shortcut menu is attached to the *firstName* and *lastName* text box controls and allows the user to clear these controls. The second shortcut menu is attached to the form and contains commands for saving the currently displayed member's information and for clearing the form.

 Note Text box controls are associated with a default shortcut menu that provides *Cut*, *Copy*, and *Paste* commands for performing text editing. The shortcut menu that you will define in the following exercise will override this default menu.

Create the *firstName* and *lastName* shortcut menu

1. In the *Design View* window displaying Window1.xaml, add the following *ContextMenu* element shown in bold type to the end of the window resources in the *XAML* pane after the style definitions:

```
<Window.Resources>
    ...
    <ContextMenu x:Key="textBoxMenu" Style="{StaticResource bellRingersFontStyle}">
    </ContextMenu>
</Window.Resources>
```

This shortcut menu will be shared by the *firstName* and *lastName* text boxes. Adding the shortcut menu to the window resources makes it available to any controls in the window.

2. Add the following *MenuItem* element shown in bold type to the *textBoxMenu* shortcut menu:

```
<Window.Resources>
    ...
    <ContextMenu x:Key="textBoxMenu" Style="{StaticResource bellRingersFontStyle}">
        <MenuItem Header="Clear Name" Name="clearName" />
    </ContextMenu>
</Window.Resources>
```

This code adds to the shortcut menu a menu item called *clearName* with the legend "Clear Name".

3. In the *XAML* pane, modify the definitions of the *firstName* and *lastName* text box controls, and add the *ContextMenu* property, shown here in bold type:

    ```
    <TextBox ... Name="firstName" ContextMenu="{StaticResource textBoxMenu}" ... />
    ...
    <TextBox ... Name="lastName" ContextMenu="{StaticResource textBoxMenu}" ... />
    ```

    The *ContextMenu* property determines which menu (if any) will be displayed when the user right-clicks the control.

4. Return to the definition of the *textBoxMenu* style, and to the *clearName* menu item add a *Click* event method called *clearName_Click*. (This is the default name generated by the *<New Event Handler>* command.)

    ```
    <MenuItem Header="Clear Name" Name="clearName" Click="clearName_Click" />
    ```

5. In the *Code and Text Editor* window displaying Window1.xaml.cs, add the following statements to the *clearName_Click* event method that the *<New Event Handler>* command generated:

    ```
    firstName.Text = String.Empty;
    lastName.Text = String.Empty;
    ```

    This code clears both text boxes when the user clicks the *Clear Name* command on the shortcut menu.

6. On the *Debug* menu, click *Start Without Debugging* to build and run the application. When the form appears, click *File*, and then click *New Member*.

7. Type a name in the *First Name* and *Last Name* text boxes. Right-click the *First Name* text box. On the shortcut menu, click the *Clear Name* command, and verify that both text boxes are cleared.

8. Type a name in the *First Name* and *Last Name* text boxes. This time, right-click the *Last Name* text box. On the shortcut menu, click the *Clear Name* command and again verify that both text boxes are cleared.

9. Right-click anywhere on the form outside the *First Name* and *Last Name* text boxes.

    Only the *First Name* and *Last Name* text boxes have shortcut menus, so no pop-up menu should appear.

10. Close the form, and return to Visual Studio 2008.

Now you can add the second shortcut menu, which contains commands that the user can use to save member information and to clear the fields on the form. To provide a bit of variation, and to show you how easy it is to create shortcut menus dynamically, in the following exercise you will create the shortcut menu by using code. The best place to put this code is in the constructor of the form. You will then add code to enable the shortcut menu for the window when the user creates a new member.

Create the window shortcut menu

1. Switch to the *Code and Text Editor* window displaying the Window1.xaml.cs file.

2. Add the following private variable shown in bold type to the *Window1* class:

```
public partial class Window1 : Window
{
    ...
    private ContextMenu windowContextMenu = null;
    ...
}
```

3. Locate the constructor for the *Window1* class. This is actually the first method in the class and is called *Window1*. Add the statements shown in bold type after the code that calls the *Reset* method to create the menu items for saving member details:

```
public Window1()
{
    InitializeComponent();
    this.Reset();

    MenuItem saveMemberMenuItem = new MenuItem();
    saveMemberMenuItem.Header = "Save Member Details";
    saveMemberMenuItem.Click += new RoutedEventHandler(saveMember_Click);
}
```

This code sets the *Header* property for the menu item and then specifies that the *Click* event should invoke the *saveMember_Click* event method; this is the same method that you wrote in an earlier exercise in this chapter. The *RoutedEventHandler* type is a delegate that represents methods for handling the events raised by many WPF controls. (For more information about delegates and events, refer to Chapter 17, "Interrupting Program Flow and Handling Events.")

4. In the *Window1* constructor, add the following statements shown in bold type to create the menu items for clearing the fields on the form and resetting them to their default values:

```
public Window1()
{
    ...
    MenuItem clearFormMenuItem = new MenuItem();
    clearFormMenuItem.Header = "Clear Form";
    clearFormMenuItem.Click += new RoutedEventHandler(clear_Click);
}
```

This menu item invokes the *clear_Click* event method when clicked by the user.

5. In the *Window1* constructor, add the following statements shown in bold type to construct the shortcut menu and populate it with the two menu items you have just created:

```
public Window1()
{
    ...
    windowContextMenu = new ContextMenu();
    windowContextMenu.Items.Add(saveMemberMenuItem);
    windowContextMenu.Items.Add(clearFormMenuItem);
}
```

The *ContextMenu* type contains a collection called *Items* that holds the menu items.

6. At the end of the *newMember_Click* event method, add the statement shown in bold type to associate the context menu with the form:

```
private void newMember_Click(object sender, RoutedEventArgs e)
{
    ...
    this.ContextMenu = windowContextMenu;
}
```

Notice that the application associates the shortcut menu with the form only when the new member functionality is available. If you were to set the *ContextMenu* property of the form in the constructor, the *Save Member Details* and *Clear Details* shortcut menu items would be available even when the controls on the form were disabled, which is not how you want this application to behave.

 Tip You can disassociate a shortcut menu from a form by setting the *ContextMenu* property of the form to *null*.

7. On the *Debug* menu, click *Start Without Debugging* to build and run the application.

8. When the form appears, right-click the form and verify that the shortcut menu does not appear.

9. On the *File* menu, click *New Member*, and then input some details for a new member.

10. Right-click the form. On the shortcut menu, click *Clear Form* and verify that the fields on the form are reset to their default values.

11. Input some more member details. Right-click the form. On the shortcut menu, click *Save Member Details*. Verify that the "Member details saved" message box appears, and then click *OK*.

12. Close the form, and return to Visual Studio 2008.

Windows Common Dialog Boxes

The Bell Ringers application now lets you save member information, but it always saves data to the same file, overwriting anything that is already there. Now is the time to address this issue.

A number of everyday tasks require the user to specify some sort of information. For example, if the user wants to open or save a file, the user is usually asked which file to open or where to save it. You might have noticed that the same dialog boxes are used by many different applications. This is not a result of a lack of imagination by applications developers; it is just that this functionality is so common that Microsoft has standardized it and made it available as a "common dialog box"—a component supplied with the Microsoft Windows operating system that you can use in your own applications. The Microsoft .NET Framework class library provides the *OpenFileDialog* and *SaveFileDialog* classes, which act as wrappers for these common dialog boxes.

Using the *SaveFileDialog* Class

In the following exercise, you will use the *SaveFileDialog* class. In the BellRingers application, when the user saves details to a file, you will prompt the user for the name and location of the file by displaying the Save File common dialog box.

Use the *SaveFileDialog* class

1. In the *Code and Text Editor* window displaying Window1.xaml.cs, add the following *using* statement to the list at the top of the file:

   ```
   using Microsoft.Win32;
   ```

   The *SaveFileDialog* class is in the *Microsoft.Win32* namespace.

2. Locate the *saveMember_Click* method, and add the code shown in bold type to the start of this method, replacing *YourName* with the name of your own account:

   ```
   private void saveMember_Click(object sender, RoutedEventArgs e)
   {
       SaveFileDialog saveDialog = new SaveFileDialog();
       saveDialog.DefaultExt = "txt";
       saveDialog.AddExtension = true;
       saveDialog.FileName = "Members";
       saveDialog.InitialDirectory = @"C:\Users\YourName\Documents\";
       saveDialog.OverwritePrompt = true;
       saveDialog.Title = "Bell Ringers";
       saveDialog.ValidateNames = true;
       ...
   }
   ```

> **Note** If you are using Windows XP, replace the statement that sets the InitialDirectory property of the saveDialog object with the following code:
>
> ```
> saveDialog.InitialDirectory = @"C:\Documents and Settings\YourName\My Documents\";
> ```

This code creates a new instance of the *SaveFileDialog* class and sets its properties. The following table describes the purpose of these properties.

Property	Description
DefaultExt	The default file name extension to use if the user does not specify the extension when providing the file name.
AddExtension	Enables the dialog box to add the file name extension indicated by the *DefaultExt* property to the name of the file specified by the user if the user omits the extension.
FileName	The name of the currently selected file. You can populate this property to specify a default file name, or clear it if you don't want a default file name.
InitialDirectory	The default directory to be used by the dialog box.
OverwritePrompt	Causes the dialog box to warn the user when an attempt is made to overwrite an existing file with the same name. For this to work, the *ValidateNames* property must also be set to *true*.
Title	A string that is displayed on the title bar of the dialog box.
ValidateNames	Indicates whether file names are validated. It is used by some other properties, such as *OverwritePrompt*. If the *ValidateNames* property is set to *true*, the dialog box also checks to verify that any file name typed by the user contains only valid characters.

3. Add the following statements shown in bold type to the *saveMember_Click* method, and enclose the previous code that creates the *StreamWriter* object and writes the member details to a file in an *if* statement:

```
if (saveDialog.ShowDialog().Value)
{
    using (StreamWriter writer = new StreamWriter("Members.txt"))
    {
        // existing code
        ...
    }
}
```

The *ShowDialog* method displays the *Save File* dialog box. The *Save File* dialog box is modal, which means that the user cannot continue using any other forms in the application until she has closed this dialog box by clicking one of its buttons. The *Save File* dialog box has a *Save* button and a *Cancel* button. If the user clicks *Save*, the value returned by the *ShowDialog* method is *true*; otherwise, it is *false*.

The *ShowDialog* method prompts the user for the name of a file to save to but does not actually do any saving—you still have to supply that code yourself. All it does is provide the name of the file that the user has selected in the *FileName* property.

4. In the *saveMember_Click* method, modify the statement that creates the *StreamWriter* object as shown in bold type here:

```
using (StreamWriter writer = new StreamWriter(saveDialog.FileName))
{
    ...
}
```

The *saveMember_Click* method will now write to the file specified by the user rather than to Members.txt.

5. On the *Debug* menu, click *Start Without Debugging* to build and run the application.

6. On the *File* menu, click *New Member*, and then add some details for a new member.

7. On the *File* menu, click *Save Member Details*.

The *Save File* dialog box should appear, with the caption "Bell Ringers." The default folder should be your Documents folder, and the default file name should be Members, as shown in the following image:

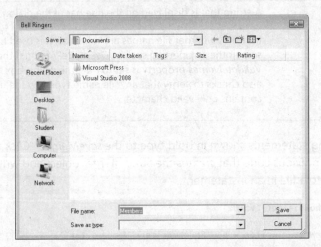

If you omit the file name extension, .txt is added automatically when the file is saved. If you pick an existing file, the dialog box warns you before it closes.

8. Change the value in the *File name* text box to **TestMember**, and then click *Save*.

9. In the Bell Ringers application, verify that the "Member details saved" message appears, click *OK*, and then close the application.

10. Using Windows Explorer, move to your Documents folder.

Verify that the TestMember.txt file has been created.

11. Double-click the file, and verify that it contains the details of the member that you added. Close Notepad when you have finished.

You can use a similar technique for opening a file: create an *OpenFileDialog* object, activate it by using the *ShowDialog* method, and retrieve the *FileName* property when the method returns if the user has clicked the *Open* button. You can then open the file, read its contents, and populate the fields on the screen. For more details on using the *OpenFileDialog* class, consult the MSDN Library for Visual Studio 2008.

- If you want to continue to the next chapter

  Keep Visual Studio 2008 running, and turn to Chapter 24.

- If you want to exit Visual Studio 2008 now

  On the *File* menu, click *Exit*. If you see a *Save* dialog box, click *Yes* (if you are using Visual Studio 2008) or *Save* (if you are using Microsoft Visual C# 2008 Express Edition) and save the project.

Chapter 23 Quick Reference

To	Do this
Create a menu for a form	Add a *DockPanel* control, and place it at the top of the form. Add a *Menu* control to the *DockPanel* control.
Add menu items to a menu	Add *MenuItem* elements to the *Menu* control. Specify the text for a menu item by setting the *Header* property, and give each menu item a name by specifying the *Name* property. You can optionally specify properties so that you can display features such as icons and child menus. You can add an access key to a menu item by prefixing the appropriate letter with an underscore character.
Create a separator bar in a menu	Add a *Seperator* element to the menu.
Enable or disable a menu item	Set the *IsEnabled* property to *True* or *False* in the *Properties* window at design time, or write code to set the *IsEnabled* property of the menu item to *true* or *false* at run time.
Perform an action when the user clicks a menu item	Select the menu item, and specify an event method for the *Click* event. Add your code to the event method.
Create a shortcut menu	Add a *ContextMenu* to the window resources. Add items to the shortcut menu just as you add items to an ordinary menu.
Associate a shortcut menu with a form or control	Set the *ContextMenu* property of the form or control to refer to the shortcut menu.

Create a shortcut menu dynamically	Create a *ContextMenu* object. Populate the *Items* collection of this object with *MenuItem* objects defining each of the menu items. Set the *ContextMenu* property of the form or control to refer to the shortcut menu.
Prompt the user for the name of a file to save	Use the *SaveFileDialog* class. Display the dialog box by using the *ShowDialog* method. When the dialog box closes, the *FileName* property of the *SaveFileDialog* instance contains the name of the file selected by the user.

Chapter 24
Performing Validation

After completing this chapter, you will be able to:

- Examine the information entered by a user to ensure that it does not violate any application or business rules.

- Use data binding validation rules to validate information entered by a user.

- Perform validation effectively but unobtrusively.

In the previous two chapters, you have seen how to create a Microsoft Windows Presentation Foundation (WPF) application that uses a variety of controls for data entry. You created menus to make the application easier to use. You have learned how to trap events raised by menus, forms, and controls so that your application can actually do something besides just look pretty. Although careful design of a form and the appropriate use of controls can help to ensure that the information entered by a user makes sense, you often need to perform additional checks. In this chapter, you will learn how to validate the data entered by a user running an application to ensure that it matches any business rules specified by the application's requirements.

Validating Data

The concept of input validation is simple enough, but it is not always easy to implement, especially if validation involves cross-checking data the user has entered into two or more controls. The underlying business rule might be relatively straightforward, but all too often, the validation is performed at an inappropriate time, making the form difficult (and infuriating) to use.

Strategies for Validating User Input

You can employ many strategies to validate the information entered by the users of your applications. A common technique that many Microsoft Windows developers familiar with previous versions of the Microsoft .NET Framework use is to handle the *LostFocus* event of controls. The *LostFocus* event is raised when the user moves away from a control. You can add code to this event to examine the data in the control that the user is vacating and ensure that it matches the requirements of the application before allowing the cursor to move away. The problem with this strategy is that often you need to cross-check data entered into one control against the values in others, and the validation logic can become quite convoluted; you frequently end up repeating similar logic in the *LostFocus* event handler for several controls. Additionally, you have no power over the sequence in which the user moves from control to control. Users can move through the controls on a form in any order, so you cannot always assume that every control contains a valid value if you are cross-checking a particular control against others on the form.

Another fundamental issue with this strategy is that it can tie the validation logic of the presentation elements of an application too closely to the business logic. If the business requirements change, you might need to modify the validation logic, and maintenance can become a complex task.

With WPF, you can define validation rules as part of the business model used by your applications. You can then reference these rules from the Extensible Application Markup Language (XAML) description of the user interface. To do this, you define the classes required by the business model and then bind properties of the user interface controls to properties exposed by these classes. At run time, WPF can create instances of these classes. When you modify the data in a control, the data can be automatically copied back to the specified property in the appropriate business model class instance and validated. You will learn more about data binding in Part V, "Managing Data," of this book. For the purposes of this chapter, we will concentrate on the validation rules that you can associate with data binding.

An Example—Customer Information Maintenance

Consider a simple scenario. You have been asked to build a Customer Information maintenance application. Part of the application needs to record the essential details of a customer, including the customer's title, name, and gender. You decide to create a form like the one shown in the following graphic.

You need to ensure that the user's input is consistent: the title (Mr, Mrs, Miss, or Ms) must match the selected gender (Male or Female), and vice versa.

Performing Validation by Using Data Binding

In the following exercises, you will examine the Customer Information application and add validation rules by using data binding. As a cautionary step, you will see how easy it is to get the validation timing wrong and render an application almost unusable!

Examine the Customer Details form

1. Start Microsoft Visual Studio 2008 if it is not already running.

2. Open the CustomerDetails project, located in the \Microsoft Press\Visual CSharp Step By Step\Chapter 24\CustomerDetails folder in your Documents folder.

3. On the *Debug* menu, click *Start Without Debugging* to build and run the application.

4. When the form appears, click the drop-down arrow in the *Title* combo box, and then click *Mr.*

5. In the *Gender* group box, click the *Female* radio button.

6. On the *File* menu, click *Save*, and verify that the "Customer saved" message box appears.

   The application does not actually save any data. The important point is that if it did, the information saved would have been inconsistent because the application does not currently perform any checking. Ideally, all customers should have a name, and the values specified for the *Title* and *Gender* controls should match.

7. Click *OK*, and then close the form and return to Visual Studio 2008.

The first step in adding the necessary validation logic is to create a class that can model a customer. You will start by learning how to use this class to ensure that the user always enters a first name and last name for the customer.

Create the *Customer* class with validation logic for enforcing entry of a name

1. In *Solution Explorer*, right-click the CustomerDetails project, point to *Add*, and then click *Class*.

2. In the *Add New Item – CustomerDetails* dialog box, in the *Name* text box, type **Customer.cs**, and then click *Add*.

3. In the *Code and Text Editor* window displaying the Customer.cs file, add to the *Customer* class the private *foreName* and *lastName* fields shown in bold type here:

```
class Customer
{
    private string foreName;
    private string lastName;
}
```

4. Add the following public *ForeName* property to the *Customer* class as shown in bold type, based on the *foreName* field you added in the preceding step:

```
class Customer
{
    ...
    public string ForeName
    {
        get { return this.foreName; }
        set
        {
            if (String.IsNullOrEmpty(value))
            {
                throw new ApplicationException
                    ("Specify a forename for the customer");
            }
            else
            {
                this.foreName = value;
            }
        }
    }
}
```

The property *set* accessor examines the value supplied for the first name, and if it is empty, it raises an exception with a suitable message.

5. Add to the *Customer* class the *LastName* property shown in bold type in the following code. This property follows a similar pattern to that of the *ForeName* property:

```
class Customer
{
    ...
    public string LastName
    {
        get { return this.lastName; }
        set
        {
            if (String.IsNullOrEmpty(value))
            {
                throw new ApplicationException
                    ("Specify a last name for the customer");
            }
            else
            {
                this.lastName = value;
            }
        }
    }
}
```

Now that you have created the *Customer* class, the next step is to bind the *foreName* and *lastName* text boxes on the form to the corresponding properties of the class.

Bind the text box controls on the form to properties in the *Customer* class

1. In *Solution Explorer*, double-click the CustomerForm.xaml file to display the form in the *Design View* window.

2. In the *XAML* pane, add the XML namespace declaration shown here in bold type to the *Window* definition:

```
<Window x:Class="CustomerDetails.CustomerForm"
    xmlns="http://schemas.microsoft.com/winfx/2006/xaml/presentation"
    xmlns:x="http://schemas.microsoft.com/winfx/2006/xaml"
    xmlns:cust="clr-namespace:CustomerDetails"
    Title="Customer Details" Height="273" Width="370" ResizeMode="NoResize">
...
```

With this declaration in place, you can reference the types in the *CustomerDetails* namespace in the XAML code for the window.

3. Add the following *Window.Resources* element shown in bold type to the window:

```
<Window x:Class="CustomerDetails.CustomerForm"
    ...
    ...ResizeMode="NoResize">
    <Window.Resources>
        <cust:Customer x:Key="customerData" />
    </Window.Resources>
    <Grid>
    ...
```

This resource creates a new instance of the *Customer* class. You can reference this instance by using the key value, *customerData*, elsewhere in the XAML definition of the window.

4. Find the definition of the *foreName* text box in the *XAML* pane, and modify it as shown here in bold type (make sure that you replace the closing delimiter tag (/>) for the *TextBox* control with an ordinary delimiter (>) and that you add a closing </TextBox> tag):

```
<TextBox Height="21" Margin="70,74,0,0" Name="foreName" VerticalAlignment="Top"
    HorizontalAlignment="Left" Width="120" >
    <TextBox.Text>
        <Binding Source="{StaticResource customerData}" Path="ForeName" />
    </TextBox.Text>
</TextBox>
```

This code binds the data displayed in the *Text* property of this text box to the value in the *ForeName* property of the *customerData* object. If the user updates the value in the *foreName* text box on the form, the new data is automatically copied to the *customer-Data* object. Remember that the *ForeName* property in the *Customer* class checks that the user has actually specified a value and not just blanked it out.

5. Modify the definition of the binding that you added in the preceding step and add a *Binding.ValidationRules* child element, as shown here in bold type:

```
<TextBox Height="21" Margin="70,74,0,0" Name="foreName" VerticalAlignment="Top"
    HorizontalAlignment="Left" Width="120" >
    <TextBox.Text>
        <Binding Source="{StaticResource customerData}" Path="ForeName" >
            <Binding.ValidationRules>
                <ExceptionValidationRule/>
            </Binding.ValidationRules>
        </Binding>
    </TextBox.Text>
</TextBox>
```

With the *ValidationRules* element of a binding, you can specify the validation that the application should perform when the user enters data in this control. The *ExceptionValidationRule* element is a built-in rule that checks for any exceptions thrown by the application when the data in this control changes. If it detects any exceptions, it highlights the control so that the user can see that there is a problem with the input.

6. Add the equivalent binding and binding rule to the *lastName* text box, associating it with the *LastName* property of the *customerData* object, as follows:

```
<TextBox Height="21" Margin="210,74,0,0" Name="lastName" VerticalAlignment="Top"
    HorizontalAlignment="Left" Width="120" >
    <TextBox.Text>
        <Binding Source="{StaticResource customerData}" Path="LastName" >
            <Binding.ValidationRules>
                <ExceptionValidationRule/>
            </Binding.ValidationRules>
        </Binding>
    </TextBox.Text>
</TextBox>
```

7. On the *Debug* menu, click *Start Without Debugging* to build and run the application.

8. When the form appears, type your name in the *foreName* and *lastName* text boxes, and then click the *title* combo box.

Nothing noteworthy should happen.

9. Click the *foreName* text box, delete the first name that you entered, and then click the *title* combo box again.

This time, the *foreName* text box is highlighted with a red border.

10. Enter a value in the *foreName* text box again, and delete the value in the *lastName* text box. On the *File* menu, click *Save*.

Notice that the red border has disappeared from the *foreName* text box but, rather surprisingly, there is no red border around the *lastName* text box.

11. In the message box, click *OK*, and then click the *title* combo box.

The red border now appears around the *lastName* text box.

12. Close the form, and return to Visual Studio 2008.

There are at least two questions that you should be asking yourself at this point:

- Why doesn't the form always detect when the user has forgotten to enter a value in a text box? The answer is that the validation occurs only when the text box loses its focus. This in turn happens only when the user moves the focus to another control on the form. Menus are not actually treated as though they are part of the form (they are handled differently), so when you select a menu item you are not moving to another control on the form, and hence the text box has not yet lost its focus. Only when you click the *title* combo box (or some other control) does the focus move and the validation occur. Additionally, the *foreName* and *lastName* text boxes are initially empty. If you move from the *foreName* text box to the *lastName* text box and then on to the *title* combo box without typing anything, the validation will not be performed. Only when you type something and then delete it does the validation run. You will address these problems later in this chapter.

- How can I get the form to display a meaningful error message rather than just highlighting that there is a problem with the input in a control? You can capture the message generated by an exception and display it elsewhere on the form. You will see how to do this in the following exercise.

Add a style to display exception messages

1. In the *Design View* window displaying the CustomerForm.xaml file, in the *XAML* pane, add the following style shown in bold type to the *Window.Resources* element:

```
<Window.Resources>
    <cust:Customer x:Key="customerData" />
    <Style x:Key="errorStyle" TargetType="Control">
        <Style.Triggers>
            <Trigger Property="Validation.HasError" Value="True">
                <Setter Property="ToolTip"
                        Value="{Binding RelativeSource={x:Static RelativeSource.
Self},Path=(Validation.Errors)[0].ErrorContent}" />
            </Trigger>
        </Style.Triggers>
    </Style>
</Window.Resources>
```

This style contains a trigger that detects when the *Validation.HasError* property of the control is set to *true*. This occurs if a binding validation rule for the control generates an exception. The trigger sets the *ToolTip* property of the current control to display the text of the exception. Detailed explanation of the binding syntax shown here is outside the scope of this book, but the binding source `{Binding RelativeSource={x:Static RelativeSource.Self}` is a reference to the current control, and the binding path `(Validation.Errors)[0].ErrorContent` associates the first exception message found in this binding source with the *ToolTip* property. (An exception could throw further exceptions, all of which generate their own messages. The first message is usually the most significant, though.)

2. Apply the *errorStyle* style to the *foreName* and *lastName* text box controls, as shown in bold type here:

```
<TextBox Style="{StaticResource errorStyle}" ... Name="foreName" ... >
  ...
</TextBox>
<TextBox Style="{StaticResource errorStyle}" ... Name="lastName" ... >
  ...
</TextBox>
```

3. On the *Debug* menu, click *Start Without Debugging* to build and run the application.

4. When the form appears, type your name in the *foreName* and *lastName* text boxes, and then click the *title* combo box.

5. Click the *foreName* text box, delete the first name that you entered, and then click the *title* combo box again.

   The *foreName* text box is highlighted with a red border.

 Note Make sure that you actually delete the contents of the *foreName* text box rather than just overtyping the text with spaces.

6. Click the *Title* combo box again to hide the list of titles, and then rest the mouse pointer on the *foreName* text box. A ScreenTip should appear, displaying the message "Specify a forename for the customer," like this:

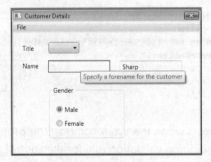

This is the message raised by the exception you added to the *ForeName* property of the *Customer* class.

7. Clear the *lastName* text box, and then click the *Title* combo box. Click the *Title* combo box again to hide the list, and then rest the mouse pointer on the *lastName* text box and verify that the tooltip "Specify a last name for the customer" appears.

8. Close the form, and return to Visual Studio 2008.

There are still some issues left to fix, but you will correct them after you have seen how to validate the title and gender of customers.

Add properties to validate the customer title and gender

1. Switch to the *Code and Text Editor* window displaying the Customer.cs file.

2. Add the *Title* and *Gender* enumerations shown here in bold type to the file above the *Customer* class.

```
enum Title { Mr, Mrs, Miss, Ms }
enum Gender { Male, Female }

class Customer
{
    ...
}
```

You will use these enumerations to specify the types of the *Title* and *Gender* properties of the *Customer* class.

3. Add the *title* and *gender* private fields to the *Customer* class, as shown in bold type here:

```
class Customer
{
    private string foreName;
    private string lastName;
    private Title title;
    private Gender gender;
    ...
}
```

4. Add the private method to the *Customer* class as shown in bold type here:

```
class Customer
{
    ...
    private bool checkTitleAndGender(Title proposedTitle, Gender proposedGender)
    {
        bool retVal = false;

        if (proposedGender.Equals(Gender.Male))
        {
            retVal = (proposedTitle.Equals(Title.Mr)) ? true : false;
        }
```

```
        if (proposedGender.Equals(Gender.Female))
        {
            retVal = (proposedTitle.Equals(Title.Mr)) ? false : true;
        }

        return retVal;
    }
}
```

This method examines the values in the *proposedTitle* and *proposedGender* parameters and tests them for consistency. If the values in *proposedTitle* and *proposedGender* are consistent, this method returns *true*; otherwise, it returns *false*.

> **Note** You might not be familiar with the ternary operator (indicated by the ? and :) used in this method. It operates like a condensed *if ... else* statement. It has the following form:
>
> *boolean expression ? true result : false result*
>
> The Boolean expression is evaluated. If it yields *true*, the expression between the question mark (?) and the colon (:) is evaluated and used as the result of the entire expression; otherwise, the expression after the colon (:) is evaluated and used as the result.

5. Add the public *Title* and *Gender* properties shown here in bold type to the *Customer* class. The type of the *Title* property is the *Title* enumeration, and the type of the *Gender* property is the *Gender* enumeration:

```
class Customer
{
    ...
    public Title Title
    {
        get { return this.title; }
        set
        {
            this.title = value;
            if (!checkTitleAndGender(value, this.gender))
            {
                throw new ApplicationException(
                    "The title must match the gender of the customer");
            }
        }
    }

    public Gender Gender
    {
        get { return this.gender; }
        set
        {
            this.gender = value;
            if (!checkTitleAndGender(this.title, value))
```

```
        {
            throw new ApplicationException(
                "The gender must match the title of the customer");
        }
    }
}
```

The *set* accessors of these properties call the *checkTitleAndGender* method to verify that the *title* and the *gender* fields match, and they raise an exception if the fields do not match.

6. Add the *ToString* method shown here in bold type to the *Customer* class:

```
class Customer
{
    ...
    public override string ToString()
    {
        return this.Title.ToString() + " " + this.ForeName + " " +
            this.LastName + " - " + this.Gender.ToString();
    }
}
```

You will use this method to display the details of customers when you save them to verify that the data is correct.

The next step is to bind the *title* combo box and the *male* and *female* radio buttons on the form to these new properties. However, if you stop and think for a moment, you will realize that there are a couple of small problems. First, you need to bind the *Text* property of the *title* combo box to the *Title* property of the *Customer* object created by the form. The type of the *Text* property is *string*. The type of the *Title* property is *Title* (an enumeration). You must convert between *string* and *Title* values for the binding to work. Fortunately, with the binding mechanism implemented by WPF, you can specify a converter class to perform actions such as this.

The second problem is similar. You need to bind the *IsChecked* property of each radio button (which is a *boolean* value) to the *Gender* property of the *Customer* object (which has the *Gender* type). Again, you can create a converter class to convert between a *boolean* value and a *Gender* value, but you also need to indicate which of the two radio buttons has been clicked. When you click either of these radio buttons, you are setting the *Gender* property, but you are setting it to a different value in each case. If the *IsChecked* property of the *male* radio button is set to *true*, you should set the *Gender* property to *Gender.Male*. If the *IsChecked* property of the *female* radio button is set to *true*, you should set the *Gender* property to *Gender.Female*. Happily, those clever people in the WPF team at Microsoft thought of this as well, and you can pass a parameter to a converter method that will let you indicate which radio button has been clicked.

Converter methods reside in their own classes that must implement the *IValueConverter* interface. This interface defines two methods: *Convert*, which converts from the type used by the property in the class that is providing the data for the binding to the type displayed on the form, and *ConvertBack*, which converts the data from the type displayed on the form to the type required by the class.

Create the converter classes and methods

1. In the Customer.cs file, add the following *using* statement to the list at the top of the file.

   ```
   using System.Windows.Data;
   ```

   The *IValueConverter* interface is defined in this namespace.

2. Add the *TitleConverter* class shown here to the file.

   ```
   [ValueConversion(typeof(string), typeof(Title))]
   public class TitleConverter : IValueConverter
   {
   }
   ```

   The text in brackets directly above the class is an example of an attribute. An attribute provides descriptive metadata for a class. The *ValueConversion* attribute is used by tools such as the WPF designer in the *Design View* window to verify that you are applying the class correctly when you reference it. The parameters to the *ValueConversion* attribute specify the type of the value displayed by the form (*string*) and the type of the value in the corresponding property in the class (*Title*). You will see more examples of attributes in later chapters in this book.

3. In the *TitleConverter* class, add the *Convert* method shown here in bold type:

   ```
   public class TitleConverter : IValueConverter
   {
       public object Convert(object value, Type targetType, object parameter,
           System.Globalization.CultureInfo culture)
       {
           Title title = (Title)value;
           return title.ToString();
       }
   }
   ```

   The signature of the *Convert* method is defined by the *IValueConverter* interface. The *value* parameter is the value in the class that you are converting from. (You can ignore the other parameters for now.) The return value from this method is the data bound to the property on the form. In this case, the *Convert* method converts a *Title* value to a *string*. Notice that the *value* parameter is passed in as an *object*, so you need to cast it to the appropriate type before attempting to use it.

4. Add the following *ConvertBack* method shown in bold type to the *TitleConverter* class:

```
public class TitleConverter : IValueConverter
{
    ...
    public object ConvertBack(object value, Type targetType, object parameter,
        System.Globalization.CultureInfo culture)
    {
        Title retVal = Title.Miss;

        switch ((string)value)
        {
            case "Mr"   : retVal = Title.Mr;
                          break;
            case "Mrs"  : retVal = Title.Mrs;
                          break;
            case "Ms"   : retVal = Title.Ms;
                          break;
            case "Miss": retVal = Title.Miss;
                          break;
        }
        return retVal;
    }
}
```

In the *ConvertBack* method, the *value* parameter is now the value from the form that you are converting back to a value of the appropriate type for the class. In this case, the *ConvertBack* method converts the data from a *string* (displayed in the *Text* property in the combo box) to the corresponding *Title* value.

5. After the *TitleConverter* class, add the *GenderConverter* class shown here to the Customer.cs file:

```
[ValueConversion(typeof(bool), typeof(Gender))]
public class GenderConverter : IValueConverter
{
}
```

This time, the class will convert between *Gender* values and the Boolean values corresponding to the radio buttons on the form.

6. Add the *Convert* method shown in bold type in the following code to the *GenderConverter* class:

```
public class GenderConverter : IValueConverter
{
    public object Convert(object value, Type targetType, object parameter,
        System.Globalization.CultureInfo culture)
    {
        string radioButtonId = (string)parameter;
        Gender gender = (Gender)value;
        bool retVal = false;
```

```
            if (String.Equals(radioButtonId, "Female") && gender.Equals(Gender.Female))
                retVal = true;

            if (String.Equals(radioButtonId, "Male") && gender.Equals(Gender.Male))
                retVal = true;

            return retVal;
        }
    }
```

On this occasion, the method makes use of the *parameter* parameter. When you reference a converter from a form, you can specify additional data to be passed in. This is useful if more than one control must bind its values to the same property in a class. You can use this parameter to determine which control is calling the converter method. When you add the binding for the radio buttons to the form in the next exercise, you will specify a parameter of "Male" for the *male* radio button and "Female" for the *female* radio button. The *Convert* method examines the data in this parameter and compares it with the data in the *value* parameter. If the *parameter* parameter is "Female", the converter has been called for the *female* radio button. If the *value* parameter contains the value *Gender.Female*, the customer object is also female, and the method returns *true*. If the *parameter* parameter is "Female" but the *value* parameter is *Gender.Male*, the customer object is male and the method returns *false*. The method uses the same logic if the *parameter* parameter is "Male".

7. Add the *ConvertBack* method shown here in bold type to the *GenderConverter* class:

```
public class GenderConverter : IValueConverter
{
    ...
    public object ConvertBack(object value, Type targetType, object parameter,
        System.Globalization.CultureInfo culture)
    {
        if (String.Equals((string)parameter, "Female"))
            return Gender.Female;
        else
            return Gender.Male;
    }
}
```

This method looks to be suspiciously simple, and it is. The important point to realize is that the converter method is called only when the user changes the value on the form. In the case of a radio button, a user can only select it; the user can never clear it. A radio button is cleared only when another radio button in the same group is selected. This means that you don't need to check the *value* parameter because it will always be *true*. You only need to return a *Gender* value based on the *parameter* parameter.

Bind the combo box and radio button controls on the form to the properties in the *Customer* class

1. Return to the *Design View* window displaying the CustomerForm.xaml file.

2. In the *XAML* pane, add a *TitleConverter* object as a resource to the window, and specify a key value of *titleConverter.* Add a *GenderConverter* object as another resource, with a key value of *genderConverter*, as shown in bold type here:

```
<Window.Resources>
    <cust:Customer x:Key="customerData" />
    <cust:TitleConverter x:Key="titleConverter" />
    <cust:GenderConverter x:Key="genderConverter" />
    ...
</Window.Resources>
```

3. Locate the definition of the *title* combo box control, and style the control by using the *errorStyle* style. After the list of combo box items, add the XAML code shown here in bold type to bind the *Text* property of the combo box to the *Title* property in the *customerData* object, specifying the *titleConverter* resource as the object providing the converter methods:

```
<ComboBox Style="{StaticResource errorStyle}" ... Name="title" ...>
    <ComboBox.Items>
        ...
    </ComboBox.Items>
    <ComboBox.Text>
        <Binding Source="{StaticResource customerData}" Path="Title"
            Converter="{StaticResource titleConverter}" >
            <Binding.ValidationRules>
                <ExceptionValidationRule />
            </Binding.ValidationRules>
        </Binding>
    </ComboBox.Text>
</ComboBox>
```

4. Modify the definition for the *male* radio button. As shown in bold type in the following code, apply the *errorStyle* style, remove the *IsChecked="True"* property from the definition of the radio button, and add XAML code to bind the *IsChecked* property to the *Gender* property of the *customerData* object. Specify the *genderConverter* object as the resource providing the converter methods, and set the *ConverterParameter* property to *"Male"*:

```
<RadioButton Style="{StaticResource errorStyle}" Height="16" Name="male"
    Width="120" Margin="0,20,0,0" >
    Male
    <RadioButton.IsChecked>
        <Binding Source="{StaticResource customerData}" Path="Gender"
         Converter="{StaticResource genderConverter}" ConverterParameter="Male">
            <Binding.ValidationRules>
                <ExceptionValidationRule />
            </Binding.ValidationRules>
```

```
        </Binding>
    </RadioButton.IsChecked>
</RadioButton>
```

5. Modify the definition for the *female* radio button in a similar manner, but set the *ConverterParameter* property to *"Female"*:

```
<RadioButton Style="{StaticResource errorStyle}" Height="16" Name="female"
    Width="120" Margin="0,10,0,0" >
    Female
    <RadioButton.IsChecked>
        <Binding Source="{StaticResource customerData}" Path="Gender"
         Converter="{StaticResource genderConverter}" ConverterParameter="Female">
            <Binding.ValidationRules>
                <ExceptionValidationRule />
            </Binding.ValidationRules>
        </Binding>
    </RadioButton.IsChecked>
</RadioButton>
```

6. On the *View* menu, click *Code* to switch to the *Code and Text Editor* window displaying the CustomerForm.xaml.cs file.

7. Change the code in the *saveCustomer_Click* method, as shown here in bold type:

```
private void saveCustomer_Click(object sender, RoutedEventArgs e)
{
    Binding customerBinding =
        BindingOperations.GetBinding(this.title, ComboBox.TextProperty);
    Customer customer = customerBinding.Source as Customer;
    MessageBox.Show(customer.ToString(), "Saved");
}
```

This code displays the details of the customer in the message box. (It still does not actually save the customer information anywhere.) The static *GetBinding* method of the *BindingOperations* class returns a reference to the object to which the specified property is bound. In this case, the *GetBinding* method retrieves the object bound to the *Text* property of the *title* combo box. This should be the same object referred to by the *customerData* resource. In fact, the code could have queried any of the bound properties of the *foreName*, *lastName*, *male*, and *female* controls to retrieve the same reference. The reference is returned as a *Binding* object. The code then casts this *Binding* object into a *Customer* object before displaying its details.

Important The remaining steps in this exercise are necessary because of a bug in the current release of the .NET Framework. When you select a radio button, any other radio buttons in the same group are automatically cleared. However, the cleared radio buttons also lose their bindings, and validation no longer works if you select them again. The fix is to rebuild and reattach the bindings for all cleared radio buttons each time a radio button is selected. In this example, whenever the user selects the *male* radio button, the application must rebuild and reattach the binding for the *female* button, and vice versa. This bug should be corrected in a future release of the .NET Framework.

8. Add the private method shown here in bold type to the *CustomerForm* class:

```
public partial class CustomerForm : Window
{
    ...
    private Binding rebuildBinding(string parameter)
    {
        Binding customerBinding =
            BindingOperations.GetBinding(this.title, ComboBox.TextProperty);
        Customer customer = customerBinding.Source as Customer;
        Binding binding = new Binding();
        binding.Source = customer;
        binding.Path = new PropertyPath("Gender");
        binding.Converter = new GenderConverter();
        binding.ConverterParameter = parameter;
        binding.ValidationRules.Add(new ExceptionValidationRule());
        return binding;
    }
}
```

This method creates a new binding for the *Gender* radio buttons. The first two statements should be familiar; they retrieve a reference to the *customerData* object created by the form. The remaining steps create a new binding object that references the *Gender* property of the *customerData* object as its source and adds a reference to a *GenderConverter* converter object, as required by the radio buttons. The *parameter* variable is a string that will be passed in to this method and will contain the text "Male" or "Female" depending on which radio button the method is re-creating the binding for. Finally, the code adds the *ExceptionValidationRule* validation rule to the binding before returning the binding back to the caller. Take the time to compare this code with the XAML description of the binding for either of the two radio buttons.

9. Return to the *Design View* window displaying the CustomerForm.xaml file. In the *XAML* pane, locate the definition of the *male* radio button and specify a *Checked* event method called *male_Checked*. (This is the default name generated by the *<New Event Handler>* command.)

```
<RadioButton ... Name="male" ... Checked="male_Checked">
```

10. In the definition of the *female* radio button, specify a *Checked* event method called *female_Checked*.

```
<RadioButton ... Name="female" ... Checked="female_Checked">
```

11. Switch to the *Code and Text Editor* window displaying the CustomerForm.xaml.cs file. Add the code shown here in bold type to the *male_Checked* and *female_Checked* methods:

```
public partial class CustomerForm : Window
{
    ...
    private void male_Checked(object sender, RoutedEventArgs e)
```

```
        {
            Binding binding = rebuildBinding("Female");
            if (this.female != null)
            {
                this.female.SetBinding(RadioButton.IsCheckedProperty, binding);
                BindingExpression femaleBe =
                    this.female.GetBindingExpression(RadioButton.IsCheckedProperty);
                femaleBe.UpdateTarget();
            }
        }

        private void female_Checked(object sender, RoutedEventArgs e)
        {
            Binding binding = rebuildBinding("Male");
            if (this.male != null)
            {
                this.male.SetBinding(RadioButton.IsCheckedProperty, binding);
                BindingExpression maleBe =
                    this.male.GetBindingExpression(RadioButton.IsCheckedProperty);
                maleBe.UpdateTarget();
            }
        }
    }
```

This *male_Checked* method rebuilds and reattaches the binding for the *female* radio button. The code then creates a *BindingExpression* object that provides a mechanism for synchronizing the state of the *female* radio button on the form with the *Gender* property of the underlying *Customer* object. The *UpdateTarget* method ensures that the *female* radio button indicates the correct value for the *Gender* property; if the *Gender* property of the *Customer* object is *Female*, the *IsChecked* property of the *female* radio button will be set to *true*; otherwise, it will be set to *false*.

The *female_Checked* method performs the same tasks for the *male* radio button.

Run the application, and test the validation

1. On the *Debug* menu, click *Start Without Debugging* to build and run the application.

   Notice that the default title is "Mr" and the default gender is "Male".

2. In the *Title* combo box, click *"Mrs"*.

   The *checkTitleAndGender* method in the *Customer* class generates an exception because the title and the gender don't agree. The *Title* box is highlighted with a red border. Rest the mouse pointer on the *Title* combo box, and verify that the ScreenTip text "The title must match the gender of the customer" appears.

3. In the *Title* combo box, click *"Mr"*.

   Verify that the error disappears.

4. In the *Gender* group box, click the *Female* radio button.

   Again, the *checkTitleAndGender* method generates an exception, and the *Female* radio button appears highlighted with a red border. Rest the mouse pointer on the *Female* combo box, and verify that the ScreenTip text "The gender must match the title of the customer" appears.

5. On the *File* menu, click *Save*.

   A message box appears, displaying the title ("Mr") and the gender ("Female") of the customer. Although the form contains erroneous and missing data (you have not entered a name), you can still save the data!

6. Click *OK*, and then type a name in the *foreName* and *lastName* text boxes, but do not click away from the *lastName* text box.

7. On the *File* menu, click *Save* again.

   The message box now includes the first name of the customer but not the last name. This happens because the *lastName* text box on the form has not lost the focus. Remember from earlier that data binding validation for a text box occurs only when the user clicks another control on the form. The same applies to the data itself; by default, it is copied to the *customerDetails* object only when the text box loses the focus. In fact, it is the act of copying the data from the form to the *customerDetails* object that triggers the validation.

8. Click *OK*, and then click the *Title* combo box. Set the title to *"Mrs"*. On the *File* menu, click *Save*.

   This time, the message box displays the first name and last name of the customer. Also, although the title ("Mrs") and gender ("Female") now match, the radio button still flags an error.

9. Click *OK*, close the application, and return to Visual Studio 2008.

Changing the Point at Which Validation Occurs

The issue with the Customer Information application is that the validation is performed at the wrong time, is inconsistently applied, and does not actually prevent the user from saving inconsistent data. You just need an alternative approach to handling the validation. The solution is to check the user's input only when the user saves the data. This way, you can ensure that the user has finished entering all the data and that it is consistent. If there are any problems, you can display an error message and prevent the data from being saved until the problems have been corrected. In the following exercise, you will modify the Customer Information application to postpone validation until the user attempts to save the customer information.

Validate data explicitly

1. Return to *the Design View* window displaying CustomerForm.xaml. In the *XAML* pane, modify the binding for the *title* combo box and set the *UpdateSourceTrigger* property to *"Explicit"*, as shown in bold type here:

```
<ComboBox ... Name="title" ...>
...
    <ComboBox.Text>
        <Binding Source="{StaticResource customerData}" Path="Title"
          Converter="{StaticResource titleConverter}" UpdateSourceTrigger="Explicit" >
            ...
        </Binding>
    </ComboBox.Text>
</ComboBox>
```

The *UpdateSourceTrigger* property governs when the information entered by the user is sent back to the underlying *Customer* object and validated. Setting this property to *"Explicit"* postpones this synchronization until your application explicitly performs it by using code.

2. Modify the bindings for the *foreName* and *lastName* text boxes to set the *UpdateSourceTrigger* property to *"Explicit"*:

```
<TextBox ... Name="foreName" ... >
    <TextBox.Text>
        <Binding Source="{StaticResource customerData}" Path="ForeName"
            UpdateSourceTrigger="Explicit" >
            ...
        </Binding>
    </TextBox.Text>
</TextBox>
...
<TextBox ... Name="lastName" ... >
    <TextBox.Text>
        <Binding Source="{StaticResource customerData}" Path="LastName"
            UpdateSourceTrigger="Explicit" >
            ...
        </Binding>
    </TextBox.Text>
</TextBox>
```

The application is not going to check consistency between the title and gender of the customer until the user saves the customer information. If the validation rule for the *Title* property passes, there is no need to validate the *Gender* property. Similarly, if the validation rule for the *Title* property fails, there is no need to examine the *Gender* property. In other words, the validation rule for the *Gender* radio buttons on the form is now redundant, so remove it. To do so, remove the *Binding.ValidationRules* and *ExceptionValidationRule* elements from the bindings for the *male* and *female* radio buttons. Do not set the *UpdateSourceTrigger* property of the bindings to *"Explicit"*.

3. Return to the *Code and Text Editor* window displaying the CustomerForm.xaml.cs file. In the *rebuildBinding* method, locate the statement that adds the validation rule to the binding and comment it out as shown in bold type here:

```
private Binding rebuildBinding(string parameter)
{
    ...
    // binding.ValidationRules.Add(new ExceptionValidationRule());
    ...
}
```

4. In the *saveCustomer_Click* method, add the statements shown here in bold type to the start of the method:

```
private void saveCustomer_Click(object sender, RoutedEventArgs e)
{
    BindingExpression titleBe =
        this.title.GetBindingExpression(ComboBox.TextProperty);
    BindingExpression foreNameBe =
        this.foreName.GetBindingExpression(TextBox.TextProperty);
    BindingExpression lastNameBe =
        this.lastName.GetBindingExpression(TextBox.TextProperty);
    ...
}
```

These statements create *BindingExpression* objects for each of the three controls with binding validation rules. In an earlier exercise in this chapter, you saw that you can use a *BindingExpression* object to ensure that the data displayed on the form is synchronized with the data in the *Customer* object by calling the *UpdateTarget* method. The *BindingExpression* class also provides the *UpdateSource* method, which synchronizes data the other way around, sending the values in the bound properties of controls on the form back to the *Customer* object. When this occurs, the data will also be validated.

5. Add the following statements shown in bold type to the *saveCustomer_Click* method after the code you added in the preceding step:

```
private void saveCustomer_Click(object sender, RoutedEventArgs e)
{
    ...
    titleBe.UpdateSource();
    foreNameBe.UpdateSource();
    lastNameBe.UpdateSource();
    ...
}
```

These statements update the properties in the *Customer* object with the values entered by the user on the form, and they validate the data as they do so. Notice that there is no need to update the *Gender* property manually because you did not set the *UpdateSourceTrigger* property to *"Explicit"* for the binding for the radio buttons; the *Gender* property is still updated automatically.

The *BindingExpression* class provides a property called *HasError* that indicates whether the *UpdateSource* method was successful or whether it caused an exception.

6. Add the code shown here in bold type to the *saveCustomer_Click* method to test the *HasError* property of each *BindingExpression* object and display a message if the validation fails. Move the original code that displays the customer details to the *else* part of the *if* statement.

```
private void saveCustomer_Click(object sender, RoutedEventArgs e)
{
    ...
    if (titleBe.HasError || foreNameBe.HasError || lastNameBe.HasError)
    {
        MessageBox.Show("Please correct errors", "Not Saved");
    }
    else
    {
        Binding customerBinding =
            BindingOperations.GetBinding(this.title, ComboBox.TextProperty);
        Customer customer = customerBinding.Source as Customer;
        MessageBox.Show(customer.ToString(), "Saved");
    }
}
```

Test the application again

1. On the *Debug* menu, click *Start Without Debugging* to build and run the application.

2. When the Customer Details form appears, set the *Title* combo box to *"Mrs"*.

   Notice that the combo box is not highlighted because it has not yet been validated.

3. On the *File* menu, click *Save*.

   A message box should appear with the message "Please correct errors," and the *Title* and *Name* fields on the form should be highlighted. This is because the value in the *Title* combo box does not match the gender, and you have left the *Name* fields blank. If you rest the mouse on the fields highlighted with the red border, the ScreenTips display the reason for the error message.

4. Click *OK* to close the message box.

5. Click the *Female* radio button, and then enter a first name and last name for the customer.

   Notice that the highlighting of the controls with errors does not disappear.

6. On the *File* menu, click *Save* again.

   This time, the data is complete and consistent. A message box should appear, displaying the full details of the customer, and the highlighting on the form disappears.

7. Click *OK*, and exit the application.

This chapter has shown you how to perform basic validation by using the default exception validation rule processing provided by using data binding. You can also define your own custom validation rules if you want to perform more complex checks. For further information, see the documentation provided with Visual Studio 2008.

- If you want to continue to the next chapter

    Keep Visual Studio 2008 running, and turn to Chapter 25.

- If you want to exit Visual Studio 2008 now

    On the *File* menu, click *Exit*. If you see a *Save* dialog box, click *Yes* (if you are using Visual Studio 2008) or *Save* (if you are using Microsoft Visual C# 2008 Express Edition) and save the project.

Chapter 24 Quick Reference

To	Do this

Use data binding to bind a property of a control on a form to a property of an object	In the XAML code for the property of the control, specify a binding source identifying the object and the name of the property in the object to bind to. For example:
	
	`<TextBox ...>`
	
	`    <TextBox.Text>`
	
	`        <Binding Source="{StaticResource customerData}"`
	`            Path="ForeName" />`
	
	`    </TextBox.Text>`
	
	`</TextBox>`

Enable a data binding to validate data entered by the user	Specify the *Binding.ValidationRules* element as part of the binding. For example:
	
	`<Binding Source="{StaticResource customerData}"`
	`    Path="ForeName" />`
	`    <Binding.ValidationRules>`
	`        <ExceptionValidationRule/>`
	`    </Binding.ValidationRules>`
	`</Binding>`

Display error information in a nonintrusive manner	Define a style that detects a change to the *Validation.HasError* property of the control, and then set the *ToolTip* property of the control to the message returned by the exception. Apply this style to all controls that require validation. For example:

```
<Style x:Key="errorStyle" TargetType="Control">
  <Style.Triggers>
    <Trigger Property="Validation.HasError"
     Value="True">
      <Setter Property="ToolTip"
        Value="{Binding RelativeSource=
          {x:Static RelativeSource.Self},
          Path=(Validation.Errors)[0].ErrorContent}" />
    </Trigger>
  </Style.Triggers>
</Style>
```

Validate all the controls on a form under programmatic control rather than when the user moves from control to control	In the XAML code for the binding, set the *UpdateSourceTrigger* property of the binding to *"Explicit"* to defer validation until the application requests it. To validate the data for all controls, create a *BindingExpression* object for each bound property of each control, and call the *UpdateSource* method. Examine the *HasError* property of each *BindingExpression* object. If this property is *true*, the validation failed.

Part V
Managing Data

Chapter 25
Querying Information in a Database

After completing this chapter, you will be able to:

- Fetch and display data from a Microsoft SQL Server database by using Microsoft ADO.NET.

- Define entity classes for holding data retrieved from a database.

- Use DLINQ to query a database and populate instances of entity classes.

- Create a custom *DataContext* class for accessing a database in a typesafe manner.

In Part IV of this book, "Working with Windows Applications," you learned how to use Microsoft Visual C# to build user interfaces and present and validate information. In Part V, you will learn about managing data by using the data access functionality available in Microsoft Visual Studio 2008 and the Microsoft .NET Framework. The chapters in this part of the book describe ADO.NET, a library of objects specifically designed to make it easy to write applications that use databases. In this chapter, you will also learn how to query data by using DLINQ—extensions to LINQ based on ADO.NET that are designed for retrieving data from a database. In Chapter 26, "Displaying and Editing Data by Using Data Binding," you will learn more about using ADO.NET and DLINQ for updating data.

 Important To perform the exercises in this chapter, you must have installed Microsoft SQL Server 2005 Express Edition, Service Pack 2. This software is available on the retail DVD with Microsoft Visual Studio 2008 and Visual C# 2008 Express Edition and is installed by default.

 Important It is recommended that you use an account that has Administrator privileges to perform the exercises in this chapter and the remainder of this book.

Querying a Database by Using ADO.NET

The ADO.NET class library contains a comprehensive framework for building applications that need to retrieve and update data held in a relational database. The model defined by ADO.NET is based on the notion of data providers. Each database management system (such as SQL Server, Oracle, IBM DB2, and so on) has its own data provider that implements an abstraction of the mechanisms for connecting to a database, issuing queries, and updating data. By using these abstractions, you can write portable code that is independent of the underlying database management system. In this chapter, you will connect to a database managed by SQL Server 2005 Express Edition, but the techniques that you will learn are equally applicable when using a different database management system.

The Northwind Database

Northwind Traders is a fictitious company that sells edible goods with exotic names. The Northwind database contains several tables with information about the goods that Northwind Traders sells, the customers they sell to, orders placed by customers, suppliers from whom Northwind Traders obtains goods to resell, shippers that they use to send goods to customers, and employees who work for Northwind Traders. Figure 25-1 shows all the tables in the Northwind database and how they are related to one another. The tables that you will be using in this chapter are *Orders* and *Products*.

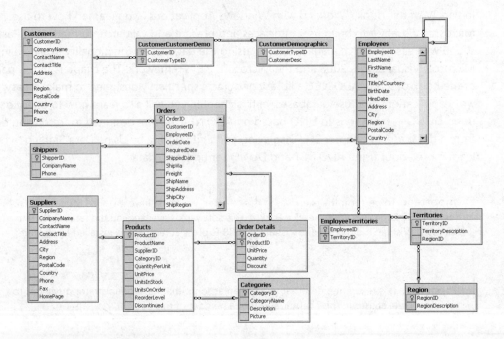

Creating the Database

Before proceeding further, you need to create the Northwind database.

Granting Permissions for Creating a SQL Server 2005 Database

You must have administrative rights for SQL Server 2005 Express before you can create a database. By default, if you are using the Windows Vista operating system, the computer *Administrator* account and members of the *Administrators* group do not have these rights. You can easily grant these permissions by using the SQL Server 2005 User Provisioning Tool for Vista, as follows:

1. Log on to your computer as an account that has administrator access.

2. Run the sqlprov.exe program, located in the folder C:\Program Files\Microsoft SQL Server\90\Shared.

3. In the *User Account Control* dialog box, click *Continue*. A console window briefly appears, and then the *SQL Server User Provisioning on Vista* window is displayed.

4. In the *User to provision* text box, type the name of the account you are using to perform the exercises. (Replace *YourComputer\YourAccount* with the name of your computer and your account.)

5. In the *Available privileges* box, click *Member of SQL Server SysAdmin role on SQLEXPRESS*, and then click the **>>** button.

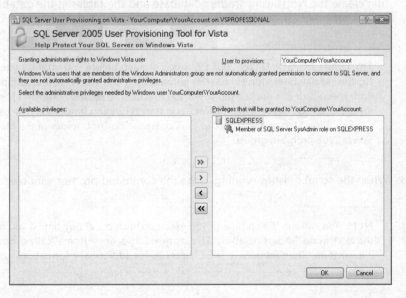

6. Click *OK*.

   The permission will be granted to the specified user, and the SQL Server 2005 User Provisioning Tool for Vista will close automatically.

Create the Northwind database

1. On the Windows *Start* menu, click *All Programs*, click *Accessories*, and then click *Command Prompt* to open a command prompt window. If you are using Windows Vista, in the command prompt window type the following command to go to the \Microsoft Press\Visual CSharp Step by Step\Chapter 25 folder under your Documents folder. Replace *Name* with your user name.

```
cd "\Users\Name\Documents\Microsoft Press\Visual CSharp Step by Step\Chapter 25"
```

If you are using Windows XP, type the following command to go to the \Microsoft Press\Visual CSharp Step by Step\Chapter 25 folder under your My Documents folder, replacing *Name* with your user name.

```
cd "\Documents and Settings\Name\My Documents\Microsoft Press\Visual CSharp Step by Step\Chapter 25"
```

2. In the command prompt window, type the following command:

```
sqlcmd -S YourComputer\SQLExpress -E -iinstnwnd.sql
```

Replace *YourComputer* with the name of your computer.

This command uses the sqlcmd utility to connect to your local instance of SQL Server 2005 Express and run the instnwnd.sql script. This script contains the SQL commands that create the Northwind Traders database and the tables in the database and fills them with some sample data.

> **Tip** Ensure that SQL Server 2005 Express is running before you attempt to create the Northwind database. (It is set to start automatically by default. You will simply receive an error message if it is not started when you execute the *sqlcmd* command.) You can check the status of SQL Server 2005 Express, and start it running if necessary, by using the SQL Configuration Manager available in the Configuration Tools folder of the Microsoft SQL Server 2005 program group.

3. When the script finishes running, close the command prompt window.

> **Note** You can run the command you executed in step 2 at any time if you need to reset the Northwind Traders database. The instnwnd.sql script automatically drops the database if it exists and then rebuilds it. See Chapter 26 for additional information.

Using ADO.NET to Query Order Information

In the next set of exercises, you will write code to access the Northwind database and display information in a simple console application. The aim of the exercise is to help you learn more about ADO.NET and understand the object model it implements. In later exercises, you will use DLINQ to query the database. In Chapter 26, you will see how to use the wizards included with Visual Studio 2008 to generate code that can retrieve and update data and display data graphically in a Windows Presentation Foundation (WPF) application.

The application you are going to create first will produce a simple report displaying information about customers' orders. The program will prompt the user for a customer ID and then display the orders for that customer.

Connect to the database

1. Start Visual Studio 2008 if it is not already running.

2. Create a new project called ReportOrders by using the Console Application template. Save it in the \Microsoft Press\Visual CSharp Step By Step\Chapter 25 folder under your Documents folder, and then click *OK*.

> **Note** Remember, if you are using Visual C# 2008 Express Edition, you can specify the location for saving your project by setting the *Visual Studio projects location* in the *Projects and Solutions* section of the *Options* dialog box on the *Tools* menu.

3. In *Solution Explorer*, change the name of the file Program.cs to Report.cs. In the *Microsoft Visual Studio* message, click *Yes* to change all references of the *Program* class to *Report*.

4. In the *Code and Text Editor* window, add the following *using* statement to the list at the top of the file:

```
using System.Data.SqlClient;
```

The *System.Data.SqlClient* namespace contains the SQL Server data provider classes for ADO.NET. These classes are specialized versions of the ADO.NET classes, optimized for working with SQL Server.

5. In the *Main* method of the *Report* class, add the following statement shown in bold type, which declares a *SqlConnection* object:

```
static void Main(string[] args)
{
    SqlConnection dataConnection = new SqlConnection();
}
```

SqlConnection is a subclass of an ADO.NET class called *Connection*. It is designed to handle connections to SQL Server databases.

6. After the variable declaration, add a *try/catch* block to the *Main* method. All the code that you will write for gaining access to the database goes inside the *try* part of this block. In the *catch* block, add a simple handler that catches *SqlException* exceptions. The new code is shown in bold type here:

```
static void Main(string[] args)
{
    ...
    try
    {
        // You will add your code here in a moment
    }
    catch(SqlException e)
    {
        Console.WriteLine("Error accessing the database: {0}", e.Message);
    }
}
```

A *SqlException* is thrown if an error occurs when accessing a SQL Server database.

7. Replace the comment in the *try* block with the code shown in bold type here:

```
try
{
    dataConnection.ConnectionString =
        "Integrated Security=true;Initial Catalog=Northwind;" +
        "Data Source=YourComputer\\SQLExpress";
    dataConnection.Open();
}
```

Important In the *ConnectionString* property, replace *YourComputer* with the name of your computer. Make sure that you type the string on a single line.

This code attempts to create a connection to the Northwind database. The contents of the *ConnectionString* property of the *SqlConnection* object contain elements that specify that the connection will use Windows Authentication to connect to the Northwind database on your local instance of SQL Server 2005 Express Edition. This is the preferred method of access because you do not have to prompt the user for any form of user name or password, and you are not tempted to hard-code user names and passwords into your application. Notice that a semicolon separates all the elements in the *ConnectionString*.

You can also encode many other elements in the connection string. See the documentation supplied with Visual Studio 2008 for details.

Using SQL Server Authentication

Windows Authentication is useful for authenticating users who are all members of a Windows domain. However, there might be occasions when the user accessing the database does not have a Windows account, for example, if you are building an application designed to be accessed by remote users over the Internet. In these cases, you can use the *User ID* and *Password* parameters instead, like this:

```
string userName = ...;
string password = ...;
// Prompt the user for their name and password, and fill these variables

string connString = String.Format(
    "User ID={0};Password={1};Initial Catalog=Northwind;" +
    "Data Source=YourComputer\\SQLExpress", username, password);

myConnection.ConnectionString = connString;
```

At this point, I should offer a sentence of advice: never hard-code user names and passwords into your applications. Anyone who obtains a copy of the source code (or who reverse-engineers the compiled code) can see this information, and this renders the whole point of security meaningless.

The next step is to prompt the user for a customer ID and then query the database to find all of the orders for that customer.

Query the *Orders* table

1. Add the statements shown here in bold type to the *try* block after the *dataConnection. Open();* statement:

```
try
{
    ...
    Console.Write("Please enter a customer ID (5 characters): ");
    string customerId = Console.ReadLine();
}
```

These statements prompt the user for a customer ID and read the user's response in the string variable *customerId*.

2. Type the following statements shown in bold type after the code you just entered:

```
try
{
    ...
    SqlCommand dataCommand = new SqlCommand();
    dataCommand.Connection = dataConnection;
```

```
dataCommand.CommandText =
    "SELECT OrderID, OrderDate, ShippedDate, ShipName, ShipAddress, " +
    "ShipCity, ShipCountry " +
    "FROM Orders WHERE CustomerID='" + customerId + "'";
Console.WriteLine("About to execute: {0}\n\n", dataCommand.CommandText);
}
```

The first statement creates a *SqlCommand* object. Like *SqlConnection*, this is a specialized version of an ADO.NET class, *Command*, that has been designed for performing queries against a SQL Server database. An ADO.NET *Command* object is used to execute a command against a data source. In the case of a relational database, the text of the command is a SQL statement.

The second line of code sets the *Connection* property of the *SqlCommand* object to the database connection you opened in the preceding exercise. The next two statements populate the *CommandText* property with a SQL SELECT statement that retrieves information from the *Orders* table for all orders that have a *CustomerID* that matches the value in the *customerId* variable. The *Console.WriteLine* statement just repeats the command about to be executed to the screen.

> **Important** If you are an experienced database developer, you will probably be about to e-mail me telling me that using string concatenation to build SQL queries is bad practice. This approach renders your application vulnerable to SQL injection attacks. However, the purpose of this code is to quickly show you how to execute queries against a SQL Server database by using ADO.NET, so I have deliberately kept it simple. Do not write code such as this in your production applications.
>
> For a description of what a SQL injection attack is, how dangerous it can be, and how you should write code to avoid such attacks, see the SQL Injection topic in SQL Server Books Online, available at *http://msdn2.microsoft.com/en-us/library/ms161953.aspx*.

3. Add the following statement shown in bold type after the code you just entered:

```
try
{
    ...
    SqlDataReader dataReader = dataCommand.ExecuteReader();
}
```

The *ExecuteReader* method of a *SqlCommand* object constructs a *SqlDataReader* object that you can use to fetch the rows identified by the SQL statement. The *SqlDataReader* class provides the fastest mechanism available (as fast as your network allows) for retrieving data from a SQL Server.

The next task is to iterate through all the orders (if there are any) and display them.

Fetch data and display orders

1. Add the *while* loop shown here in bold type after the statement that creates the *SqlDataReader* object:

```
try
{
    ...
    while (dataReader.Read())
    {
        // Code to display the current row
    }
}
```

The *Read* method of the *SqlDataReader* class fetches the next row from the database. It returns *true* if another row was retrieved successfully; otherwise, it returns *false*, usually because there are no more rows. The *while* loop you have just entered keeps reading rows from the *dataReader* variable and finishes when there are no more rows.

2. Add the statements shown here in bold type to the body of the *while* loop you created in the preceding step:

```
while (dataReader.Read())
{
    int orderId = dataReader.GetInt32(0);
    DateTime orderDate = dataReader.GetDateTime(1);
    DateTime shipDate = dataReader.GetDateTime(2);
    string shipName = dataReader.GetString(3);
    string shipAddress = dataReader.GetString(4);
    string shipCity = dataReader.GetString(5);
    string shipCountry = dataReader.GetString(6);
    Console.WriteLine(
        "Order: {0}\nPlaced: {1}\nShipped: {2}\n" +
        "To Address: {3}\n{4}\n{5}\n{6}\n\n", orderId, orderDate,
        shipDate, shipName, shipAddress, shipCity, shipCountry);
}
```

This block of code shows how you read the data from the database by using a *SqlDataReader* object. A *SqlDataReader* object contains the most recent row retrieved from the database. You can use the *GetXXX* methods to extract the information from each column in the row—there is a *GetXXX* method for each common type of data. For example, to read an *int* value, you use the *GetInt32* method; to read a string, you use the *GetString* method; and you can probably guess how to read a *DateTime* value. The *GetXXX* methods take a parameter indicating which column to read: 0 is the first column, 1 is the second column, and so on. The preceding code reads the various columns from the current *Orders* row, stores the values in a set of variables, and then prints out the values of these variables.

Firehose Cursors

One of the major drawbacks in a multiuser database application is locked data. Unfortunately, it is common to see applications retrieve rows from a database and keep those rows locked to prevent another user from changing the data while the application is using them. In some extreme circumstances, an application can even prevent other users from reading data that it has locked. If the application retrieves a large number of rows, it locks a large proportion of the table. If there are many users running the same application at the same time, they can end up waiting for one another to release locks and it all leads to a slow-running and frustrating mess.

The *SqlDataReader* class has been designed to remove this drawback. It fetches rows one at a time and does not retain any locks on a row after it has been retrieved. It is wonderful for improving concurrency in your applications. The *SqlDataReader* class is sometimes referred to as a "firehose cursor." (The term *cursor* is an acronym that stands for "current set of rows.")

When you have finished using a database, it's good practice to close your connection and release any resources you have been using.

Disconnect from the database, and test the application

1. Add the statement shown here in bold type after the *while* loop in the *try* block:

   ```
   try
   {
       ...
       while(dataReader.Read())
       {
           ...
       }
       dataReader.Close();
   }
   ```

   This statement closes the *SqlDataReader* object. You should always close a *SqlDataReader* object when you have finished with it because you will not able to use the current *SqlConnection* object to run any more commands until you do. It is also considered good practice to do it even if all you are going to do next is close the *SqlConnection*.

Note If you activate multiple active result sets (MARS) with SQL Server 2005, you can open more than one *SqlDataReader* object against the same *SqlConnection* object and process multiple sets of data. MARS is disabled by default. To learn more about MARS and how you can activate and use it, consult SQL Server 2005 Books Online.

2. After the *catch* block, add the following *finally* block:

```
catch(SqlException e)
{
    ...
}
finally
{
    dataConnection.Close();
}
```

Database connections are scarce resources. You need to ensure that they are closed when you have finished with them. Putting this statement in a *finally* block guarantees that the *SqlConnection* will be closed, even if an exception occurs; remember that the code in the *finally* block will be executed after the *catch* handler has finished.

> **Tip** An alternative approach to using a *finally* block is to wrap the code that creates the *SqlDataConnection* object in a *using* statement, as shown in the following code. At the end of the block defined by the *using* statement, the *SqlConnection* object is closed automatically, even if an exception occurs:
>
> ```
> using (SqlConnection dataConnection = new SqlConnection())
> {
>     try
>     {
>         dataConnection.ConnectionString = "...";
>         ...
>     }
>     catch (SqlException e)
>     {
>         Console.WriteLine("Error accessing the database: {0}", e.Message);
>     }
> }
> ```

3. On the *Debug* menu, click *Start Without Debugging* to build and run the application.

4. At the customer ID prompt, type the customer ID **VINET**, and press Enter.

The SQL SELECT statement appears, followed by the orders for this customer, as shown in the following image:

```
C:\Windows\system32\cmd.exe                                              _ □ X
Please enter a customer ID (5 characters): VINET
About to execute: SELECT OrderID, OrderDate, ShippedDate, ShipName, ShipAddress,
 ShipCity, ShipCountry FROM Orders WHERE CustomerID='VINET'


Order: 10248
Placed: 04/07/1996 00:00:00
Shipped: 16/07/1996 00:00:00
To Address: Vins et alcools Chevalier
59 rue de l'Abbaye
Reims
France


Order: 10274
Placed: 06/08/1996 00:00:00
Shipped: 16/08/1996 00:00:00
To Address: Vins et alcools Chevalier
59 rue de l'Abbaye
Reims
France


Order: 10295
Placed: 02/09/1996 00:00:00
```

You can scroll back through the console window to view all the data. Press the Enter key to close the console window when you have finished.

5. Run the application again, and then type **BONAP** when prompted for the customer ID.

Some rows appear, but then an error occurs. If you are using Windows Vista, a message box appears with the message "ReportOrders has stopped working." Click *Close program* (or *Close the program* if you are using Visual C# Express). If you are using Windows XP, a message box appears with the message "ReportOrders has encountered a problem and needs to close. We are sorry for the inconvenience." Click *Don't Send*.

An error message containing the text "Data is Null. This method or property cannot be called on Null values" appears in the console window.

The problem is that relational databases allow some columns to contain null values. A null value is a bit like a null variable in C#: It doesn't have a value, but if you try to read it, you get an error. In the *Orders* table, the *ShippedDate* column can contain a null value if the order has not yet been shipped. You should also note that this is a *SqlNullValueException* and consequently is not caught by the *SqlException* handler.

6. Press Enter to close the console window and return to Visual Studio 2008.

Closing Connections

In many older applications, you might notice a tendency to open a connection when the application starts and not close the connection until the application terminates. The rationale behind this strategy was that opening and closing database connections were expensive and time-consuming operations. This strategy had an impact on the scalability of applications because each user running the application had a connection to the database open while the application was running, even if the user went to lunch for a few hours. Most databases limit the number of concurrent connections that they allow. (Sometimes this is because of licensing, but usually it's because each connection consumes resources on the database server that are not infinite.) Eventually, the database would hit a limit on the number of users that could operate concurrently.

Most .NET Framework data providers (including the SQL Server provider) implement *connection pooling*. Database connections are created and held in a pool. When an application requires a connection, the data access provider extracts the next available connection from the pool. When the application closes the connection, it is returned to the pool and made available for the next application that wants a connection. This means that opening and closing database connections are no longer expensive operations. Closing a connection does not disconnect from the database; it just returns the connection to the pool. Opening a connection is simply a matter of obtaining an already-open connection from the pool. Therefore, you should not hold on to connections longer than you need to—open a connection when you need it, and close it as soon as you have finished with it.

You should note that the *ExecuteReader* method of the *SqlCommand* class, which creates a *SqlDataReader*, is overloaded. You can specify a *System.Data. CommandBehavior* parameter that automatically closes the connection used by the *SqlDataReader* when the *SqlDataReader* is closed, like this:

```
SqlDataReader dataReader =
    dataCommand.ExecuteReader(System.Data.CommandBehavior.CloseConnection);
```

When you read the data from the *SqlDataReader* object, you should check that the data you are reading is not null. You'll see how to do this next.

Handle null database values

1. In the *Main* method, change the code in the body of the *while* loop to contain an *if ... else* block, as shown here in bold type:

```
while (dataReader.Read())
{
    int orderId = dataReader.GetInt32(0);
    if (dataReader.IsDBNull(2))
    {
        Console.WriteLine("Order {0} not yet shipped\n\n", orderId);
    }
    else
    {
        DateTime orderDate = dataReader.GetDateTime(1);
        DateTime shipDate = dataReader.GetDateTime(2);
        string shipName = dataReader.GetString(3);
        string shipAddress = dataReader.GetString(4);
        string shipCity = dataReader.GetString(5);
        string shipCountry = dataReader.GetString(6);
        Console.WriteLine(
            "Order {0}\nPlaced {1}\nShipped{2}\n" +
            "To Address {3}\n{4}\n{5}\n{6}\n\n", orderId, orderDate,
            shipDate, shipName, shipAddress, shipCity, shipCountry);
    }
}
```

The *if* statement uses the *IsDBNull* method to determine whether the *ShippedDate* column (column 2 in the table) is null. If it is null, no attempt is made to fetch it (or any of the other columns, which should also be null if there is no *ShippedDate* value); otherwise, the columns are read and printed as before.

2. Build and run the application again.

3. Type **BONAP** for the customer ID when prompted.

   This time you do not get any errors, but you receive a list of orders that have not yet been shipped.

4. When the application finishes, press Enter and return to Visual Studio 2008.

Querying a Database by Using DLINQ

In Chapter 20, "Querying In-Memory Data by Using Query Expressions," you saw how to use LINQ to examine the contents of enumerable collections held in memory. LINQ provides query expressions, which use SQL-like syntax for performing queries and generating a result set that you can then step through. It should come as no surprise that you can use an extended form of LINQ, called DLINQ, for querying and manipulating the contents of a database. DLINQ is built on top of ADO.NET. DLINQ provides a high level of abstraction, removing the need for you to worry about the details of constructing an ADO.NET *Command* object, iterating through a result set returned by a *DataReader* object, or fetching data column by column by using the various *GetXXX* methods.

Defining an Entity Class

You saw in Chapter 20 that using LINQ requires the objects that you are querying be enumerable; they must be collections that implement the *IEnumerable* interface. DLINQ can create its own enumerable collections of objects based on classes you define and that map directly to tables in a database. These classes are called *entity classes*. When you connect to a database and perform a query, DLINQ can retrieve the data identified by your query and create an instance of an entity class for each row fetched.

The best way to explain DLINQ is to see an example. The *Products* table in the Northwind database contains columns that contain information about the different aspects of the various products that Northwind Traders sells. The part of the instnwnd.sql script that you ran in the first exercise in this chapter contains a *CREATE TABLE* statement that looks similar to this (some of the columns, constraints, and other details have been omitted):

```
CREATE TABLE "Products" (
    "ProductID" "int" NOT NULL ,
    "ProductName" nvarchar (40) NOT NULL ,
    "SupplierID" "int" NULL ,
    "UnitPrice" "money" NULL,
    CONSTRAINT "PK_Products" PRIMARY KEY CLUSTERED ("ProductID"),
    CONSTRAINT "FK_Products_Suppliers" FOREIGN KEY ("SupplierID")
        REFERENCES "dbo"."Suppliers" ("SupplierID")
)
```

You can define an entity class that corresponds to the *Products* table like this:

```
[Table(Name = "Products")]
public class Product
{
    [Column(IsPrimaryKey = true, CanBeNull = false)]
    public int ProductID { get; set; }

    [Column(CanBeNull = false)]
    public string ProductName { get; set; }
```

```
    [Column]
    public int? SupplierID { get; set; }

    [Column(DbType = "money")]
    public decimal? UnitPrice { get; set; }
}
```

The *Product* class contains a property for each of the columns in which you are interested in the *Products* table. You don't have to specify every column from the underlying table, but any columns that you omit will not be retrieved when you execute a query based on this entity class. The important points to note are the *Table* and *Column* attributes.

The *Table* attribute identifies this class as an entity class. The *Name* parameter specifies the name of the corresponding table in the database. If you omit the *Name* parameter, DLINQ assumes that the entity class name is the same as the name of the corresponding table in the database.

The *Column* attribute describes how a column in the *Products* table maps to a property in the *Product* class. The *Column* attribute can take a number of parameters. The ones shown in this example and described in the following list are the most common:

- The *IsPrimaryKey* parameter specifies that the property makes up part of the primary key. (If the table has a composite primary key spanning multiple columns, you should specify the *IsPrimaryKey* parameter for each corresponding property in the entity class.)

- The *DbType* parameter specifies the type of the underlying column in the database. In many cases, DLINQ can detect and convert data in a column in the database to the type of the corresponding property in the entity class, but in some situations you need to specify the data type mapping yourself. For example, the *UnitPrice* column in the *Products* table uses the SQL Server *money* type. The entity class specifies the corresponding property as a *decimal* value.

> **Note** The default mapping of *money* data in SQL Server is to the *decimal* type in an entity class, so the *DbType* parameter shown here is actually redundant. However, I wanted to show you the syntax.

- The *CanBeNull* parameter indicates whether the column in the database can contain a null value. The default value for the *CanBeNull* parameter is *true*. Notice that the two properties in the *Product* class that correspond to columns that permit null values in the database (*SupplierID* and *UnitPrice*) are defined as nullable types in the entity class.

Note You can also use DLINQ to create new databases and tables based on the definitions of your entity classes by using the *CreateDatabase* method of the *DataContext* object. In the current version of DLINQ, the part of the library that creates tables uses the definition of the *DbType* parameter to specify whether a column should allow null values. If you are using DLINQ to create a new database, you should specify the nullability of each column in each table in the *DbType* parameter, like this:

```
[Column(DbType = "NVarChar(40) NOT NULL", CanBeNull = false)]
public string ProductName { get; set; }
...
[Column(DbType = "Int NULL", CanBeNull = true)]
public int? SupplierID { get; set; }
```

Like the *Table* attribute, the *Column* attribute provides a *Name* parameter that you can use to specify the name of the underlying column in the database. If you omit this parameter, DLINQ assumes that the name of the column is the same as the name of the property in the entity class.

Creating and Running a DLINQ Query

Having defined an entity class, you can use it to fetch and display data from the *Products* table. The following code shows the basic steps for doing this:

```
DataContext db = new DataContext("Integrated Security=true;" +
    "Initial Catalog=Northwind;Data Source=YourComputer\\SQLExpress");

Table<Product> products = db.GetTable<Product>();
var productsQuery = from p in products
                    select p;

foreach (var product in productsQuery)
{
    Console.WriteLine("ID: {0}, Name: {1}, Supplier: {2}, Price: {3:C}",
                    product.ProductID, product.ProductName,
                    product.SupplierID, product.UnitPrice);
}
```

Note Remember that the keywords *from*, *in*, and *select* in this context are C# identifiers. You must type them in lowercase.

The *DataContext* class is responsible for managing the relationship between your entity classes and the tables in the database. You use it to establish a connection to the database and create collections of the entity classes. The *DataContext* constructor expects a connection string as a parameter, specifying the database that you want to use. This connection string is exactly the same as the connection string that you would use when connecting through an ADO.NET *Connection* object. (The *DataContext* class actually creates an ADO.NET connection behind the scenes.)

The generic *GetTable<TEntity>* method of the *DataContext* class expects an entity class as its *IEntity* type parameter. This method constructs an enumerable collection based on this type and returns the collection as a *Table<TEntity>* type. You can perform DLINQ queries over this collection. The query shown in this example simply retrieves every object from the *Products* table.

Note If you need to recap your knowledge of LINQ query expressions, turn back to Chapter 20.

The *foreach* statement iterates through the results of this query and displays the details of each product. The following image shows the results of running this code. (The prices shown are per case, not per individual item.)

```
C:\Windows\system32\cmd.exe                                    _ □ X
ID: 1, Name: Chai, Supplier: 1, Price: £18.00
ID: 2, Name: Chang, Supplier: 1, Price: £19.00
ID: 3, Name: Aniseed Syrup, Supplier: 1, Price: £10.00
ID: 4, Name: Chef Anton's Cajun Seasoning, Supplier: 2, Price: £22.00
ID: 5, Name: Chef Anton's Gumbo Mix, Supplier: 2, Price: £21.35
ID: 6, Name: Grandma's Boysenberry Spread, Supplier: 3, Price: £25.00
ID: 7, Name: Uncle Bob's Organic Dried Pears, Supplier: 3, Price: £30.00
ID: 8, Name: Northwoods Cranberry Sauce, Supplier: 3, Price: £40.00
ID: 9, Name: Mishi Kobe Niku, Supplier: 4, Price: £97.00
ID: 10, Name: Ikura, Supplier: 4, Price: £31.00
ID: 11, Name: Queso Cabrales, Supplier: 5, Price: £21.00
ID: 12, Name: Queso Manchego La Pastora, Supplier: 5, Price: £38.00
ID: 13, Name: Konbu, Supplier: 6, Price: £6.00
ID: 14, Name: Tofu, Supplier: 6, Price: £23.25
ID: 15, Name: Genen Shouyu, Supplier: 6, Price: £15.50
ID: 16, Name: Pavlova, Supplier: 7, Price: £17.45
ID: 17, Name: Alice Mutton, Supplier: 7, Price: £39.00
ID: 18, Name: Carnarvon Tigers, Supplier: 7, Price: £62.50
ID: 19, Name: Teatime Chocolate Biscuits, Supplier: 8, Price: £9.20
ID: 20, Name: Sir Rodney's Marmalade, Supplier: 8, Price: £81.00
ID: 21, Name: Sir Rodney's Scones, Supplier: 8, Price: £10.00
ID: 22, Name: Gustaf's Knäckebröd, Supplier: 9, Price: £21.00
ID: 23, Name: Tunnbröd, Supplier: 9, Price: £9.00
ID: 24, Name: Guaraná Fantástica, Supplier: 10, Price: £4.50
ID: 25, Name: NuNuCa Nuß-Nougat-Creme, Supplier: 11, Price: £14.00
```

The *DataContext* object controls the database connection automatically; it opens the connection immediately prior to fetching the first row of data in the *foreach* statement and then closes the connection after the last row has been retrieved.

The DLINQ query shown in the preceding example retrieves every column for every row in the *Products* table. In this case, you can actually iterate through the *products* collection directly, like this:

```
Table<Product> products = db.GetTable<Product>();

foreach (Product product in products)
{
    ...
}
```

When the *foreach* statement runs, the *DataContext* object constructs a SQL SELECT statement that simply retrieves all the data from the *Products* table. If you want to retrieve a single row in the *Products* table, you can call the *Single* method of the *Products* entity class.

Single is an extension method that itself takes a method that identifies the row you want to find and returns this row as an instance of the entity class (as opposed to a collection of rows in a *Table* collection). You can specify the method parameter as a lambda expression. If the lambda expression does not identify exactly one row, the *Single* method returns an *InvalidOperationException*. The following code example queries the Northwind database for the product with the *ProductID* value of 27. The value returned is an instance of the *Product* class, and the *Console.WriteLine* statement prints the name of the product. As before, the database connection is opened and closed automatically by the *DataContext* object.

```
Product singleProduct = products.Single(p => p.ProductID == 27);
Console.WriteLine("Name: {0}", singleProduct.ProductName);
```

Deferred and Immediate Fetching

An important point to emphasize is that by default, DLINQ retrieves the data from the database only when you request it and not when you define a DLINQ query or create a *Table* collection. This is known as deferred fetching. In the example shown earlier that displays all of the products from the *Products* table, the *productsQuery* collection is populated only when the *foreach* loop runs. This mode of operation matches that of LINQ when querying in-memory objects; you will always see the most up-to-date version of the data, even if the data changes after you have run the statement that creates the *productsQuery* enumerable collection.

When the *foreach* loop starts, DLINQ creates and runs a SQL SELECT statement derived from the DLINQ query to create an ADO.NET *DataReader* object. Each iteration of the *foreach* loop performs the necessary *GetXXX* methods to fetch the data for that row. After the final row has been fetched and processed by the *foreach* loop, DLINQ closes the database connection.

Deferred fetching ensures that only the data an application actually uses is retrieved from the database. However, if you are accessing a database running on a remote instance of SQL Server, fetching data row by row does not make the best use of network bandwidth. In this scenario, you can fetch and cache all the data in a single network request by forcing immediate evaluation of the DLINQ query. You can do this by calling the *ToList* or *ToArray* extension methods, which fetch the data into a list or array when you define the DLINQ query, like this:

```
var productsQuery = from p in products.ToList()
                    select p;
```

In this code example, *productsQuery* is now an enumerable list, populated with information from the *Products* table. When you iterate over the data, DLINQ retrieves it from this list rather than sending fetch requests to the database.

Joining Tables and Creating Relationships

DLINQ supports the *join* query operator for combining and retrieving related data held in multiple tables. For example, the *Products* table in the Northwind database holds the ID of the supplier for each product. If you want to know the name of each supplier, you have to query the *Suppliers* table. The *Suppliers* table contains the *CompanyName* column, which specifies the name of the supplier company, and the *ContactName* column, which contains the name of the person in the supplier company that handles orders from Northwind Traders. You can define an entity class containing the relevant supplier information like this (the *SupplierName* column in the database is mandatory, but the *ContactName* allows null values):

```
[Table(Name = "Suppliers")]
public class Supplier
{
    [Column(IsPrimaryKey = true, CanBeNull = false)]
    public int SupplierID { get; set; }

    [Column(CanBeNull = false)]
    public string CompanyName { get; set; }

    [Column]
    public string ContactName { get; set; }
}
```

You can then instantiate *Table<Product>* and *Table<Supplier>* collections and define a DLINQ query to join these tables together, like this:

```
DataContext db = new DataContext(...);
Table<Product> products = db.GetTable<Product>();
Table<Supplier> suppliers = db.GetTable<Supplier>();
var productsAndSuppliers = from p in products
                           join s in suppliers
                           on p.SupplierID equals s.SupplierID
                           select new { p.ProductName, s.CompanyName, s.ContactName };
```

When you iterate through the *productsAndSuppliers* collection, DLINQ will execute a SQL SELECT statement that joins the *Products* and *Suppliers* tables in the database over the *SupplierID* column in both tables and fetches the data.

However, with DLINQ you can specify the relationships between tables as part of the definition of the entity classes. DLINQ can then fetch the supplier information for each product automatically without requiring that you code a potentially complex and error-prone *join* statement. Returning to the products and suppliers example, these tables have a many-to-one relationship in the Northwind database; each product is supplied by a single supplier, but a single supplier can supply several products. Phrasing this relationship slightly differently, a row in the *Products* table can reference a single row in the *Suppliers* table through the *SupplierID* columns in both tables, but a row in the *Suppliers* table can reference a whole set of rows in the *Products* table. DLINQ provides the *EntityRef<TEntity>* and *EntitySet<TEntity>* generic types to model this type of relationship. Taking the *Product* entity class first, you can define the "one" side of the relationship with the *Supplier* entity class by using the *EntityRef<Supplier>* type, as shown here in bold type:

```
[Table(Name = "Products")]
public class Product
{
    [Column(IsPrimaryKey = true, CanBeNull = false)]
    public int ProductID { get; set; }
    ...
    [Column]
    public int? SupplierID { get; set; }
    ...
    private EntityRef<Supplier> supplier;
    [Association(Storage = "supplier", ThisKey = "SupplierID", OtherKey = "SupplierID")]
    public Supplier Supplier
    {
        get { return this.supplier.Entity; }
        set { this.supplier.Entity = value; }
    }
}
```

The private *supplier* field is a reference to an instance of the *Supplier* entity class. The public *Supplier* property provides access to this reference. The *Association* attribute specifies how DLINQ locates and populates the data for this property. The *Storage* parameter identifies the *private* field used to store the reference to the *Supplier* object. The *ThisKey* parameter indicates which property in the *Product* entity class DLINQ should use to locate the *Supplier* to reference for this product, and the *OtherKey* parameter specifies which property in the *Supplier* table DLINQ should match against the value for the *ThisKey* parameter. In this example, The *Product* and *Supplier* tables are joined across the *SupplierID* property in both entities.

> **Note** The *Storage* parameter is actually optional. If you specify it, DLINQ accesses the corresponding data member directly when populating it rather than going through the *set* accessor. The *set* accessor is required for applications that manually fill or change the entity object referenced by the *EntityRef<TEntity>* property. Although the *Storage* parameter is actually redundant in this example, it is recommended practice to include it.

The *get* accessor in the *Supplier* property returns a reference to the *Supplier* entity by using the *Entity* property of the *EntityRef<Supplier>* type. The *set* accessor populates this property with a reference to a *Supplier* entity.

You can define the "many" side of the relationship in the *Supplier* class with the *EntitySet<Product>* type, like this:

```
[Table(Name = "Suppliers")]
public class Supplier
{
    [Column(IsPrimaryKey = true, CanBeNull = false)]
    public int SupplierID { get; set; }
    ...
    private EntitySet<Product> products = null;
    [Association(Storage = "products", OtherKey = "SupplierID", ThisKey = "SupplierID")]
    public EntitySet<Product> Products
    {
        get { return this.products; }
        set { this.products.Assign(value); }
    }
}
```

> **Tip** It is conventional to use a singular noun for the name of an entity class and its properties. The exception to this rule is that *EntitySet<TEntity>* properties typically take the plural form because they represent a collection rather than a single entity.

This time, notice that the *Storage* parameter of the *Association* attribute specifies the private *EntitySet<Product>* field. An *EntitySet<TEntity>* object holds a collection of references to entities. The *get* accessor of the public *Products* property returns this collection. The *set* accessor uses the *Assign* method of the *EntitySet<Product>* class to populate this collection.

So, by using the *EntityRef<TEntity>* and *EntitySet<TEntity>* types you can define properties that can model a one-to-many relationship, but how do you actually fill these properties with data? The answer is that DLINQ fills them for you when it fetches the data. The following code creates an instance of the *Table<Product>* class and issues a DLINQ query to fetch the details of all products. This code is similar to the first DLINQ example you saw earlier. The difference is in the *foreach* loop that displays the data.

```
DataContext db = new DataContext(...);
Table<Product> products = db.GetTable<Product>();

var productsAndSuppliers = from p in products
                           select p;

foreach (var product in productsAndSuppliers)
{
    Console.WriteLine("Product {0} supplied by {1}",
        product.ProductName, product.Supplier.CompanyName);
}
```

The *Console.WriteLine* statement reads the value in the *ProductName* property of the product entity as before, but it also accesses the *Supplier* entity and displays the *CompanyName* property from this entity. If you run this code, the output looks like this:

As the code fetches each *Product* entity, DLINQ executes a second, deferred, query to retrieve the details of the supplier for that product so that it can populate the *Supplier* property, based on the relationship specified by the *Association* attribute of this property in the *Product* entity class.

When you have defined the *Product* and *Supplier* entities as having a one-to-many relationship, similar logic applies if you execute a DLINQ query over the *Table<Supplier>* collection, like this:

```
DataContext db = new DataContext(...);
Table<Supplier> suppliers = db.GetTable<Supplier>();
var suppliersAndProducts = from s in suppliers
                           select s;

foreach (var supplier in suppliersAndProducts)
{
    Console.WriteLine("Supplier name: {0}", supplier.CompanyName);
    Console.WriteLine("Products supplied");
    foreach (var product in supplier.Products)
    {
        Console.WriteLine("\t{0}", product.ProductName);
    }
    Console.WriteLine();
}
```

In this case, when the *foreach* loop fetches a supplier, it runs a second query (again deferred) to retrieve all the products for that supplier and populate the *Products* property. This time, however, the property is a collection (an *EntitySet<Product>*), so you can code a nested *foreach* statement to iterate through the set, displaying the name of each product. The output of this code looks like this:

Deferred and Immediate Fetching Revisited

Earlier in this chapter, I mentioned that DLINQ defers fetching data until the data is actually requested but that you could apply the *ToList* or *ToArray* extension method to retrieve data immediately. This technique does not apply to data referenced as *EntitySet<TEntity>* or *EntityRef<TEntity>* properties; even if you use *ToList* or *ToArray*, the data will still be fetched only when accessed. If you want to force DLINQ to query and fetch referenced data immediately, you can set the *LoadOptions* property of the *DataContext* object as follows:

```
DataContext db = new DataContext(...);
Table<Supplier> suppliers = db.GetTable<Supplier>();
DataLoadOptions loadOptions = new DataLoadOptions();
loadOptions.LoadWith<Supplier>(s => s.Products);
db.LoadOptions = loadOptions;
var suppliersAndProducts = from s in suppliers
                           select s;
```

The *DataLoadOptions* class provides the generic *LoadWith* method. By using this method, you can specify whether an *EntitySet<TEntity>* property in an instance should be loaded when the instance is populated. The parameter to the *LoadWith* method is another method, which you can supply as a lambda expression. The example shown here causes the *Products* property of each *Supplier* entity to be populated as soon as the data for each *Product* entity is fetched rather than being deferred. If you specify the *LoadOptions* property of the *DataContext* object together with the *ToList* or *ToArray* extension method of a *Table* collection, DLINQ will load the entire collection as well as the data for the referenced properties for the entities in that collection into memory as soon as the DLINQ query is evaluated.

> **Tip** If you have several *EntitySet<TEntity>* properties, you can call the *LoadWith* method of the same *LoadOptions* object several times, each time specifying the *EntitySet<TEntity>* to load.

Defining a Custom *DataContext* Class

The *DataContext* class provides functionality for managing databases and database connections, creating entity classes, and executing commands to retrieve and update data in a database. Although you can use the raw *DataContext* class provided with the .NET Framework, it is better practice to use inheritance and define your own specialized version that declares the various *Table<TEntity>* collections as public members. For example, here is a specialized *DataContext* class that exposes the *Products* and *Suppliers Table* collections as public members:

```
public class Northwind : DataContext
{
    public Table<Product> Products;
    public Table<Supplier> Suppliers;

    public Northwind(string connectionInfo) : base(connectionInfo)
    {
    }
}
```

Notice that the *Northwind* class also provides a constructor that takes a connection string as a parameter. You can create a new instance of the *Northwind* class and then define and run DLINQ queries over the *Table* collection classes it exposes like this:

```
Northwind nwindDB = new Northwind(...);

var suppliersQuery = from s in nwindDB.Suppliers
                     select s;

foreach (var supplier in suppliersQuery)
{
    ...
}
```

This practice makes your code easier to maintain, especially if you are retrieving data from multiple databases. Using an ordinary *DataContext* object, you can instantiate any entity class by using the *GetTable* method, regardless of the database to which the *DataContext* object connects. You find out that you have used the wrong *DataContext* object and have connected to the wrong database only at run time, when you try to retrieve data. With a custom *DataContext* class, you reference the *Table* collections through the *DataContext* object. (The base *DataContext* constructor uses a mechanism called *reflection* to examine its members, and it automatically instantiates any members that are *Table* collections—the details of how reflection works are outside the scope of this book.) It is obvious to which database you need to connect to retrieve data for a specific table; if IntelliSense does not display your table when you define the DLINQ query, you have picked the wrong *DataContext* class, and your code will not compile.

Using DLINQ to Query Order Information

In the following exercise, you will write a version of the console application that you developed in the preceding exercise that prompts the user for a customer ID and displays the details of any orders placed by that customer. You will use DLINQ to retrieve the data. You will then be able to compare DLINQ with the equivalent code written by using ADO.NET.

Define the *Order* entity class

1. Using Visual Studio 2008, create a new project called DLINQOrders by using the Console Application template. Save it in the \Microsoft Press\Visual CSharp Step By Step\Chapter 25 folder under your Documents folder, and then click *OK*.

2. In *Solution Explorer*, change the name of the file Program.cs to DLINQReport.cs. In the *Microsoft Visual Studio* message, click *Yes* to change all references of the *Program* class to *DLINQReport*.

3. On the *Project* menu, click *Add Reference*. In the *Add Reference* dialog box, click the *.NET* tab, select the *System.Data.Linq* assembly, and then click *OK*.

   This assembly holds the DLINQ types and attributes.

4. In the *Code and Text Editor* window, add the following *using* statements to the list at the top of the file:

   ```
   using System.Data.Linq;
   using System.Data.Linq.Mapping;
   using System.Data.SqlClient;
   ```

5. Add the *Order* entity class to the DLINQReport.cs file after the *DLINQReport* class, as follows:

   ```
   [Table(Name = "Orders")]
   public class Order
   {
   }
   ```

   The table is called *Orders* in the Northwind database. Remember that it is common practice to use the singular noun for the name of an entity class because an entity object represents one row from the database.

6. Add the property shown here in bold type to the *Order* class:

```
[Table(Name = "Orders")]
public class Order
{
    [Column(IsPrimaryKey = true, CanBeNull = false)]
    public int OrderID { get; set; }
}
```

The *OrderID* column is the primary key for this table in the Northwind database.

7. Add the following properties shown in bold type to the *Order* class:

```
[Table(Name = "Orders")]
public class Order
{
    ...
    [Column]
    public string CustomerID { get; set; }

    [Column]
    public DateTime? OrderDate { get; set; }

    [Column]
    public DateTime? ShippedDate { get; set; }

    [Column]
    public string ShipName { get; set; }

    [Column]
    public string ShipAddress { get; set; }

    [Column]
    public string ShipCity { get; set; }

    [Column]
    public string ShipCountry { get; set; }
}
```

These properties hold the customer ID, order date, and shipping information for an order. In the database, all of these columns allow null values, so it is important to use the nullable version of the *DateTime* type for the *OrderDate* and *ShippedDate* properties (*string* is a reference type that automatically allows null values). Notice that DLINQ automatically maps the SQL Server *NVarChar* type to the .NET Framework *string* type and the SQL Server *DateTime* type to the .NET Framework *DateTime* type.

8. Add the following *Northwind* class to the DLINQReport.cs file after the *Order* entity class:

```
public class Northwind : DataContext
{
    public Table<Order> Orders;
```

```
    public Northwind(string connectionInfo) : base (connectionInfo)
    {
    }
}
```

The *Northwind* class is a *DataContext* class that exposes a *Table* property based on the *Order* entity class. In the next exercise, you will use this specialized version of the *DataContext* class to access the *Orders* table in the database.

Retrieve order information by using a DLINQ query

1. In the *Main* method of the *DLINQReport* class, add the statement shown here in bold type, which creates a *Northwind* object. Be sure to replace *YourComputer* with the name of your computer:

```
static void Main(string[] args)
{
    Northwind northwindDB = new Northwind("Integrated Security=true;" +
        "Initial Catalog=Northwind;Data Source=YourComputer\\SQLExpress");
}
```

The connection string specified here is exactly the same as in the earlier exercise. The *northwindDB* object uses this string to connect to the Northwind database.

2. After the variable declaration, add a *try/catch* block to the *Main* method:

```
static void Main(string[] args)
{
    ...
    try
    {
        // You will add your code here in a moment
    }
    catch(SqlException e)
    {
        Console.WriteLine("Error accessing the database: {0}", e.Message);
    }
}
```

As when using ordinary ADO.NET code, DLINQ raises a *SqlException* if an error occurs when accessing a SQL Server database.

3. Replace the comment in the *try* block with the following code shown in bold type:

```
try
{
    Console.Write("Please enter a customer ID (5 characters): ");
    string customerId = Console.ReadLine();
}
```

These statements prompt the user for a customer ID and save the user's response in the string variable *customerId*.

4. Type the statement shown here in bold type after the code you just entered:

```
try
{
    ...
    var ordersQuery = from o in northwindDB.Orders
                      where String.Equals(o.CustomerID, customerId)
                      select o;
}
```

This statement defines the DLINQ query that will retrieve the orders for the specified customer.

5. Add the *foreach* statement and *if...else* block shown here in bold type after the code you added in the preceding step:

```
try
{
    ...
    foreach (var order in ordersQuery)
    {
        if (order.ShippedDate == null)
        {
            Console.WriteLine("Order {0} not yet shipped\n\n", order.OrderID);
        }
        else
        {
            // Display the order details
        }
    }
}
```

The *foreach* statement iterates through the orders for the customer. If the value in the *ShippedDate* column in the database is *null*, the corresponding property in the *Order* entity object is also *null*, and then the *if* statement outputs a suitable message.

6. Replace the comment in the *else* part of the *if* statement you added in the preceding step with the code shown here in bold type:

```
if (order.ShippedDate == null)
{
    ...
}
else
{
    Console.WriteLine("Order: {0}\nPlaced: {1}\nShipped: {2}\n" +
                      "To Address: {3}\n{4}\n{5}\n{6}\n\n", order.OrderID,
                      order.OrderDate, order.ShippedDate, order.ShipName,
                      order.ShipAddress, order.ShipCity,
                      order.ShipCountry);
}
```

7. On the *Debug* menu, click *Start Without Debugging* to build and run the application.

8. In the console window displaying the message "Please enter a customer ID (5 characters):", type **VINET**.

The application should display a list of orders for this customer. When the application has finished, press Enter to return to Visual Studio 2008.

9. Run the application again. This time type **BONAP** when prompted for a customer ID.

The final order for this customer has not yet shipped and contains a null value for the *ShippedDate* column. Verify that the application detects and handles this null value. When the application has finished, press Enter to return to Visual Studio 2008.

You have now seen the basic elements that DLINQ provides for querying information from a database. DLINQ has many more features that you can employ in your applications, including the ability to modify data and update a database. You will look briefly at some of these aspects of DLINQ in the next chapter.

- If you want to continue to the next chapter

  Keep Visual Studio 2008 running, and turn to Chapter 26.

- If you want to exit Visual Studio 2008 now

  On the *File* menu, click *Exit*. If you see a *Save* dialog box, click *Yes* (if you are using Visual Studio 2008) or *Save* (if you are using Visual C# 2008 Express Edition) and save the project.

Chapter 25 Quick Reference

To	Do this
Connect to a SQL Server database by using ADO.NET	Create a *SqlConnection* object, set its *ConnectionString* property with details specifying the database to use, and call the *Open* method.
Create and execute a database query by using ADO.NET	Create a *SqlCommand* object. Set its *Connection* property to a valid *SqlConnection* object. Set its *CommandText* property to a valid SQL SELECT statement. Call the *ExecuteReader* method to run the query and create a *SqlDataReader* object.
Fetch data by using an ADO.NET *SqlDataReader* object	Ensure that the data is not null by using the *IsDBNull* method. If the data is not null, use the appropriate *GetXXX* method (such as *GetString* or *GetInt32*) to retrieve the data.

Define an entity class	Define a class with public properties for each column. Prefix the class definition with the *Table* attribute, specifying the name of the table in the underlying database. Prefix each property with the *Column* attribute, and specify parameters indicating the name, type, and nullability of the corresponding column in the database.
Create and execute a query by using DLINQ	Create a *DataContext* variable, and specify a connection string for the database. Create a *Table* collection variable based on the entity class corresponding to the table you want to query. Define a DLINQ query that identifies the data to be retrieved from the database and returns an enumerable collection of entities. Iterate through the enumerable collection to retrieve the data for each row and process the results.

Chapter 26
Displaying and Editing Data by Using Data Binding

After completing this chapter, you will be able to:

- Use the Object Relational Designer to generate entity classes.

- Use data binding in a Microsoft Windows Presentation Foundation (WPF) application to display and maintain data retrieved from a database.

- Update a database by using DLINQ.

- Detect and resolve conflicting updates made by multiple users.

In Chapter 25, "Querying Information in a Database," you learned the essentials of using Microsoft ADO.NET and DLINQ for executing queries against a database. In this chapter, you will learn how to write applications that use DLINQ to modify data. You will see how to use data binding in a WPF application to present to a user data retrieved from a database and to enable the user to update that data. You will then learn how to propagate these updates back to the database.

Using Data Binding with DLINQ

You first encountered the idea of data binding in a WPF application in Chapter 24, "Performing Validation," when you used this technique to associate the properties of controls on a WPF form with properties in an instance of a class. You can adopt a similar strategy and bind properties of controls to entity objects so that you can display and maintain data held in a database by using a graphical user interface. First, however, you need to define the entity classes required by DLINQ. You saw how to do this manually in Chapter 25, and by now you should understand how entity classes work. You will be pleased to know that Microsoft Visual Studio 2008 provides the Object Relational Designer, which can connect to a database and generate entity classes for you. The Object Relational Designer can even generate the appropriate *EntityRef<TEntity>* and *EntitySet<TEntity>* relationship properties. You will use this tool in the following exercises.

Granting Access to a SQL Server 2005 Database File—Visual C# 2008 Express Edition

If you are using Microsoft Visual C# 2008 Express Edition, when you define a Microsoft SQL Server database connection for the entity wizard, you connect directly to the SQL Server database file. Visual C# 2008 Express Edition starts its own instance of SQL Server Express, called a *user instance* for accessing the database. The user instance runs using the credentials of the user executing the application. If you are using Visual C# 2008 Express Edition, you must detach the database from the SQL Server Express default instance because it will not allow a user instance to connect to a database that it is currently using. The following procedure describes how to perform this task.

Detach the Northwind database

1. On the Windows *Start* menu, click *All Programs*, click *Accessories*, and then click *Command Prompt* to open a command prompt window. If you are using Windows Vista, in the command prompt window, type the following command to move to the \Microsoft Press\Visual CSharp Step by Step\Chapter 26 folder under your Documents folder. Replace *Name* with your user name.

   ```
   cd "\Users\Name\Documents\Microsoft Press\Visual CSharp Step by Step\Chapter 26"
   ```

   If you are using Windows XP, type the following command to go to the \ Microsoft Press\Visual CSharp Step by Step\Chapter 26 folder under your My Documents folder, replacing *Name* with your user name.

   ```
   cd "\Documents and Settings\Name\My Documents\Microsoft Press\Visual CSharp Step by Step\Chapter 26"
   ```

2. In the command prompt window, type the following command:

   ```
   sqlcmd -S YourComputer\SQLExpress -E -idetach.sql
   ```

   Replace *YourComputer* with the name of your computer.

   The detach.sql script contains the following SQL Server command, which detaches the Northwind database from the SQL Server instance:

   ```
   sp_detach_db 'Northwind'
   ```

3. When the script finishes running, close the command prompt window.

 Note If you need to rebuild the Northwind database, you can run the instnwnd.sql script as described in Chapter 25. However, if you have detached the Northwind database you must first delete the Northwind.mdf and Northwind_log.ldf files in the C:\Program Files\ Microsoft SQL Server\MSSQL.1\MSSQL\Data folder; otherwise, the script will fail.

If you are running under the Windows Vista operating system, you must grant this user access to the folder holding the database and grant Full Control over the database files themselves. The next procedure shows how to do this.

Grant access to the Northwind database file under Windows Vista

1. Log on to your computer using an account that has administrator access.

2. Using Windows Explorer, move to the folder C:\Program Files\Microsoft SQL Server\MSSQL.1\MSSQL.

3. In the message box that appears, displaying the message "You don't currently have permission to access this folder," click *Continue*. In the *User Account Control* message that follows, click *Continue* again.

4. Move to the Data folder, right-click the Northwind file, and then click *Properties*.

5. In the *Northwind Properties* dialog box, click the *Security* tab.

6. If the *Security* page contains the message "Do you want to continue?" click *Continue*. In the *User Account Control* message box, click *Continue*.

   If the *Security* page contains the message "To change permissions, click Edit" click *Edit*. If a *User Account Control* message box appears, click *Continue*.

7. If your user account is not listed in the *Group or user names* list box, in the *Permissions for Northwind* dialog box, click *Add*. In the *Select Users or Groups* dialog box, enter the name of your user account, and then click *OK*.

8. In the *Permissions for Northwind* dialog box, in the *Group or user names* list box, click your user account.

9. In the *Permissions for **Account*** list box (where *Account* is your user account name), select the *Allow* checkbox for the *Full Control* entry, and then click *OK*.

10. In the *Northwind Properties* dialog box, click *OK*.

11. Repeat steps 4 through 10 for the Northwind_log file in the Data folder.

Generate entity classes for the *Suppliers* and *Products* tables

1. Start Visual Studio 2008 if it is not already running.

2. Create a new project by using the WPF Application template. Name the project *Suppliers*, and save it in the \Microsoft Press\Visual CSharp Step by Step\Chapter 26 folder in your Documents folder.

> **Note** If you are using Visual C# 2008 Express Edition, you can specify the location for saving your project by setting the *Visual Studio projects location* in the *Projects and Solutions* section of the *Options* dialog box on the *Tools* menu.

3. On the *Project* menu, click *Add Class*.

4. In the *Add New Item – Suppliers* dialog box, select the *LINQ to SQL Classes* template, type **Northwind.dbml** in the *Name* box, and then click *Add*.

   The *Object Relational Designer* window appears. You can use this window to specify the tables in the Northwind database for which you want to create entity classes, select the columns that you want to include, and define the relationships between them.

   The Object Relational Designer requires you to configure a connection to a database. The steps for performing this task are slightly different depending on whether you are using Visual Studio 2008 Professional Edition or Enterprise Edition, or Visual C# 2008 Express Edition.

5. If you are using Visual Studio 2008 Professional Edition or Enterprise Edition, perform the following tasks:

   5.1. On the *View* menu, click *Server Explorer*.

   5.2. In the *Server Explorer* window, right-click *Data Connections*, and then click *Add Connection*.

   5.3. If the *Choose Data Source* dialog box appears, *click Microsoft SQL Server*, and then click *Continue*.

   5.4. In the *Add Connection* dialog box, click the *Change* button adjacent to the *Data source* box.

   5.5. In the *Change Data Source* dialog box, click the *Microsoft SQL Server* data source, make sure the *.NET Framework Data Provider for SQL Server* is selected as the data provider, and then click *OK*.

   5.6. In the *Add Connection* dialog box, type **YourServer\SQLExpress** in the *Server name* box, where *YourServer* is the name of your computer.

5.7. Select the *Use Windows Authentication* radio button. This option uses your Microsoft Windows account name to connect to the database and is the recommended way to log on to SQL Server.

5.8. In the *Connect to a database* section of the dialog box, click *Select or enter a database name*, select the *Northwind* database, and then click *OK*.

6. If you are using Visual C# 2008 Express Edition, perform the following tasks:

6.1. On the *View* menu, point to *Other Windows*, and then click *Database Explorer*.

6.2. In the *Database Explorer* window, right-click *Data Connections*, and then click *Add Connection*.

6.3. If the *Choose Data Source* dialog box appears, click the *Microsoft SQL Server Database File* data source, make sure the *.NET Framework Data Provider for SQL Server* is selected as the data provider, and then click *Continue*.

6.4. In the *Add Connection* dialog box, verify that the *Data source* box displays *Microsoft SQL Server Database File (SqlClient)*. If it does not, click *Change*, and in the *Change Data Source* dialog box, click the *Microsoft SQL Server Database File* data source, make sure the *.NET Framework Data Provider for SQL Server* is selected as the data provider, and then click *OK*.

6.5. In the *Add Connection* dialog box, to the right of the *Database file name* text box, click *Browse*.

6.6. In the *Select SQL Server Database File* dialog box, move to the folder C:\Program Files\Microsoft SQL Server\MSSQL.1\MSSQL\Data, click the Northwind database file, and then click *Open*.

6.7. Select the *Use Windows Authentication* option to log on to the server, and then click *OK*.

> **Note** Some data sources can be accessed by using more than one data provider. For example, if you are using Visual Studio 2008 Professional Edition or Enterprise Edition, you can connect to SQL Server by using the Microsoft .NET Framework Data Provider for SQL Server or the .NET Framework Data Provider for OLE DB. The .NET Data Provider for SQL Server is optimized for connecting to SQL Server databases, whereas the .NET Framework Data Provider for OLE DB is a more generic provider that can be used to connect to a variety of data sources, not just SQL Server.

7. In *Server Explorer* or *Database Explorer*, expand the new data connection (*YourComputer*\sqlexpress.Northwind.dbo if you are running Visual Studio 2008 or Northwind.mdf if you are running Visual C# 2008 Express Edition), and then expand *Tables*.

8. Click the *Suppliers* table, and drag it onto the *Object Relational Designer* window.

   The Object Relational Designer generates an entity class called *Supplier* based on the *Suppliers* table, with properties for each column in the table.

    Note If you are using Visual C# 2008 Express Edition, a message box appears, asking you whether you want to add the data file for the Northwind database to your project. Click *No*.

9. In the *Supplier* class, click the *HomePage* column, and then press Delete.

   The Object Relational Designer removes the *HomePage* property from the *Supplier* class.

10. Using the same technique, remove all the remaining columns from the *Supplier* class except for *SupplierID* , *CompanyName*, and *ContactName*.

11. In *Server Explorer* or *Database Explorer*, click the *Products* table and drag it onto the *Object Relational Designer* window.

   The Object Relational Designer generates an entity class called *Product*, based on the *Products* table. Notice that the Object Relational Designer detects the relationship between the *Suppliers* and *Products* tables.

12. Remove the *Discontinued, ReorderLevel, UnitsOnOrder, UnitsInStock*, and *CategoryID* properties from the *Product* class. The complete classes should look like the following image.

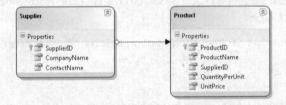

    Tip You can modify the attributes of an entity class and any of its properties by selecting the class or property and changing the values in the *Properties* window.

13. In *Solution Explorer*, expand the Northwind.dbml folder, and then double-click *Northwind.designer.cs*.

   The code generated by the Object Relational Designer appears in the *Code and Text Editor* window. If you examine this code, you will see that it contains a *DataContext* class called *NorthwindDataContext* and the two entity classes. These entity classes are a little more complicated than are the classes that you created manually in Chapter 25, but the general principles are the same. The additional complexity is the result of the entity classes implementing the *INotifyPropertyChanging* and *INotifyPropertyChanged* interfaces. These interfaces define events that the entity classes raise when their property values change. The various user interface controls in the WPF library subscribe to these events to detect any changes to data and ensure that the information displayed on a WPF form is up-to-date.

The information concerning the connection you specified before creating the two entity classes is saved in an application configuration file. Storing the connection string in a configuration file enables you to modify the connection string without rebuilding the application; you simply edit the application configuration file. It is useful if you envisage ever needing to relocate or rename the database, or switch from using a local development database to a production database that has the same set of tables.

Using an Application Configuration File

An application configuration file provides a very useful mechanism enabling a user to modify some of the resources used by an application without rebuilding the application itself. The connection string used for connecting to a database is an example of just such a resource.

When you use the Object Relational Designer to generate entity classes, a new file is added to your project called app.config. This is the source for the application configuration file, and it appears in the *Solution Explorer* window. You can examine the contents of the app.config file by double-clicking it. You will see that it is an XML file, as shown here (the text has been reformatted to fit on the printed page):

```xml
<?xml version="1.0" encoding="utf-8" ?>
<configuration>
    <configSections>
    </configSections>
    <connectionStrings>
        <add name="DisplayProducts.Properties.Settings.NorthwindConnectionString"
            connectionString="Data Source=YourComputer\SQLExpress;
                Initial Catalog=Northwind;Integrated Security=True"
            providerName="System.Data.SqlClient" />
    </connectionStrings>
</configuration>
```

The connection string is held in the *<connectionStrings>* element of the file. When you build the application, the C# compiler copies the app.config file to the folder holding the compiled code and renames it as *application*.exe.config, where *application* is the name of your application. When your application connects to the database, it should read the connection string value from the configuration file rather than using a connection string that is hard-coded in your C# code. You will see how to do this when using generated entity classes later in this chapter.

You should deploy the application configuration file (the *application*.exe.config file) with the executable code for the application. If the user needs to connect to a different database, she can edit the configuration file by using a text editor to modify the *<connectionString>* attribute of the *<connectionStrings>* element. When the application runs, it will use the new value automatically.

Be aware that you should take steps to protect the application configuration file and prevent a user from making inappropriate changes.

Create the user interface for the Suppliers application

1. In *Solution Explorer*, right-click the Window1.xaml file, click *Rename*, and rename the file SupplierInfo.xaml.

2. Double-click the App.xaml file to display it in the *Design View* window. In the *XAML* pane, change the *StartupUri* element to "SupplierInfo.xaml", as shown here in bold type:

```
<Application x:Class="Suppliers.App"
    xmlns="http://schemas.microsoft.com/winfx/2006/xaml/presentation"
    xmlns:x="http://schemas.microsoft.com/winfx/2006/xaml"
    StartupUri="SupplierInfo.xaml">
    ...
</Application>
```

3. In *Solution Explorer*, double-click the SupplierInfo.xaml file to display it in the *Design View* window. In the *XAML* pane, as shown in bold type below, change the value of the *x:Class* element to "Suppliers.SupplierInfo", set the *Title* to "Supplier Information", set the *Height* to"362", and set the *Width* to "614":

```
<Window x:Class="Suppliers.SupplierInfo"
    xmlns="http://schemas.microsoft.com/winfx/2006/xaml/presentation"
    xmlns:x="http://schemas.microsoft.com/winfx/2006/xaml"
    Title="Supplier Information" Height="362" Width="614">
    ...
</Window>
```

4. Display the SupplierInfo.xaml.cs file in the *Code and Text Editor* window. Change the name of the *Window1* class to *SupplierInfo*, and change the name of the constructor, as shown here in bold type:

```
public partial class SupplierInfo : Window
{
    public SupplierInfo()
    {
        InitializeComponent();
    }
}
```

5. In *Solution Explorer*, double-click the SupplierInfo.xaml file to display it in the *Design View* window. From the *Toolbox*, add a *ComboBox* control, a *ListView* control, and a *Button* control to the Supplier Information form.

6. Using the *Properties* window, set the properties of these controls to the values specified in the following table.

Control	Property	Value
comboBox1	Name	suppliersList
	Height	21
	Width	Auto
	Margin	40,16,42,0
	VerticalAlignment	Top
	HorizontalAlignment	Stretch
listView1	Name	productsList
	Height	Auto
	Width	Auto
	Margin	40,44,40,60
	VerticalAlignment	Stretch
	HorizontalAlignment	Stretch
button1	Name	saveChanges
	Content	Save Changes
	IsEnabled	False (clear the check box)
	Height	23
	Width	90
	Margin	40,0,0,10
	VerticalAlignment	Bottom
	HorizontalAlignment	Left

The Supplier Information form should look like this in the *Design View* window:

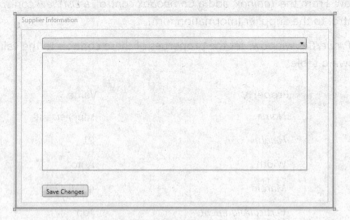

7. In the *XAML* pane, add the following Window resource shown in bold type to the *Window* element:

```
<Window x:Class="Suppliers.SupplierInfo"
...>
    <Window.Resources>
        <DataTemplate x:Key="SuppliersTemplate">
            <StackPanel Orientation="Horizontal">
                <TextBlock Text="{Binding Path=SupplierID}" />
                <TextBlock Text=" : " />
                <TextBlock Text="{Binding Path=CompanyName}" />
                <TextBlock Text=" : " />
                <TextBlock Text="{Binding Path=ContactName}" />
            </StackPanel>
        </DataTemplate>
    </Window.Resources>
    <Grid>
    ...
    </Grid>
</Window>
```

You can use a *DataTemplate* to specify how to display data in a control. You will apply this template to the *suppliersList* combo box in the next step. This template contains three *TextBlock* controls organized horizontally by using a *StackPanel*. The first, third, and fifth *TextBlock* controls will display the data in the *SupplierID*, *CompanyName*, and *ContactName* properties of the *Supplier* entity object to which you will bind later. The other *TextBlock* controls just display a ":" separator.

8. In the *XAML* pane, modify the definition of the *suppliersList* combo box and specify the *IsSynchronizedWithCurrentItem*, *ItemsSource*, and *ItemTemplate* properties, as follows in bold type:

```
<ComboBox ... Name="suppliersList" IsSynchronizedWithCurrentItem="True"
    ItemsSource="{Binding}" ItemTemplate="{StaticResource SuppliersTemplate}" />
```

You will display the data in a *Table<Supplier>* collection in the *suppliersList* control. Setting the *IsSynchronizedWithCurrentItem* property ensures that the *SelectedItem* property of the control is kept synchronized with the current item in the collection. If you don't set this property to *True*, when the application starts up and establishes the binding with the collection, the combo box will not automatically display the first item in this collection.

ItemsSource currently has an empty binding. In Chapter 24, you defined an instance of a class as a static resource and specified that resource as the binding source. If you do not specify a binding source, WPF binds to an object specified in the *DataContext* property of the control. (Do not confuse the *DataContext* property of a control with a *DataContext* object used to communicate with a database; it is unfortunate that they happen to have the same name.) You will set the *DataContext* property of the control to a *Tables<Supplier>* collection object in code.

The *ItemTemplate* property specifies the template to use to display data retrieved from the binding source. In this case, the *suppliersList* control will display the *SupplierID*, *CompanyName*, and *ContactName* fields from the binding source.

9. Modify the definition of the *productsList* list box, and specify the *IsSynchronizedWithCurrentItem* and *ItemsSource* properties:

```
<ListView ... Name="productsList" IsSynchronizedWithCurrentItem="True"
    ItemsSource="{Binding}" />
```

The *Supplier* entity class contains an *EntitySet<Product>* property that references the products the supplier can provide. You will set the *DataContext* property of the *productsList* control to the *Products* property of the currently selected *Supplier* object in code. In a later exercise, you will also provide functionality enabling the user to add and remove products. This code will modify the list of products acting as the binding source. Setting the *IsSynchronizedWithCurrentItem* property to *True* ensures that the newly created product is selected in the list when the user adds a new one or that an existing item is selected if the user deletes one. (If you set this property to *False*, when you delete a product, no item in the list will be selected afterward, which can cause problems in your application if your code attempts to access the currently selected item.)

10. Add the following *ListView.View* child element containing a *GridView* and column definitions to the *productsList* control. Be sure to replace the closing delimiter (/>) of the *ListView* element with an ordinary delimiter (>) and add a terminating </ListView> element.

```
<ListView ... Name="productsList" ...>
    <ListView.View>
        <GridView>
            <GridView.Columns>
                <GridViewColumn Width="75" Header="Product ID"
```

```
                              DisplayMemberBinding="{Binding Path=ProductID}" />
                <GridViewColumn Width="225" Header="Name"
                        DisplayMemberBinding="{Binding Path=ProductName}" />
                <GridViewColumn Width="135" Header="Quantity Per Unit"
                        DisplayMemberBinding="{Binding Path=QuantityPerUnit}" />
                <GridViewColumn Width="75" Header ="Unit Price"
                        DisplayMemberBinding="{Binding Path=UnitPrice}" />
            </GridView.Columns>
          </GridView>
        </ListView.View>
    </ListView>
```

You can make a *ListView* control display data in various formats by setting the *View* property. This Extensible Application Markup Language (XAML) code uses a *GridView* component. A *GridView* displays data in a tabular format; each row in the table has a fixed set of columns defined by the *GridViewColumn* properties. Each column has its own header that displays the name of the column. The *DisplayMemberBinding* property of each column specifies the data that the column should display from the binding source.

The data for the *UnitPrice* column is a *decimal?* property. WPF will convert this information to a string and apply a default numeric format. Ideally, the data in this column should be displayed as a currency value. You can reformat the data in a *GridView* column by creating a *converter* class. You first encountered converter classes in Chapter 24 when converting a Boolean value represented by the state of a radio button into an enumeration. This time, the converter class will convert a *decimal?* value to a *string* containing a representation of a currency value.

11. Switch to the *Code and Text Editor* window displaying the SupplierInfo.xaml.cs file. Add the following *PriceConverter* class to this file after the *SupplierInfo* class:

```
[ValueConversion(typeof(string), typeof(decimal?))]
class PriceConverter : IValueConverter
{
    public object Convert(object value, Type targetType, object parameter,
                    System.Globalization.CultureInfo culture)
    {
        if (value != null)
            return String.Format("{0:C}", value);
        else
            return "";
    }

    public object ConvertBack(object value, Type targetType, object parameter,
                        System.Globalization.CultureInfo culture)
    {
        throw new NotImplementedException();
    }
}
```

The *Convert* method calls the *String.Format* method to create a string that uses the local currency format of your computer. The user will not actually modify the unit price in the list view, so there is no need to implement the *ConvertBack* method to convert a *string* back to a *decimal*? value.

12. Return to the *Design View* window displaying the SupplierInfo.xaml form. Add the following XML namespace declaration to the *Window* element, and define an instance of the *PriceConverter* class as a Window resource, as shown here in bold type:

```
<Window x:Class="Suppliers.SupplierInfo"
...
xmlns:app="clr-namespace:Suppliers"
...>
    <Window.Resources>
        <app:PriceConverter x:Key="priceConverter" />
        ...
    </Window.Resources>
    ...
</Window>
```

13. Modify the definition of the Unit Price *GridViewColumn*, and apply the converter class to the binding, like this:

```
<GridViewColumn ... Header ="Unit Price" DisplayMemberBinding=
    "{Binding Path=UnitPrice, Converter={StaticResource priceConverter}}" />
```

You have now laid out the form. Next, you need to write some code to retrieve the data displayed by the form, and you must set the *DataContext* properties of the *suppliersList* and *productsList* controls so that the bindings function correctly.

Write code to retrieve supplier information and establish the data bindings

1. Change the definition of the *Window* element, and specify a *Loaded* event method called *Window_Loaded*. (This is the default name of this method, generated when you click *<New Event Handler>*.) The XAML code for the Window element should look like this:

```
<Window x:Class="Suppliers.SupplierInfo"
    ...
    Title="Supplier Information" ... Loaded="Window_Loaded">
    ...
</Window>
```

2. In the *Code and Text Editor* window displaying the SupplierInfo.xaml.cs file, add the following *using* statements to the list at the top of the file:

```
using System.ComponentModel;
using System.Collections;
```

3. Add the following three private fields shown here in bold type to the *SupplierInfo* class.

```
public partial class SupplierInfo : Window
{
    private NorthwindDataContext ndc = null;
    private Supplier supplier = null;
    private BindingList<Product> productsInfo = null;
    ...
}
```

You will use the *ndc* variable to connect to the Northwind database and retrieve the data from the *Suppliers* table. The *supplier* variable will hold the data for the current supplier displayed in the *suppliersList* control. The *productsInfo* variable will hold the products provided by the currently displayed supplier. It will be bound to the *productsList* control.

You might be wondering about this definition of the *productsInfo* variable; after all, the *Supplier* class has an *EntitySet<Product>* property that references the products supplied by a supplier. You could actually bind this *EntitySet<Product>* property to the *productsList* control, but there is one important problem with this approach. I mentioned earlier that the *Supplier* and *Product* entity classes implement the *INotifyPropertyChanging* and *INotifyPropertyChanged* interfaces. When you bind a WPF control to a data source, the control automatically subscribes to the events exposed by these interfaces to update the display when the data changes. However, the *EntitySet<Product>* class does not implement these interfaces, so the list view control will not be updated if any products are added to, or removed from, the supplier. (It will be updated if an existing product changes, however, because each item in *EntitySet<Product>* is a *Product* object, which does send the appropriate notifications to the WPF controls to which it is bound.)

4. Add the following code to the *Window_Loaded* method:

```
private void Window_Loaded(object sender, RoutedEventArgs e)
{
    ndc = new NorthwindDataContext();
    this.suppliersList.DataContext = ndc.Suppliers;
}
```

When the application starts and loads the window, this code creates a *NorthwindDataContext* variable that connects to the Northwind database. Remember that the Object Relational Designer created this class earlier. The default constructor for this class reads the database connection string from the application configuration file. The method then sets the *DataContext* property of the *suppliersList* combo box to the *Suppliers Table* collection property of the *ndc* variable. This action resolves the binding for the combo box, and the data template used by this combo box displays the values in the *SupplierID*, *CompanyName*, and *ContactName* for each *Supplier* object in the collection.

> **Note** If a control is a child of another control, for example, a *GridViewColumn* in a *ListView*, you need to set the *DataContext* property only of the parent control. If the *DataContext* property of a child control is not set, the WPF runtime will use the *DataContext* of the parent control instead. This technique makes it possible for you to share a data context between several child controls and a parent control.
>
> If the immediate parent control does not have a data context, the WPF runtime will examine the grandparent control, and so on, all the way up to the *Window* control defining the form. If no data context is available, any data bindings for a control are ignored.

5. Return to the *Design View* window. Double-click the *suppliersList* combo box to create the *suppliersList_SelectionChanged* event method. (If you are unable to click on the *suppliersList* combo box, try closing and reopening the SupplierInfo.xaml file in the *Design View* window.) This method runs whenever the user selects a different item in the combo box.

6. In the *Code and Text Editor* window, add the following statements shown in bold type to the *suppliersList_SelectionChanged* method:

```
private void suppliersList_SelectionChanged(object sender,
SelectionChangedEventArgs e)
{
    supplier = this.suppliersList.SelectedItem as Supplier;
    IList list = ((IListSource)supplier.Products).GetList();
    productsInfo = list as BindingList<Product>;
    this.productsList.DataContext = productsInfo;
}
```

This method obtains the currently selected supplier and copies the data in the *EntitySet<Product>* property for this supplier to the *productsInfo* variable after converting it to a *BindingList<Product>* collection. Notice that the *EntitySet<Product>* class implements the *IListSource* interface, which provides the *GetList* method for copying the data in the entity set into an *IList* object. Finally, the method sets the *DataContext* property of the *productsList* control to this list of products.

7. On the *Debug* menu, click *Start Without Debugging* to build and run the application.

When the form runs, it should display the products for the first supplier—Exotic Liquids. The form should look like the following image.

> **Note** Under some circumstances, the application can fail with a timeout exception if SQL Server does not respond within a reasonable time. (It can take SQL Server a few seconds to open a connection to the database.) If this happens, simply run the application again.

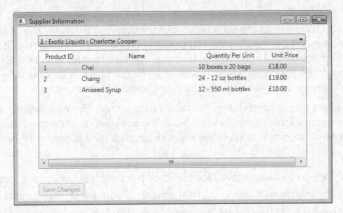

8. Select a different supplier from the combo box, and verify that the list view displays the products for that supplier. When you have finished browsing the data, close the form and return to Visual Studio 2008.

The final step is to provide functionality enabling the user to modify the details of products, remove products, and create new products. Before you can do that, you need to learn how to use DLINQ to update data.

Using DLINQ to Modify Data

DLINQ provides a two-way communication channel with a database. You have seen how to use DLINQ to fetch data, but you can also modify the information you have retrieved and send these changes back to the database.

Updating Existing Data

You can change the values in the *Product* objects in the *Table<Product>* collection in exactly the same way that you change the values in any ordinary object—by setting its properties. However, updating an object in memory does not update the database. To persist changes to the database, you need to generate the appropriate SQL UPDATE commands and arrange for them to be executed by the database server. You can do this quite easily with DLINQ. The following code fragment fetches product number 14 and changes its name to "Bean Curd" (product 14 was originally named "Tofu" in the Northwind database), and then sends the change back to the database:

```
NorthwindDataContext ndc = new NorthwindDataContext();
Product product = ndc.Products.Single(p => p.ProductID == 14);
product.ProductName = "Bean Curd";
ndc.SubmitChanges();
```

The key statement in this code example is the call to the *SubmitChanges* method of the *DataContext* object. When you modify the information in a DLINQ entity object that was populated by running a query, the *DataContext* object managing the connection that was used to run the original query tracks the changes you make to the data. The *SubmitChanges* method propagates these changes back to the database. Behind the scenes, the *DataContext* object constructs and executes a SQL UPDATE statement.

If you fetch and modify several products, you need to call *SubmitChanges* only once, after the final modification. The *SubmitChanges* method batches all of the updates together. The *DataContext* object creates a database transaction and performs all of the SQL UPDATE statements within this transaction. If any of the updates fail, the transaction is aborted, all the changes made by the *SubmitChanges* method are rolled back in the database, and the *SubmitChanges* method throws an exception. If all the updates succeed, the transaction is committed, and the changes become permanent in the database. You should note that if the *SubmitChanges* method fails, only the database is rolled back; your changes are still present in the entity objects in memory. The exception thrown when the *SubmitChanges* method fails provides some information on the reason for the failure. You can attempt to rectify the problem and call *SubmitChanges* again.

The *DataContext* class also provides the *Refresh* method. With this method, you can repopulate *Table* collections from the database and discard any changes you have made. You use it like this:

```
ndc.Refresh(RefreshMode.OverwriteCurrentValues, ndc.Products);
```

The first parameter is a member of the *System.Data.Linq.RefreshMode* enumeration. Specifying the value *RefreshMode.OverwriteCurrentValues* forces the data to be refreshed from the database. (This enumeration contains other values, as you will see in the next section.) The second parameter is the table to be refreshed. Actually, the *Refresh* method can take a *params* array as its second parameter, so you can provide a whole list of tables if you need to refresh more than one.

> **Tip** Change tracking is a potentially expensive operation for a *DataContext* object to perform. If you know that you are not going to modify data (if for example your application generates a read-only report), you can disable change tracking by setting the *ObjectTrackingEnabled* property to *false*. You must set this property *before* fetching any data. Any attempt to call *SubmitChanges* on a read-only *DataContext* object will raise an *InvalidOperationException*.

Handling Conflicting Updates

There could be any number of reasons why an update operation fails, but one of the most common causes is conflicts occurring when two users attempt to update the same data simultaneously. If you think about what happens when you run an application that uses DLINQ, you can see that there is plenty of scope for conflict. When you retrieve data through a *DataContext* object, it is buffered in the memory of your application in a collection of entity objects. Another user could perform the same query and retrieve the same data. If you both modify the data and then you both call the *SubmitChanges* method, one of you will overwrite the changes made by the other in the database. This phenomenon is known as a *lost update*. The *SubmitChanges* method detects this condition and raises a *ChangeConflictException*, which you should be prepared to handle.

When a *ChangeConflictException* arises, you can ascertain the reason for the conflict by examining the *ChangeConflicts* property of the *DataContext* object. This property is a collection containing *ObjectChangeConflict* objects, which contain information about the reason for each conflict. The important properties in the *ObjectChangeConflict* class are *IsDeleted*, which is a Boolean value indicating whether the conflict was caused by another user deleting the row that you were attempting to update, and *MemberConflicts*, which is a read-only collection of *MemberChangeConflict* objects. The *MemberChangeConflict* class contains a further set of properties, including the current value of the data in your application, the current value of the data in the database, and the original value you retrieved from the database. If you detect conflicts when performing the *SubmitChanges* method, your application can examine the reason for each conflict and determine how to handle it. Depending on the nature of the application, you could even present information about the conflict to the user and let the user choose. To help you correct the problems caused by a conflict, the *ObjectChangeConflict* class contains a method called *Resolve*. For each conflict in the *ChangeConflicts* collection property of the *DataContext* object, you can call the *Resolve* method and pass in a parameter indicating your preferred resolution strategy. This parameter should be a member of the *RefreshMode* enumeration. You can specify the following values:

- **RefreshMode.KeepCurrentValues**   This value indicates that the data in memory should overwrite the conflicting changes in the database—the current user is the winner of the conflict.

- **RefreshMode.OverwriteCurrentValue**   This value indicates that the data in the database should be used. The conflicting changes in memory will be overwritten with the values from the database—the current user is the loser of the conflict.

- **RefreshMode.KeepChanges**   This value specifies what happens if two users update different columns in the same row. In this case, the changes made by the other user to the other columns are merged with the changes the current user has made in memory—both users are winners of the conflict.

The following code shows a *ChangeConflictException* handler that displays conflicting data and resolves the conflict by using the *RefreshMode.OverwriteCurrentValues* option.

```
try
{
    ndc.SubmitChanges();
}
catch (ChangeConflictException)
{
    foreach (ObjectChangeConflict conflict in ndc.ChangeConflicts)
    {
        foreach (MemberChangeConflict changeConflict in conflict.MemberConflicts)
        {
            Console.WriteLine("Conflict Details");
            Console.WriteLine("Original value retrieved from database: {0}",
                changeConflict.OriginalValue.ToString());
            Console.WriteLine("Current value in database: {0}",
                changeConflict.DatabaseValue.ToString());
            Console.WriteLine("Current value in memory: {0}",
                changeConflict.CurrentValue.ToString());
        }
        conflict.Resolve(RefreshMode.OverwriteCurrentValues);
    }
}
```

Note The *ChangeConflicts* collection of the *DataContext* class provides the *ResolveAll* method that lets you apply the same *RefreshMode* value to resolve all conflicts.

When you have resolved the conflicts, you should call the *SubmitChanges* method again to resubmit your changes. There is one potential issue with this technique as it currently stands: if the user has updated several rows, there could be more than one conflict. The *ChangeConflictException* is thrown the first time a conflict is detected, and you can handle it in the manner just described, but only one *ObjectChangeConflict* object will be set in the *ChangeConflicts* collection. When you call *SubmitChanges* again to send the resolved update to the database, another *ChangeConflictException* for the next conflict will arise, which you have to detect and handle. To help you, the *SubmitChanges* method is overloaded, so you can specify how to handle the *ChangeConflictException*. Calling *SubmitChanges* with a parameter value of *ConflictMode.ContinueOnConflict* indicates that the *SubmitChanges* method should try to perform all the updates and only throw the *ChangeConflictException* at the end if one or more conflicts have occurred. Call the overloaded method like this:

```
ndc.SubmitChanges(ConflictMode.ContinueOnConflict);
```

The code in your *ChangeConflictException* handler can then iterate through all the items in the *ObjectChangeConflict* property of the *DataContext* object and resolve them all (the example shown earlier already does this) before calling *SubmitChanges* again.

When you call *SubmitChanges*, you can also specify the parameter value of *ConflictMode. FailOnFirstConflict*. This is the default behavior and raises a *ChangeConflictException* as soon as the first conflict is detected.

Adding and Deleting Data

As well as modifying existing data, with DLINQ you can add new items to a *Table* collection and remove items from a *Table* collection. To add a new item, call the *Add* method and provide an entity object with the new information, like this:

```
NorthwindDataContext ndc = new NorthwindDataContext(...);
Table<Product> products = ndc.Products;
Product newProduct = new Product() {ProductName = "New Product", ... };
products.Add(newProduct);
```

When you call *SubmitChanges*, the *DataContext* object will generate a SQL INSERT statement for each new item in the *Table* collection.

Note When you add a new entity object to the *Table* collection, you must provide values for every column that does not allow a null value in the database. The exception to this rule is for primary key columns that are designated as IDENTITY columns in the database—SQL Server will generate values for these columns and will raise an error if you try to specify a value of your own.

Deleting an entity object from a *Table* collection is equally straightforward. You call the *Remove* method and specify the entity to be deleted. The following code deletes product 14 from the *products* collection.

```
Product product = products.Single(p => p.ProductID == 14);
products.Remove(product);
```

When you call *SubmitChanges*, the *DataContext* object will generate a SQL DELETE statement for each row that has been removed from the *Table* collection.

Note Be careful when deleting rows in tables that have relationships to other tables because such deletions can cause referential integrity errors when you update the database. For example, in the Northwind database, if you attempt to delete a supplier that currently supplies products, the update will fail. You must first change the *SupplierID* column in the *Products* table for all products available from that supplier to *null* or to a different supplier.

You now have enough knowledge to complete the Suppliers application.

Write code to modify, delete, and create products

1. Return to the Visual Studio 2008 window in which you were editing the Suppliers application.

2. In the *Design View* window, in the *XAML* pane, modify the definition of the *productsList* control to trap the *KeyDown* event and invoke an event method called *productsList_KeyDown*. (This is the default name of the event method.) If IntelliSense does not recognize the *KeyDown* keyword, try closing and reopening the SupplierInfo. xaml file.

3. In the *Code and Text Editor* window, add the following code shown in bold type to the *productsList_KeyDown* method.

```
private void productsList_KeyDown(object sender, KeyEventArgs e)
{
    switch (e.Key)
    {
        case Key.Enter: editProduct(this.productsList.SelectedItem as Product);
            break;

        case Key.Insert: addNewProduct();
            break;

        case Key.Delete: deleteProduct(this.productsList.SelectedItem as Product);
            break;
    }
}
```

This method examines the key pressed by the user. If the user presses the Enter key, the code calls the *editProduct* method, passing in the details of the product as a parameter. If the user presses the Insert key, the code calls the *addNewProduct* method to create and add a new product to the list for the current supplier, and if the user presses the Delete key, the code calls the *deleteProduct* method to delete the product. You will write the *editProduct*, *addNewProduct*, and *deleteProduct* methods in the next few steps.

4. Add the *deleteProduct* method to the *SupplierInfo* class, as follows:

```
private void deleteProduct(Product prod)
{
    MessageBoxResult response = MessageBox.Show("Delete " + prod.ProductName,
        "Confirm", MessageBoxButton.YesNo, MessageBoxImage.Question,
        MessageBoxResult.No);
    if (response == MessageBoxResult.Yes)
    {
        supplier.Products.Remove(prod);
        productsInfo.Remove(prod);
        this.saveChanges.IsEnabled = true;
    }
}
```

This method prompts the user to confirm that the user really does want to delete the currently selected product. The *if* statement calls the *Remove* method of the *Products EntitySet<TEntity>* property to delete the product from this collection and also removes it from the *productsInfo* binding list. (This step is necessary to ensure that the display is kept synchronized with the changes.) Finally, the method activates the *saveChanges* button. You will add functionality to this button to send the changes made to the *Products EntitySet<TEntity>* back to the database in a later step.

There are several approaches you can use for adding and editing products; the columns in the *ListView* control are read-only text items, but you can create a customized list view that contains text boxes or other controls that enable user input. However, the simplest strategy is to create another form that enables the user to edit or add the details of a product.

5. On the *Project* menu, click *Add Class*. In the *Add New Items – Suppliers* dialog box, select the *Window (WPF)* template, type **ProductForm.xaml** in the *Name* box, and then click *Add*.

6. In the *Design View* window, click the *ProductForm* form, and in the *Properties* window, set the *ResizeMode* property to *NoResize*, set the *Height* property to *225*, and set the *Width* property to *515*.

7. Add three *Label* controls, three *TextBox* controls, and two *Button* controls to the form. Using the *Properties* window, set the properties of these controls to the values shown in the following table.

Control	Property	Value
*label*1	*Content*	Product Name
	Height	23
	Width	120
	Margin	17,20,0,0
	VerticalAlignment	Top
	HorizontalAlignment	Left
*label*2	*Content*	Quantity Per Unit
	Height	23
	Width	120
	Margin	17,60,0,0
	VerticalAlignment	Top
	HorizontalAlignment	Left

Control	Property	Value
*label*3	*Content*	Unit Price
	Height	23
	Width	120
	Margin	17,100,0,0
	VerticalAlignment	Top
	HorizontalAlignment	Left
*textBox*1	*Name*	productName
	Height	21
	Width	340
	Margin	130,24,0,0
	VerticalAlignment	Top
	HorizontalAlignment	Left
*textBox*2	*Name*	quantityPerUnit
	Height	21
	Width	340
	Margin	130,64,0,0
	VerticalAlignment	Top
	HorizontalAlignment	Left
*textBox*3	*Name*	unitPrice
	Height	21
	Width	120
	Margin	130,104,0,0
	VerticalAlignment	Top
	HorizontalAlignment	Left
*button*1	*Name*	ok
	Content	OK
	Height	23
	Width	75
	Margin	130,150,0,0
	VerticalAlignment	Top
	HorizontalAlignment	Left

Control	Property	Value
button2	*Name*	cancel
	Content	Cancel
	Height	23
	Width	75
	Margin	300,150,0,0
	VerticalAlignment	Top
	HorizontalAlignment	Left

The Supplier Information form should look like this in the *Design View* window:

8. Double-click the *OK* button to create an event handler for the *click* event. In the *Code and Text Editor* window displaying the ProductForm.xaml.cs file, add the following code shown in bold type.

```
private void ok_Click(object sender, RoutedEventArgs e)
{
    if (String.IsNullOrEmpty(this.productName.Text))
    {
        MessageBox.Show("The product must have a name", "Error",
            MessageBoxButton.OK, MessageBoxImage.Error);
        return;
    }

    decimal result;
    if (!Decimal.TryParse(this.unitPrice.Text, out result))
    {
        MessageBox.Show("The price must be a valid number", "Error",
            MessageBoxButton.OK, MessageBoxImage.Error);
        return;
    }

    if (result < 0)
    {
```

```
MessageBox.Show("The price must not be less than zero", "Error",
    MessageBoxButton.OK, MessageBoxImage.Error);
return;
}

this.DialogResult = true;
}
```

The application will display this form by calling the *ShowDialog* method. This method displays the form as a modal dialog box. When the user clicks a button on the form, it will close automatically if the code for the *click* event sets the *DialogResult* property. If the user clicks *OK*, this method performs some simple validation of the information entered by the user. The *Quantity Per Unit* column in the database accepts *null* values, so the user can leave this field on the form empty. If the user enters a valid product name and price, the method sets the *DialogResult* property of the form to *true*. This value is passed back to the *ShowDialog* method call.

9. Return to the *Design View* window displaying the ProductForm.xaml file. Select the *Cancel* button, and in the *Properties* window, set the *IsCancel* property to *true* (select the check box).

If the user clicks the *Cancel* button, it will automatically close the form and return a *DialogResult* value of *false* to the *ShowDialog* method.

10. Switch to the *Code and Text Editor* window displaying the SupplierInfo.xaml.cs file. Add the *addNewProduct* method shown here to the *SupplierInfo* class.

```
private void addNewProduct()
{
    ProductForm pf = new ProductForm();
    pf.Title = "New Product for " + supplier.CompanyName;
    if (pf.ShowDialog().Value)
    {
        Product newProd = new Product();
        newProd.SupplierID = supplier.SupplierID;
        newProd.ProductName = pf.productName.Text;
        newProd.QuantityPerUnit = pf.quantityPerUnit.Text;
        newProd.UnitPrice = Decimal.Parse(pf.unitPrice.Text);
        supplier.Products.Add(newProd);
        productsInfo.Add(newProd);
        this.saveChanges.IsEnabled = true;
    }
}
```

The *addNewProduct* method creates a new instance of the *ProductForm* form, sets the *Title* property of this form to contain the name of the supplier, and then calls the *ShowDialog* method to display the form as a modal dialog box. If the user enters some valid data and clicks the *OK* button on the form, the code in the *if* block creates a new *Product* object and populates it with the information from the *ProductForm* instance.

The method then adds it to the *Products EntitySet<TEntity>* for the current supplier and also adds it to the list displayed in the list view control on the form. Finally, the code activates the *Save Changes* button. In a later step, you will add code to the *click* event handler for this button so that the user can save changes back to the database.

11. Add the *editProduct* method shown here to the *SupplierInfo* class.

```
private void editProduct(Product prod)
{
    ProductForm pf = new ProductForm();
    pf.Title = "Edit Product Details";
    pf.productName.Text = prod.ProductName;
    pf.quantityPerUnit.Text = prod.QuantityPerUnit;
    pf.unitPrice.Text = prod.UnitPrice.ToString();

    if (pf.ShowDialog().Value)
    {
        prod.ProductName = pf.productName.Text;
        prod.QuantityPerUnit = pf.quantityPerUnit.Text;
        prod.UnitPrice = Decimal.Parse(pf.unitPrice.Text);
        this.saveChanges.IsEnabled = true;
    }
}
```

The *editProduct* method also creates an instance of the *ProductForm* form. This time, as well as setting the *Title* property, the code also populates the fields on the form with the information from the currently selected product. When the form is displayed, the user can edit these values. If the user clicks the *OK* button to close the form, the code in the *if* block copies the new values back to the currently selected product before activating the *Save Changes* button. Notice that this time you do not need to update the current item manually in the *productsInfo* list because the *Product* class notifies the list view control of changes to its data automatically.

12. Return to the *Design View* window displaying the SupplierInfo.xaml file. Double-click the *Save Changes* button to create the *click* event handler method.

13. In the *Code and Text Editor* window, add the following code shown in bold to the *saveChanges_Click* method:

```
private void saveChanges_Click(object sender, RoutedEventArgs e)
{
    try
    {
        ndc.SubmitChanges();
        saveChanges.IsEnabled = false;
    }
    catch (Exception ex)
    {
        MessageBox.Show(ex.Message, "Error saving changes");
    }
}
```

This method calls the *SubmitChanges* method of the *DataContext* object to send all the changes back to the database. For simplicity, this method performs only very rudimentary exception handling and does not attempt to resolve errors caused by conflicting updates made by other users.

Test the Suppliers application

1. On the *Debug* menu, click *Start Without Debugging* to build and run the application. When the form appears displaying the products supplied by Exotic Liquids, click product 3 (Aniseed Syrup), and then press Enter. The *Edit Product Details* form should appear. Change the value in the *Unit Price* field to **12.5**, and then click *OK*. Verify that the new price is copied back to the list view.

2. Press the Insert key. The *New Product for Exotic Liquids* form should appear. Enter a product name, quantity per unit, and price, and then click *OK*. Verify that the new product is added to the list view.

   The value in the *Product ID* column should be 0. This value is an identity column in the database, so SQL Server will generate its own unique value for this column when you save the changes.

3. Click *Save Changes*. After the data is saved, the ID for the new product is displayed in the list view.

4. Click the new product, and then press the Delete key. In the *Confirm* dialog box, click *Yes*. Verify that the product disappears from the form. Click *Save Changes* again, and verify that the operation completes without any errors.

   Feel free to experiment by adding, removing, and editing products for other suppliers. You can make several modifications before clicking *Save Changes*—the *SubmitChanges* method saves all changes made since the data was retrieved or last saved.

> **Tip** If you accidentally delete or overwrite the data for a product that you want to keep, close the application without clicking *Save Changes*. Note that the application as written does not warn the user if the user tries to exit without first saving changes.
>
> Alternatively, you can add a *Discard Changes* button to the application that calls the *Refresh* method of the *ndc DataContext* object to repopulate its tables from the database. You would also then need to rebuild the *productsInfo* binding list for the currently selected product.
>
> However, if you are handling a relatively small number of rows, as is the case in the Suppliers application, a simpler technique is to discard the current *DataContext* object and create a new one, and then reapply the binding for the *suppliersList* combo box, like this:
>
> ```
> ndc = new NorthwindDataContext();
> this.suppliersList.DataContext = ndc.Suppliers;
> ```

5. Close the form, and return to Visual Studio 2008.

- If you want to continue to the next chapter

  Keep Visual Studio 2008 running, and turn to Chapter 27.

- If you want to exit Visual Studio 2008 now

  On the *File* menu, click *Exit*. If you see a *Save* dialog box, click *Yes* (if you are using Visual Studio 2008) or *Save* (if you are using Visual C# 2008 Express Edition) and save the project.

Chapter 26 Quick Reference

To	Do this
Create entity classes by using the Object Relational Designer	Add a new class to the project by using the LINQ to SQL Classes template. Connect to the database by using Server Explorer (Visual Studio 2008 Professional Edition or Enterprise Edition) or Database Explorer (Visual C# 2008 Express Edition). Drag tables from the database to the Object Relational Designer.
Display data from an entity object or collection in a WPF control	Define a binding for the appropriate property of the control. If the control displays a list of objects, set the *DataContext* property of the control to a collection of entity objects. If the control displays the data for a single object, set the *DataContext* property of the control to an entity object and specify the property of the entity object to display in the *Path* attribute of the binding.
Modify information in a database by using DLINQ	First do one of the following: ■ To update a row in a table in the database, fetch the data for the row into an entity object, and assign the new values to the appropriate properties of the entity object. ■ To insert a new row into a table in the database, create a new instance of the corresponding entity object, set its properties, and then call the *Add* method of the appropriate *Table* collection, specifying the new entity object as the parameter. ■ To remove a row from a table in the database, call the *Remove* method of the appropriate *Table* collection, specifying the entity object to be removed as the parameter. Then, after making all your changes, call the *SubmitChanges* method of the *DataContext* object to propagate the modifications to the database.
Detect conflicts when updating a database by using DLINQ	Provide a handler for the *ChangeConflictException*. In the exception handler, examine the *ObjectChangeConflict* objects in the *ChangeConflicts* property of the *DataContext* object. For each conflict, determine the most suitable resolution, and call the *Resolve* method with the appropriate *RefreshMode* parameter.

Part VI
Building Web Applications

Chapter 27
Introducing ASP.NET

After completing this chapter, you will be able to:

- Create simple Microsoft ASP.NET pages.

- Build applications that run in a Web browser.

- Use ASP.NET Server controls efficiently.

- Create and apply ASP.NET themes.

In the previous sections of this book, you have seen how to build Microsoft Visual C# applications that run in the Microsoft Windows environment on the desktop. These applications typically allow a user to gain access to a database by using ADO.NET and DLINQ. In this final part of the book, you will consider the world of Web applications. These are applications that are accessed over the Internet. Rather than using the desktop, Web applications rely on a Web browser to provide the user interface.

In the first three chapters of this part, you will examine the classes provided by the Microsoft .NET Framework for building Web applications. You will learn about the architecture of ASP. NET, Web forms, and Server controls. You will see that the structure of applications that execute over the Web is different from those that run on the desktop, and you will be shown some best practices for building efficient, scalable, and easily maintainable Web sites.

In the final chapter in this part, you'll learn about Web services. With Web services, you can build distributed applications composed of components and services that can be spread across the Internet (or an intranet). You will learn how to create a Web service and understand how Web services are built on the Simple Object Access Protocol (SOAP). You will also study the techniques that a desktop application can use to connect to a Web service.

> **Important** You cannot build Web applications or Web services with Microsoft Visual C# 2008 Express Edition. If you have been using Visual C# 2008 Express Edition, you can perform the exercises in the remaining chapters of this book by using Microsoft Visual Web Developer 2008 Express Edition. You can download Visual Web Developer 2008 Express Edition free of charge from the Microsoft Web site.

Understanding the Internet as an Infrastructure

The Internet is a big network (all right—a *really* big network), and, as a result, the information and data that you can access over it can be quite remote. This should have an impact on the way you design your applications. For example, you might get away with repeatedly querying and fetching individual rows of data held in a database while a user browses it in a small, local desktop application, but this strategy will not be feasible for an application that runs over the Internet. Resource use affects scalability much more for the Internet than it does for local applications.

Network bandwidth is a scarce resource that should be used sparingly. You might notice variations in the performance of your own local network according to the time of day (networks always seem to slow down on a Friday afternoon just when you are trying to get everything done before the weekend), the applications that users in your company are running, and many other factors. But no matter how variable the performance of your own local network is, the Internet is far less predictable. You are dependent on any number of servers routing your requests from your Web browser to the site you are trying to access, and the replies can get passed back along an equally tortuous route. The network protocols and data presentation mechanisms that underpin the Internet reflect the fact that networks can be (and at times most certainly will be) unreliable and that a Web application can be accessed concurrently from many different Web browsers running on many different operating systems.

Understanding Web Server Requests and Responses

A Web browser communicates with a Web application over the Internet by using the Hypertext Transfer Protocol (HTTP). Web applications are usually hosted by some sort of Web server that reads HTTP requests and determines which application should be used to respond to the request. The term *application* in this sense is a very loose term—the Web server might invoke an executable program to perform an action, or it might process the request itself by using its own internal logic or other means. However the request is processed, the Web server will send a response to the client, again by using HTTP. The content of an HTTP response is usually presented as a Hypertext Markup Language (HTML) page; this is the language that most browsers understand and know how to render.

 Note Applications run by users that access Web applications over the Internet are often referred to as clients or client applications.

Managing State

HTTP is a connectionless protocol. This means that a request (or a response) is a stand-alone packet of data. A typical exchange between a client and a Web application might involve several requests. For example, the Web application might send the client application an HTML page. The user might enter data onto this page, click some buttons, and expect the display to change as a result so that the user can enter more data, and so on. Each request sent by the client to the Web application is separate from any other requests sent both by this client and by any other clients using the same Web application simultaneously.

A client request often requires some sort of context or state. For example, consider the following common scenario. The user can browse goods for sale by using a Web application. The user might want to buy several items and places each one in a virtual shopping cart. A useful feature of such a Web application is the ability to display the current contents of the shopping cart. Where should the contents of the shopping cart (the client's state) be held? If this information is held on the Web server, the Web server must be able to piece together the different HTTP requests and determine which requests come from one client and which come from others. This is feasible but might require additional processing to reconcile client requests against state information, and, of course, it would require some sort of database to persist that state information between client requests. A complication with this technique is that the Web server has no guarantee, after the state information has been preserved, that the client will submit another request that uses or removes the information. If the Web server saved every bit of state information for every client that accessed it, it would need a very big database indeed!

An alternative strategy is to store state information on the client machine. The *Cookie Protocol* was developed so that Web servers can cache information in cookies (small files) on the client computer. The disadvantage of this approach is that the application has to arrange for the data in the cookie to be transmitted over the Web as part of every HTTP request so that the Web server can access it. The application also has to ensure that cookies are of a limited size. Perhaps the most significant drawback of cookies is that users can disable them and prevent the Web browser from storing them on user computers, causing the Web application to lose all of its state information.

Understanding ASP.NET

From the discussion in the preceding section, you can see that a framework for building and running Web applications has a number of items that it should address. It must do the following:

- Support HTTP
- Manage client state efficiently

- Provide tools allowing for the easy development of Web applications

- Generate applications that can be accessed from any browser that supports HTML

- Be responsive and scalable

Microsoft originally developed the Active Server Pages (ASP) model in response to many of these issues. By using ASP, developers can embed application code in HTML pages. A Web server such as Microsoft Internet Information Services (IIS) could execute the application code and use it to generate an HTML response. However, ASP did have its problems: you had to write a lot of application code to do relatively simple things, such as display a page of data from a database; mixing application code and HTML caused readability and maintenance issues; and performance was not always what it could be because ASP pages had to interpret application code in an HTML request every time the request was submitted, even if it was the same code each time.

With the advent of the .NET Framework, Microsoft updated the ASP model and created ASP.NET. The main features of the latest release of ASP.NET include the following:

- A rationalized program model using Web forms that contain presentation logic and code files that separate out the business logic. You can write code in any of the languages supported by the .NET Framework, including C#. ASP.NET Web forms are compiled and cached on the Web server to improve performance.

- Server controls that support server-side events but that are rendered as HTML so that they can operate correctly in any HTML-compliant browser. Microsoft has extended many of the standard HTML controls as well so that you can manipulate them in your code.

- Powerful controls for displaying, editing, and maintaining data from a database.

- Options for caching client state using cookies on the client's computer, in a special service (the ASP.NET State service) on the Web server, or in a Microsoft SQL Server database. The cache is easily programmable by using code.

- Enhanced page design and layout by using Master Pages, themes, and Web Parts. You can use Master Pages to quickly provide a common layout for all Web pages in an application. Themes help you implement a consistent look and feel across the Web site, ensuring that all controls appear in the same way if required. With Web Parts, you can create modular Web pages that users can customize to their own requirements. You will use themes later in this chapter. Using Master Pages and Web Parts is outside the scope of this book.

- Data source controls for binding data to Web pages. By using these new controls, you can build applications that can display and edit data quickly and easily. The data source controls can operate with a variety of data sources, such as DLINQ entity objects, SQL Server databases, Microsoft Access databases, XML files, Web services, and other business objects. The data source controls provide you with a consistent mechanism for working with data, independent from the source of that data. You will make use of data source controls in Chapter 29, "Protecting a Web Site and Accessing Data with Web Forms."

- Powerful controls for displaying and editing data. Microsoft provides the FormView control for displaying data and editing data one record at a time, and the GridView control is provided for presenting information in a tabular format. You can use the TreeView control to display hierarchical data, and you can use the SiteMapPath and Menu controls to assist in user navigation through your Web application. You will use the GridView control in Chapter 29.

- AJAX extensions so that you can build highly interactive and responsive Web applications that can minimize the network bandwidth required to transmit data between the client application and the Web server. By using AJAX, you can define parts of a Web page as being updatable. When information displayed in an updatable region of a page changes, only the information required for that part of the page is transmitted by the Web server.

- Security features with built-in support for authenticating and authorizing users. You can easily grant permissions to users to allow them to access your Web application, validate users when they attempt to log in, and query user information so that you know who is accessing your Web site. You can use the Login control to prompt the user for credentials and validate the user and the PasswordRecovery control for helping users remember or reset their password. You will use these security controls in Chapter 29.

- Web site configuration and management by using the ASP.NET Web Site Administration Tool. This tool provides wizards for configuring and securing ASP.NET Web applications. You will use the ASP.NET Web Site Administration Tool in Chapter 29.

In the remainder of this chapter, you will learn more about the structure of an ASP.NET application.

Creating Web Applications with ASP.NET

A Web application that uses ASP.NET typically consists of one or more ASP.NET pages or Web forms, code files, and configuration files.

A Web form is held in an .aspx file, which is essentially an HTML file with some Microsoft .NET–specific tags. An .aspx file defines the layout and appearance of a page. Often each .aspx file has an associated code file containing the application logic for the components in the .aspx file, such as event handlers and utility methods. A directive (a special tag) at the start of each .aspx file specifies the name and location of the corresponding code file. ASP. NET also supports application-level events, which are defined in Global.asax files.

Each Web application can also have a configuration file called web.config. This file, which is in XML format, contains information regarding security, cache management, page compilation, and so on.

Building an ASP.NET Application

In the following exercise, you will build a simple ASP.NET application that uses Server controls to gather input from the user about the details of the employees of a fictitious software company called Litware, Inc. The application will show you the structure of a simple Web application.

Note You do not need to have IIS running on your computer to develop Web applications. Microsoft Visual Studio 2008 includes its own Development Server. When you build and run a Web application, by default Visual Studio 2008 will run the application using this Web server. However, you should still use IIS for hosting production Web applications after you have finished developing and testing them.

Create the Web application

1. Start Visual Studio 2008 or Visual Web Developer 2008 Express Edition if it is not already running.

Note In the remainder of the book, I simply state, "Start Visual Studio 2008" when you need to open Visual Studio 2008 Standard Edition, Visual Studio 2008 Professional Edition, or Visual Web Developer 2008 Express Edition. Additionally, unless explicitly stated, all further references to Visual Studio 2008 also apply to Visual Web Developer 2008 Express Edition.

2. If you are using Visual Studio 2008, on the *File* menu, point to *New*, and then click *Web Site*.

3. If you are using Visual Web Developer 2008 Express Edition, on the *File* menu, click *New Web Site*.

4. In the *New Web Site* dialog box, click the *ASP.NET Web Site* template. Select *File System* in the *Location* drop-down list box, and specify the \Microsoft Press\Visual CSharp Step By Step\Chapter 27\Litware folder under your Documents folder. Set the *Language* to *Visual C#*, and then click *OK*.

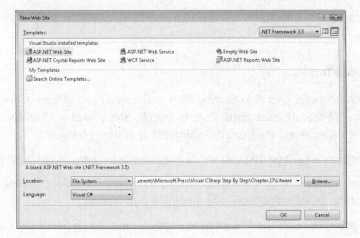

Note Setting the *Location* to *File System* creates the Web site by using the Development Server. You can use IIS by setting the *Location* to *HTTP* and specifying the URL of the Web site you want to create rather than a file name.

Visual Studio 2008 creates an application consisting of a Web folder called App_Data and a Web form called Default.aspx. The HTML code for the default page appears in the *Code and Text Editor* window.

5. In *Solution Explorer*, select the Default.aspx file. In the *Properties* window, change the *File Name* property of Default.aspx to *EmployeeForm.aspx*.

Note The *Properties* window shares the same pane as the *CSS Properties* window, the *Manage Styles* window, and the *Apply Styles* window, in the lower-right corner of Visual Studio. The *CSS Properties* window is displayed by default. To view the *Properties* window, click the *Properties* tab at the bottom of this pane.

6. Click the *Design* button at the bottom of the *Code and Text Editor* window to display the *Design View* window for the form. The *Design View* window is currently nearly empty. (There is a blank *<DIV>* element at the top of the form.)

In the *Design View* window, you can drag controls onto the Web form from the *Toolbox*, and Visual Studio 2008 will generate the appropriate HTML for you. This is the HTML that you see when you view the form in the *Source View* window. You can also edit the HTML directly if you want.

In the next exercise, you will define a style to be used by the form and then add controls to the form to make it functional. By defining a style, you can ensure that all controls on the form share a common look and feel (such as color and font), as well as set items such as a background image of the form.

Lay out the Web form

1. On the *Website* menu, click *Add Existing Item*. In the *Add Existing Item* dialog box, move to the \Microsoft Press\Visual CSharp Step By Step\Chapter 27 folder under your Documents folder, select the Computer.bmp file, and then click *Add*.

   This file contains an image that you will display on the background of your Web form.

2. Click the form in the *Design View* window. In the *Properties* window, change the *Title* property of the *DOCUMENT* object to *Employee Information*.

   The value you specify for the *Title* property appears in the title bar of the Web browser when you run the Web application.

> **Note** If the *Properties* window displays the properties for the *<DIV>* element rather than *DOCUMENT*, select *DOCUMENT* from the drop-down list at the top of the *Properties* window.

3. Click the *Manage Styles* tab underneath the *Properties* window. In the *Manage Styles* window, click the *New Style* link.

   The *New Style* dialog box opens. You can use this dialog box to create a style for the form.

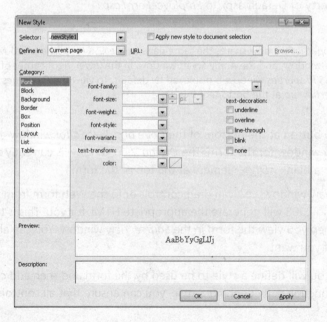

4. In the *font-family* drop-down list box, click *Arial*.

5. In the *color* drop-down list, select the dark blue square on the second row.

    The value #0000FF should appear in the color box.

6. In the *Category* list box, click *Background*.

7. Click the *Browse* button adjacent to the *background-image* combo box. In the *Picture* dialog box, click the computer.bmp file, and then click *OK*.

    The *background-image* combo box is populated with the value *url('computer.bmp')*.

8. In the *Category* list box, click *Position*.

9. In the *height* combo box, type **500**.

10. At the top of the dialog box in the *Selector* combo box, type **.employeeFormStyle** (be sure to include the leading period in this name), select the *Apply new style to document selection* check box, and then click *OK*.

    In the *Design View* window, the Web form displays the image in the background.

11. Display the *Toolbox*, and ensure that the *Standard* category of controls is expanded.

    The *Toolbox* contains controls that you can drop onto ASP.NET forms. These controls are similar, in many cases, to the controls you have been using to build Microsoft Windows Presentation Foundation (WPF) applications. The difference is that these controls have been specifically designed to operate in an HTML environment, and they are rendered by using HTML at run time.

12. From the *Toolbox*, drag four *Label* controls and three *TextBox* controls onto the Web form. Notice how the controls pick up the font and color specified by the Web form's style.

Note The controls will be automatically positioned using a left-to-right flow layout in the *Design View* window. Do not worry about their location just yet because you will move them after setting their properties.

Note As well as using a *Label* control, you can type text directly onto a Web page. However, you cannot format this text so easily, set properties, or apply themes to it. If you are building a Web site that has to support different languages (such as French or German), use *Label* controls because you can more easily localize the text they display by using Resource files. For more information, see "Resources in ASP.NET Applications" in the Microsoft Visual Studio 2008 documentation.

13. Using the *Properties* window, set the properties of these controls to the values shown in the following table.

Control	Property	Value
*Label*1	*Font Bold* (expand the Font property)	True
	Font Name	Arial Black
	Font Size	X-Large
	Text	Litware, Inc. Software Developers
	Height	36px
	Width	630px
*Label*2	*Text*	First Name
*Label*3	*Text*	Last Name
*Label*4	*Text*	Employee Id
*TextBox*1	*(ID)*	firstName
	Height	24px
	Width	230px
*TextBox*2	*(ID)*	lastName
	Height	24px
	Width	230px
*TextBox*3	*(ID)*	employeeID
	Height	24px
	Width	230px

14. Click the *Source* button at the bottom of the *Design View* window. You should see the HTML description of the form and the style in the *Code and Text Editor* window, like this (some lines have been split and reformatted to fit this code in a readable format on the printed page):

```
<%@ Page Language="C#" AutoEventWireup="true"
    CodeFile="EmployeeForm.aspx.cs" Inherits="_Default" %>

<!DOCTYPE html PUBLIC "-//W3C//DTD XHTML 1.0 Transitional//EN"
    "http://www.w3.org/TR/xhtml1/DTD/xhtml1-transitional.dtd">

<html xmlns="http://www.w3.org/1999/xhtml">
<head runat="server">
    <title>Employee Information</title>
    <style type="text/css">
        .employeeFormStyle
        {
            font-family: Arial;
            color: #0000FF;
```

```
                    background-image: url('computer.bmp');
                    height: 500px;
            }
        </style>
</head>
<body>
        <form id="form1" runat="server">
        <div class="employeeFormStyle">

            <asp:Label ID="Label1" runat="server" Font-Bold="True"
                Font-Names="Arial Black"
                Font-Size="X-Large" Height="36px"
                Text="Litware, Inc. Software Developers"
                Width="630px"></asp:Label>
            <asp:Label ID="Label2" runat="server" Text="First Name"></asp:Label>
            <asp:Label ID="Label3" runat="server" Text="Last Name"></asp:Label>
            <asp:Label ID="Label4" runat="server" Text="Employee Id"></asp:Label>
            <asp:TextBox ID="firstName" runat="server" Height="24px"
                Width="230px"></asp:TextBox>
            <asp:TextBox ID="lastName" runat="server" Height="24px"
                Width="230px"></asp:TextBox>
            <asp:TextBox ID="employeeID" runat="server" Height="24px"
                Width="230px"></asp:TextBox>

        </div>
        </form>
</body>
</html>
```

15. Modify the HTML code for the *Label1* control, and add a *Style* attribute to specify its location on the form, as shown here in bold type:

```
<asp:Label ID="Label1" ... Style="position: absolute; left: 96px; top: 24px"></
asp:Label>
```

By setting the *position* property of the *Style* attribute to *absolute*, you can specify the position of controls yourself, rather than letting Visual Studio 2008 lay them out automatically.

> **Tip** You can also specify the layout, alignment, and spacing of controls by using the commands on the *Format* menu when using the *Design View* window.

16. Edit the HTML code for the remaining label and text box controls, and add *Style* attributes to set their locations on the Web form, as shown here in bold type:

```
<asp:Label ID="Label2" ... Style="position: absolute; left: 62px; top: 104px"></
asp:Label>
<asp:Label ID="Label3" ... Style="position: absolute; left: 414px; top: 104px"></
asp:Label>
<asp:Label ID="Label4" ... Style="position: absolute; left: 62px; top: 168px"></
asp:Label>
<asp:TextBox ID="firstName" ... Style="position: absolute; left: 166px; top: 102px"></
asp:TextBox>
```

```
<asp:TextBox ID="lastName" ... Style="position: absolute; left: 508px; top: 102px"></
asp:TextBox>
<asp:TextBox ID="employeeID" ... Style="position: absolute; left: 166px; top:
166px"></asp:TextBox>
```

17. Click the *Design* button at the bottom of the window. The Web form should look like this in the *Design View* window:

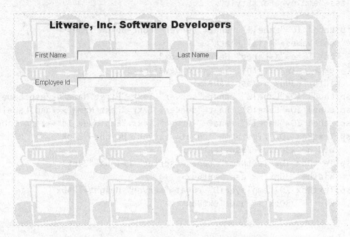

18. Add another *Label* control and four *RadioButton* controls to the Web form. Using the *Properties* window, set the properties of these controls to the values listed in the following table. Note that the controls will appear in a line across the top of the form. You will set their positions in the next step.

Control	Property	Value
*Label*5	*Text*	Position
*RadioButton*1	*(ID)*	workerButton
	Text	Worker
	TextAlign	Left
	GroupName	positionGroup
	Checked	True
*RadioButton*2	*(ID)*	bossButton
	Text	Boss
	TextAlign	Left
	GroupName	positionGroup
	Checked	False
*RadioButton*3	*(ID)*	vpButton
	Text	Vice President
	TextAlign	Left

Control	Property	Value
	GroupName	positionGroup
	Checked	False
RadioButton4	(ID)	presidentButton
	Text	President
	TextAlign	Left
	GroupName	positionGroup
	Checked	False

The *GroupName* property determines how a set of radio buttons is grouped. All buttons with the same value for *GroupName* are in the same group and are mutually exclusive—only one can be selected at a time.

19. Click the *Source* button at the bottom of the *Design View* window, and set the positions of these controls as shown in bold type here:

```
<asp:Label ID="Label5" ... Style="position: absolute; left: 86px; top: 224px">
</asp:Label>
<asp:RadioButton ID="workerButton" ... Style="position: absolute; left: 192px; top:
224px"/>
<asp:RadioButton ID="bossButton" ... Style="position: absolute; left: 206px; top:
260px"/>
<asp:RadioButton ID="presidentButton" ... Style="position: absolute; left: 174px; top:
332px"/>
<asp:RadioButton ID="vpButton" ... Style="position: absolute; left: 138px; top:
296px"/>
```

20. Click the *Design* button, and then add another *Label* control and a *DropDownList* control to the Web form. Set their properties to the values shown in the following table.

Control	Property	Value
Label6	Text	Role
DropDownList1	(ID)	positionRole
	Width	230px

The *positionRole* drop-down list will display the different positions that an employee can have within the company. This list will vary according to the position of the employee in the company. You will write code to populate this list dynamically.

21. Click the *Source* button, and add the HTML code shown here in bold type to set the position of these controls:

```
<asp:Label ID="Label6" ... Style="position: absolute; left: 456px; top: 224px">
</asp:Label>
<asp:DropDownList ID="positionRole" ... Style="position: absolute; left: 512px;
top: 224px"></asp:DropDownList>
```

22. Click the *Design* button, and add two *Button* controls and another *Label* control to the form. Set their properties to the values shown in the following table.

Control	Property	Value
*Button*1	*(ID)*	saveButton
	Text	Save
	Width	75px
*Button*2	*(ID)*	clearButton
	Text	Clear
	Width	75px
*Label*7	*(ID)*	infoLabel
	Text	*leave blank*
	Height	48px
	Width	680px

You will write event handlers for the buttons in a later exercise. The *Save* button will collate the information entered by the user and display it in the *InfoLabel* control at the bottom of the form. The *Clear* button will clear the text boxes and set other controls to their default values.

23. Click the *Source* button, and add the HTML code shown here in bold type to each of these controls:

```
<asp:Button ID="saveButton" ... Style="position: absolute; left: 328px; top: 408px"/>
<asp:Button ID="clearButton" ... Style="position: absolute; left: 424px; top: 408px"/>
<asp:Label ID="infoLabel" ... Style="position: absolute; left: 62px; top: 454px">
</asp:Label>
```

24. Click the *Design* button. The completed form should look like the following image:

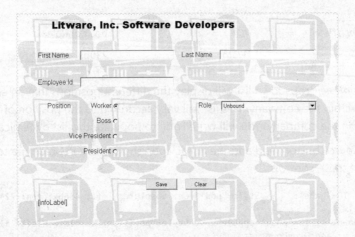

Test the Web form

1. On the *Debug* menu, click *Start Debugging*. In the *Debugging Not Enabled* message box, click *Modify the Web.config file to enable debugging*, and then click *OK*. If the *Script Debugging Disabled* message box appears, click *Yes*.

Visual Studio 2008 builds the application, the ASP.NET Development Server starts, and then Windows Internet Explorer starts and displays the form.

Tip If Internet Explorer displays a list of files rather than the Web form, close Internet Explorer and return to Visual Studio 2008. In *Solution Explorer*, right-click EmployeeForm. aspx, and then click *Set As Start Page*. Run the Web application again.

Note The first time you run a Web application by using the *Start Debugging* command, you will be prompted with a message box stating that debugging is not enabled. You can select either *Run without debugging* or *Modify the Web.config file to enable debugging*. Running in debug mode is useful initially because you can set breakpoints and single-step through the code using the debugger, as described in Chapter 3, "Writing Methods and Applying Scope." However, enabling debugging will slow the application, and debugging should be disabled before the application is deployed to a production Web site. You can do this by editing the web.config file and setting the *debug* attribute of the *compilation* element to *false*, like this:

```
<compilation debug="false">
    <assemblies>
    ...
    </assemblies>
</compilation>
```

2. Enter some information for a fictitious employee. Test the radio buttons to verify that they are all mutually exclusive. Click the drop-down arrow in the *Role* list box; the list will be empty. Click *Save* and *Clear*, and verify that they currently do nothing other than cause the form to be redisplayed.

3. Close Internet Explorer, and return to Visual Studio 2008.

Deploying a Web Site to IIS

A useful feature available in Visual Studio 2008 and Visual Web Developer 2008 Express Edition is the *Copy Web Site* command on the *Website* menu that you can use for copying a Web site from one location to another. You can use this feature to quickly deploy a Web site built and tested using the ASP.NET Development Server to a production IIS site. (You should create a new Web site or an empty virtual directory by using the Internet Information Services management console first.) The following image shows this feature in action.

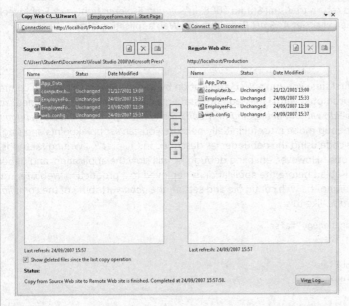

You can connect to the virtual directory on the production IIS site and then selectively copy individual files to or from the production Web site, or synchronize files between Web sites.

> **Note** If you are using the Windows Vista operating system, you must run Visual Studio 2008 using the Administrator account to connect to IIS in the *Copy Web Site* window.

For more information, see the topics "Walkthrough: Copying a Web Site Using the Copy Web Site Tool" and "How to Copy Web Site Files with the Copy Web Site Tool" in the Microsoft Visual Studio 2008 documentation.

Understanding Server Controls

The Web forms controls you added to the form are collectively known as Server controls. Server controls are similar to the standard HTML items that you can use on an ordinary Web page except that they are more programmable. Most Server controls expose event handlers, methods, and properties that code running on the server can execute and modify dynamically at run time. In the following exercises, you will learn more about programming Server controls.

Examine a Server control

1. In the *Design View* window displaying EmployeeForm.aspx, click the *Source* button.

2. Examine the HTML code for the form. Look at the definition of the first *Label* control in more detail (the following code has been laid out to make it easier to read):

```
<asp:Label ID="Label1" runat="server"
    Font-Bold="True" Font-Names="Arial Black"
    Font-Size="X-Large" Height="36px"
    Text="Litware, Inc. Software Developers" Width="630px"
    Style="position: absolute; left: 96px; top: 24px"></asp:Label>
```

There are a couple of things to observe. First, look at the type the control is, *asp:Label*. All Web forms controls live in the *asp* namespace because this is the way they are defined by Microsoft. The second noteworthy item is the *runat="server"* attribute. This attribute indicates that the control can be accessed by code running on the Web server. This code can query and change the values of any of the properties of this control (for example, change its text).

HTML Controls

ASP.NET also supports HTML controls. If you expand the HTML category in the *Toolbox*, you are presented with a list of controls. These are the controls that Microsoft supplied with the original ASP model. They are provided so that you can port existing ASP pages into ASP.NET more easily. However, if you are building a Web application from scratch, you should use the Standard Web Forms controls instead.

HTML controls also have a *runat* attribute so that you can specify where event handling code should be executed for these controls. Unlike Web forms controls, the default location for HTML controls to execute code is in the browser rather than on the server—assuming that the user's browser supports this functionality.

The EmployeeForm.aspx page requires you to add the following functionality:

■ Populate the PositionRole drop-down list when the user selects a position (Worker, Boss, Vice President, President).

■ Save the information entered when the user clicks the Save button.

■ Clear the form when the user clicks the Clear button.

You will implement this functionality by writing event handlers.

> **Note** The methods you will add in the following exercise use hard-coded values for the various roles and the jobs that they can perform. In a professional application, you should store this type of information in a database and use a technology such as ADO.NET or DLINQ to retrieve the various roles and their associated jobs from the database. You will see how to use DLINQ with an ASP.NET Web application in Chapter 29.

Handle Server control events

1. In *Solution Explorer*, expand the file EmployeeForm.aspx.

   The file EmployeeForm.aspx.cs will appear. This is the file that will actually contain the C# code for the event handlers that you write. This file is known as a *code-behind file*. You can separate the C# code from the display logic for a Web application by using this feature of ASP.NET. (You can actually write C# code and event handlers in the EmployeeForm.aspx file by using the *Source View* window, but this approach is not recommended.)

2. In the *Code and Text Editor* window displaying the source view for EmployeeForm.aspx, examine the first line of the file. It contains the following text:

   ```
   <%@ Page Language="C#" ... CodeFile="EmployeeForm.aspx.cs ... %>
   ```

   The *CodeFile* directive specifies the file containing the program code for the Web form and the language in which it is written, in this case, C#. The other supported languages include Microsoft Visual Basic and JScript.

3. In *Solution Explorer*, double-click the EmployeeForm.aspx.cs file.

   The file appears in the *Code and Text Editor* window. At the top of the file, there is a set of *using* statements. Note that this file makes heavy use of the *System.Web* namespace and its subnamespaces—this is where the ASP.NET classes reside. Also, notice that the code itself is in a class called *_Default* that descends from *System.Web.UI.Page*; this is the class from which all Web forms descend. Currently, it contains a single empty method called *Page_Load*. This method runs when the page is displayed. You can write code in this method to initialize any data required by the form.

4. Add a method called *initPositionRole* to the *_Default* class after the *Page_Load* method:

```
private void initPositionRole()
{
}
```

You will invoke this method to initialize the *positionRole* drop-down list to its default set of values.

5. Add the following statements shown in bold type to the *initPositionRole* method:

```
private void initPositionRole()
{
    positionRole.Items.Clear();
    positionRole.Enabled = true;
    positionRole.Items.Add("Analyst");
    positionRole.Items.Add("Designer");
    positionRole.Items.Add("Developer");
}
```

The first statement clears the items from the drop-down list box. The second statement activates the list box. (You will write some code shortly that disables it under certain circumstances.) The remaining statements add the three roles that are applicable to workers.

6. Add the statements shown here in bold type to the *Page_Load* method:

```
protected void Page_Load(object sender, EventArgs e)
{
    if (!IsPostBack)
    {
        initPositionRole();
    }
}
```

This block of code causes the *positionRole* drop-down list to be populated when the form appears in the user's browser. However, it is important to understand that the *Page_Load* method runs every time the Web server sends the form to the user's browser. For example, when the user clicks a button the form can be sent back to the Web server for processing; the Web server then responds by sending the form back to the browser for displaying when the processing has completed. You don't want the initialization to be performed every time the page appears because it is a waste of processing and can lead to performance problems if you are building a commercial Web site. You can determine whether the *Page_Load* method is running because this is the first time the page is being displayed by querying the *IsPostBack* property of the Web page. This property returns *false* the first time the page is displayed and *true* if the page is being redisplayed because the user has clicked a control. In the code you added, you call the *initPositionRole* method only when the form is first displayed.

7. Switch to the EmployeeForm.aspx file, and click the *Design* button. Select the *Worker* radio button. In the *Properties* window toolbar, click the *Events* toolbar button. (This button has a little lightning icon.) Double-click the *CheckedChanged* event. This event occurs when the user clicks the radio button and its value changes. Visual Studio 2008 generates the method *workerButton_CheckedChanged* to handle this event.

> **Note** The *Properties* window of an ASP.NET Web application provides additional features not currently available when you build a WPF application. These features include being able to list the events available for a control and specify an event handler. When you create a WPF application, this functionality is available only when you edit the Extensible Application Markup Language (XAML) code for a control.

8. In the *Code and Text Editor* window, add the statement shown here in bold type to the *workerButton_CheckedChanged* event method:

```
protected void workerButton_CheckedChanged(object sender, EventArgs e)
{
    initPositionRole();
}
```

Remember that the default values for the *positionRole* drop-down list are those for a worker, so the same method can be reused to initialize the list.

9. Switch to the *Design View* window displaying the EmployeeForm.aspx form. Select the *Boss* radio button, and use the *Properties* window to create an event method called *bossButton_CheckedChanged* for the *CheckedChanged* event. When the form is displayed in the *Code and Text Editor* window, type the following statements in the *BossCheckedChanged* method:

```
protected void bossButton_CheckedChanged(object sender, EventArgs e)
{
    positionRole.Items.Clear();
    positionRole.Enabled = true;
    positionRole.Items.Add("General Manager");
    positionRole.Items.Add("Project Manager");
}
```

These are the roles that a manager can fulfill.

10. Return to the *Design View* window displaying the EmployeeForm.aspx form, and create an event handler for the *CheckedChanged* event for the *Vice President* radio button. In the *Code and Text Editor* window, add the following statements shown in bold type to the *vpButton_CheckedChanged* event method:

```
protected void vpButton_CheckedChanged(object sender, EventArgs e)
{
    positionRole.Items.Clear();
    positionRole.Enabled = true;
```

```
positionRole.Items.Add("VP Sales");
positionRole.Items.Add("VP Marketing");
positionRole.Items.Add("VP Production");
positionRole.Items.Add("VP Human Resources");
}
```

11. Switch to the *Design View* window displaying the EmployeeForm.aspx form, and create an event handler for the *CheckedChanged* event for the *President* radio button. Add the code shown here in bold type to the *presidentButton_CheckedChanged* event method:

```
protected void presidentButton_CheckedChanged(object sender, EventArgs e)
{
    positionRole.Items.Clear();
    positionRole.Enabled = false;
}
```

Roles do not apply to the president of the company, so the drop-down list is cleared and disabled.

12. Return to the *Design View* window displaying the EmployeeForm.aspx form, and create an event handler for the *Click* event of the *Save* button. The method would usually save the information to a database, but to keep this application simple, the method will just echo some of the data in the *InfoLabel* control instead. Add the following statements shown in bold type to the *saveButton_Click* method:

```
protected void saveButton_Click(object sender, EventArgs e)
{
    String position = "";

    if (workerButton.Checked)
        position = "Worker";
    if (bossButton.Checked)
        position = "Manager";
    if (vpButton.Checked)
        position = "Vice President";
    if (presidentButton.Checked)
        position = "President";

    infoLabel.Text = "Employee: " + firstName.Text + " " +
        lastName.Text + "    Id: " +
        employeeID.Text + "    Position: " +
        position;
}
```

The * * character is a nonbreaking space in HTML; ordinary white-space characters after the first white-space character will usually be ignored by the browser.

13. Using the same technique, create an event method for the *Click* event of the *Clear* button. Add the following block of code shown in bold type to this method:

```
protected void clearButton_Click(object sender, EventArgs e)
{
    firstName.Text = "";
    lastName.Text = "";
```

```
employeeID.Text = "";
workerButton.Checked = true;
bossButton.Checked = false;
vpButton.Checked = false;
presidentButton.Checked = false;
initPositionRole();
infoLabel.Text = "";
}
```

This code clears the information entered by the user and then resets the role to Worker (the default value).

> **Note** Although only one radio button in a group can have its *Checked* property set to *true*, it is necessary to set the *Checked* property of the remaining radio buttons to *false* to ensure that the correct button is displayed as being selected when ASP.NET refreshes the form in the user's Web browser.

Test the Web form again

1. On the *Debug* menu, click *Start Debugging* to run the Web form again.

2. When the Web form appears in Internet Explorer, type an employee's name, enter an ID number (make them up), and then click the *Role* drop-down list.

   The list of roles for a worker is displayed.

3. Change the position of your fictitious employee to *Vice President*, and then click the *Role* drop-down list box.

   Notice that the list has not changed and still displays the roles for a worker. The list hasn't changed because the *CheckedChanged* event for the *Vice President* radio button has not been raised.

4. Close Internet Explorer, and return to Visual Studio 2008.

5. Display the EmployeeForm.aspx Web form in the *Design View* window, and then select the *worker-Button* radio button. In the *Properties* window, set the *AutoPostBack* property to *True*.

> **Tip** If the *Properties* window is still displaying the list of events for the radio button, click the *Properties* button next to the *Events* button on the *Properties* window toolbar.

When the user clicks this radio button, the form will be sent back to the server for processing, the *CheckedChanged* event will fire, and the form can be updated to display the roles for this radio button. By default, the *AutoPostBack* property is set to *False* to avoid unnecessary network traffic.

6. Set the *AutoPostBack* property to *True* for the other radio buttons: *bossButton*, *vpButton*, and *presidentButton*.

7. Run the Web form again.

This time you will find that when you click the radio buttons, there is a slight flicker while the form is submitted to the server, the event handler runs, the drop-down list is populated, and the form is displayed again.

8. On the Internet Explorer toolbar, click the *Page* drop-down list, and then click *View Source* to display the source of the HTML page being displayed in the browser.

> **Note** If the *Internet Explorer Security* message box appears, click *Allow* so that you can view the source file for the page.

Notepad starts and displays the HTML source for the page. Notice that there is no mention of any "asp:" Server controls in this file and no C# code. Instead, the Server controls and their contents have been converted to the equivalent HTML controls (and some JavaScript). This is one of the basic features of the Server controls—you access them programmatically like ordinary .NET Framework objects, with methods, properties, and events. When they are rendered by the Web server, they are converted to HTML so that you can display the form in any HTML-compliant browser.

9. When you have finished examining the file, close Notepad.

10. On the Web form, click *Save*.

The *InfoLabel* control displays the details of the new employee. If you examine the source, you will see that the HTML for the *InfoLabel* control (rendered as an HTML span with an ID of "InfoLabel") contains this text.

11. Click *Clear*.

The form resets to its default values.

12. Close Internet Explorer, and return to Visual Studio 2008.

Event Processing and Roundtrips

Server controls are undoubtedly a powerful feature of ASP.NET, but they come with a price. You should remember that although events are raised by the Web client, the event code is executed on the Web server, and that each time an event is raised, an HTTP request (or postback) is sent over the network to the Web server. The task of the Web server is to process this request and send a reply containing an HTML page to be displayed. In the case of many events, this page is the same as the one that issued the original request. However, the Web server also needs to know what other data the user has entered on the page so that when the server generates the HTML response, it can preserve these values in the display. (If the Web server sent back only the HTML that composed the original page, any data entered by the user would disappear.) If you look at the HTML source of a page generated by a Web form, you will notice a hidden input field in the form. The example shown previously had this hidden field:

```
<input type="hidden" name="__VIEWSTATE"
value="/WEPdDwxNDkOMzA1NzEOO3Q8O2w8aTwxPjs+O2w8bDxpPDE3PjtpPDE5
PjtpP DIxPjtpPDI3PjtpPDMzPjs+O2w8dDxwPHA8bDxDaGVja2VkOz47bDxvPH
Q+Oz4+Oz 47Oz47dDxwPHA8bDxDaGVja2VkOz47bDxvPGY+Oz4+Oz47Oz47dDxw
PHA8bDxDaGVja2a2 VkOz47bDxvPGY+Oz4+Oz47Oz47dDxOPDtOPGk8Mz47QDxBbm
FseXNOOORlc2lnbmVyOO RldmVsb3Blcjs+OOA8QW5hbHlzdDtEZXNpZ251lcjtE
ZXZlbG9wZXI7Pj47Pjs7Pj tOPHA8cDxsPFRleHQ7PjtsPFxlOz4+Oz47Oz47Pj
47Pj47bDxQZW9uQnVOdG9uO1BIQ kJ1dHRvbjtQSEJCdXROb247VlBCdXROb247
VlBCdXROb247UHJlc21kZW50QnVOdG9uO 1ByZXNpZGVudEJ1dHRvbjs+Pg==" />
```

This information is the content of the controls, or view state, in an encoded form. It is sent to the Web server whenever any event causes a postback. The Web server uses this information to repopulate the fields on the page when the HTML response is generated.

All of this data has an impact on scalability. The more controls you have on a form, the more state information has to be passed between the browser and Web server during the postback processing, and the more events you use, the more frequently this will happen. In general, to reduce network overhead, you should keep your Web forms relatively simple, avoid excessive use of server events, and be selective with view state to avoid sending unnecessary information across the network. You can disable the view state for a control by setting the *EnableViewState* property of the control to *False* (the default setting is *True*).

Creating and Using a Theme

When you first created the Web site, you defined a style for the form. This style determined the default font and color for controls on the form and could also be used to specify default values for other attributes, such as the way in which lists are formatted and numbered. (You can edit a style by right-clicking the style in the *Manage Styles* window and then by clicking *Modify Style*.) However, a style defined in this way applies only to a single form. Commercial Web sites typically contains tens, or maybe hundreds, of forms. Keeping all of these forms consistently formatted can be a time-consuming task; if the company you work for decided to change the font on all of its Web pages, imagine how many forms you would need to update and rebuild! This is where *themes* can be very useful. A theme is a set of properties, styles, and images that you can apply to the controls on a page or globally across all pages in a Web site.

> **Note** If you are familiar with cascading style sheets (.css files), the concept of themes might be familiar to you. However, there are some differences between cascading style sheets and themes. In particular, themes do not cascade in the same way as cascading style sheets, and properties defined in a theme applied to a control always override any local property values defined for the control.

Defining a Theme

A theme is made up of a set of skin files located in a named subfolder in the App_Themes folder for a Web site. A skin file is a text file that has the file name extension .skin. Each skin file specifies the default properties for a particular type of control using syntax very similar to that which is displayed when you view a Web form in the *Source View* window. For example, the following skin file specifies the default properties for *TextBox* and *Label* controls:

```
<asp:TextBox BackColor="Blue" ForeColor="White" Runat="Server" />
<asp:Label BackColor="White" ForeColor="Blue" Runat="Server" Font-Bold="True" />
```

You can specify many properties of a control in a skin file, but not all of them. For example, you cannot specify a value for the *AutoPostBack* property. Additionally, you cannot create skin files for every type of control, but most commonly used controls can be configured in this way.

Applying a Theme

After you have created a set of skin files for a theme, you can apply the theme to a page by modifying the *@Page* attribute that occurs at the start of the page in the *Source View* window. For example, if the skin files for a theme are located in the App_Themes\BlueTheme folder under the Web site, you can apply the theme to a page like this:

```
<%@Page Theme="BlueTheme" ...%>
```

If you want to apply the theme to all pages in the Web site, you can modify the web.config file and specify the theme in the *pages* element, like this:

```
<configuration>
    <system.web>
        <pages theme="BlueTheme" />
    </system.web>
</configuration>
```

If you modify the definition of a theme, all controls and pages that use the theme will pick up the changes automatically when they are next displayed.

In the final set of exercises in this chapter, you will create a theme for the Litware Web site and then apply this theme to all pages in the Web site.

Create a new theme

1. In *Solution Explorer*, right-click the C:\...\Litware project folder. Point to *Add ASP.NET Folder*, and then click *Theme*.

    A new folder called App_Themes is added to the project, and a subfolder is created called Theme1.

2. Change the name of the Theme1 folder to LitTheme.

3. In *Solution Explorer*, right-click the LitTheme folder, and then click *Add New Item*.

    The *Add New Item* dialog box appears, displaying the types of file that can be stored in a themes folder.

4. Click the *Skin File* template, type **Lit.skin** in the *Name* text box, and then click *Add*.

    The skin file Lit.skin is added to the LitTheme folder, and the file is displayed in the *Code and Text Editor* window.

5. Append the following lines to the end of the Lit.skin file in the *Code and Text Editor* window (this file contains a comment with some very brief instructions):

```
<asp:TextBox BackColor="Red" ForeColor="White" Runat="Server" />
<asp:Label BackColor="White" ForeColor="Red" Runat="Server" Font-Bold="True" />
<asp:RadioButton BackColor="White" ForeColor="Red" Runat="Server"/>
<asp:Button BackColor="Red" ForeColor="White" Runat="Server" Font-Bold="True"/>
<asp:DropDownList BackColor="Red" ForeColor="White" Runat="Server"/>
```

    This simple set of properties displays *TextBox*, *Button*, and *DropDownListBox* controls as white text on a red background, and *Label* and *RadioButton* controls as red text on a white background. The text on *Label* and *Button* controls is displayed using the bold font version of the current font.

Important The skin file editor is very basic and does not provide any IntelliSense to help you. If you make a mistake in this file, the application will run, but entries in this file might be ignored. When you run the application later, if any of the controls do not appear as expected, ensure that you have not mistyped anything in this file.

As mentioned previously, there are at least two ways you can apply a theme to a Web form: you can set the *@Page* attribute for each page, or you can specify the theme globally across all pages by using a Web configuration file. You are going to use the latter approach in the next exercise. This mechanism causes all pages for the Web site to apply the same theme automatically.

Create a Web configuration file, and apply the theme

1. In *Solution Explorer*, double-click the web.config file to display it in the *Code and Text Editor* window.

2. Locate the *<pages>* line, and modify it as shown here in bold type:

```
<pages theme="LitTheme">
```

3. On the *Debug* menu, click *Start Without Debugging*.

   Internet Explorer appears and displays the Web form. Verify that the style of the controls on the form have changed as expected, although any text in the text boxes might be a little hard to read (you will fix this shortly). Close Internet Explorer when you have finished.

4. In *Solution Explorer*, double-click the Lit.skin file to display it in the *Code and Text Editor* window. Modify the element defining the appearance of *TextBox* and *DropDownList* controls, as shown here in bold type:

```
<asp:TextBox BackColor="White" ForeColor="Red" Font-Bold="True" Runat="Server" />
...
<asp:DropDownList BackColor="White" ForeColor="Red" Runat="Server" />
```

5. Run the form again. Notice how the style of the *First Name*, *Last Name*, and *Employee Id TextBox* controls, and the *Role* drop-down list have changed; hopefully, they are easier to read.

6. Close Internet Explorer when you have finished.

- If you want to continue to the next chapter

  Keep Visual Studio 2008 running, and turn to Chapter 28.

- If you want to exit Visual Studio 2008 now

  On the *File* menu, click *Exit*. If you see a *Save* dialog box, click *Yes* and save the project.

Chapter 27 Quick Reference

To	Do this
Create a Web application	Create a new Web site using the ASP.NET Web Site template. Specify whether you want to use the Development Server (specify a file system location and file name) or IIS (specify an HTTP location and URL).
View and edit the HTML definition of a Web form	Click the *Source* button in the *Design View* window.
Create a style for a Web form	In the *Manage Styles* window, click *New Style*. Use the *New Style* dialog box to define the style for the form.
Add ASP.NET Server controls to a Web form	Click the *Design* button in the *Design View* window. In the *Toolbox*, expand the *Standard* category. Drag controls onto the Web form.
Add HTML controls to a Web form (with HTML controls, you can more easily port existing ASP pages into ASP.NET)	In the *Toolbox*, click the HTML category. Drag controls onto the Web form.
Create an event handler for an ASP. NET Server control	In the *Design View* window, select the control on the Web form. In the *Properties* window, click the *Events* button. Choose the event you want to handle and type the name of an event handler method or double-click the event name to select the default name. In the *Code and Text Editor* window, write the code to handle the event.
Create a theme	Add an App_Themes folder to the Web site. Create a subfolder for the theme. Create a skin file defining the properties of controls in this folder.

Apply a theme to a Web site	Either specify the theme using the @*Page* attribute of each page, like this:
	
	`<%@Page Theme="BlueTheme" ...%>`
	
	or modify the web.config file and specify the theme in the *pages* element, like this:
	
	`<pages theme="BlueTheme">`

Chapter 28

Understanding Web Forms Validation Controls

After completing this chapter, you will be able to:

- Validate user input in a Microsoft ASP.NET Web form by using the ASP.NET validation controls.

- Determine whether to perform user input validation in the user's Web browser or at the Web server.

As with a Microsoft Windows Presentation Foundation (WPF) application, validating user input is an important part of any Web application. With WPF, you can check that the user's input makes sense by binding controls to properties of business objects and letting the code in these business objects validate the data, or by writing code to validate the contents of these fields in response to events that occur when the user moves from field to field on a form. ASP.NET Web forms do not support binding to business objects for validation purposes, so at first glance it appears that your only option might be to use events. However, there is one fundamental consideration that you should think about. Web applications are distributed in their nature: the presentation logic runs in the Web browser on the user's computer, while the code for the application runs on the Web server. With this in mind, should you perform user input validation at the client (the Web browser) or at the Web server? In this chapter, you will examine this question and discover the options that are available to you.

 Note As you read this chapter, you might be surprised to discover that it contains no C# code. This is intentional. You could validate data by using C# methods, but sometimes it is equally instructive to see situations where you do not actually need to write C# code to perform potentially complex tasks.

Comparing Server and Client Validations

Consider the EmployeeForm.aspx page of the Litware Web site again. The user is expected to enter the details of an employee: name, employee ID, position, and role. All the text boxes should be mandatory. Additionally, the employee ID should be a positive integer.

Validating Data at the Web Server

If you examine the *TextBox* class, you will notice that it provides the *TextChanged* event. After the user changes the text in the text box, this event runs the next time the form is posted back to the server. As with all Web Server control events, the *TextChanged* event handler runs at the Web server. Validating data at the server involves transmitting data from the Web browser to the server, processing the event at the server to validate the data, and then packaging up any validation errors as part of the HTML response sent back to the client so that the browser can display these errors. If the validation being performed is complex or requires processing that can be performed only at the Web server (such as ensuring that an employee ID the user enters exists in a database), this is an acceptable technique. But if you are simply inspecting the data in a single text box in isolation (such as making sure that the user types a positive integer into an Employee ID text box), performing this type of validation on the Web server could impose unacceptable overhead; why not perform this check in the browser on the client computer and save a network round-trip?

Validating Data in the Web Browser

The ASP.NET Web Forms model facilitates performing client-side validation in a Web browser through the use of validation controls. If the user is running a browser (such as Microsoft Internet Explorer 4 or later) that supports dynamic HTML, the validation controls generate JavaScript code that runs in the browser and avoids the need to perform a network round-trip to the server. If the user is running an older browser, the validation controls generate server-side code instead. The key point is that the developer creating the Web form does not have to worry about checking for browser capabilities; all the browser detection and code generation features are built into the ASP.NET validation controls. The developer simply drops an ASP.NET validation control onto the Web form, sets its properties (either by using the *Properties* window or by writing code), and specifies the validation rules to be performed and any error messages to be displayed.

ASP.NET provides the following validation controls:

- *RequiredFieldValidator*   Use this control to ensure that the user has entered data into a control.

- *CompareValidator*   Use this control to compare the data entered with a constant value, the value of a property of another control, or a value retrieved from a database.

- *RangeValidator*   Use this control to check the data entered by a user against a range of values, checking that the data falls either inside or outside a given range.

- *RegularExpressionValidator*   Use this control to check that the data input by the user matches a specified regular expression, pattern, or format (such as a telephone number, for example).

- *CustomValidator*   Use this control to define your own custom validation logic and attach it to a control to be validated.

> **Note** You should be aware that if a user can type unrestricted text into a text box and send it to the Web server, the user could type text that looks like HTML tags (<b> for example). Hackers sometimes use this technique to inject HTML into a client request in an attempt to cause damage to the Web server or to try to break in. (I am not going to go into the details here!) By default, if you try this trick with an ASP.NET Web page, the request will be aborted and the user is shown the message "A potentially dangerous Request.Form value was detected from the client." You can disable this check, although that is not recommended. A better approach is to use a *RegularExpressionValidator* control to verify that the user input in a text box does not constitute an HTML tag (or anything that looks like it). For more information about regular expressions and how to use them, see the topic ".NET Framework Regular Expressions" in the Microsoft Visual Studio 2008 documentation.

Although each control performs a single well-defined type of validation, you can use several of them in combination. For example, if you want to ensure that the user enters a value in a text box and that this value falls in a particular range, you can attach a *RequiredFieldValidator* control and a *RangeValidator* control to the text box.

These controls can work in conjunction with a *ValidationSummary* control to display error messages. You will use some of these controls in the following exercises.

Implementing Client Validation

Returning to the EmployeeForm.aspx Web form, you can probably see that *RequiredFieldValidator* controls will be required for the *First Name*, *Last Name*, and *Employee Id* text boxes. The employee ID must also be numeric and should be a positive integer. In this application, you will specify that the employee ID must be between 1 and 5000. This is where a *RangeValidator* control is useful.

Add *RequiredFieldValidator* controls

1. Start Microsoft Visual Studio 2008 if it is not already running.

2. If you are using Visual Studio 2008 Professional Edition or Enterprise Edition, on the *File* menu, point to *Open*, and then click *Web Site*.

3. If you are using Microsoft Visual Web Developer 2008 Express Edition, on the *File* menu, click *Open Web Site*.

4. In the *Open Web Site* dialog box, ensure that the *File System* option is selected, browse to Microsoft Press\Visual CSharp Step by Step\Chapter 28\Litware under your Documents folder, and then click *Open*.

Note When you create a new Web site, Visual Studio 2008 creates a solution file in a solution folder in the Visual Studio 2008 folder under your Documents folder. However, you do not need to select a Microsoft Visual C# solution or project file to open a Web site for editing; just move to the folder containing the Web site files and subfolders. If you do want to open a Web site by using the solution file, on the *File* menu, point to *Open*, and click *Project/Solution* (instead of *Web Site*), move to the solution folder, and then click the solution file.

5. In *Solution Explorer*, right-click *EmployeeForm.aspx*, and then click *Set As Start Page*.

6. Right-click *EmployeeForm.aspx* again, and then click *View Designer* to display the Web form in the *Design View* window.

7. In the *Toolbox*, expand the *Validation* category.

8. Add a *RequiredFieldValidator* control to the form.

   The control appears in the upper-left part of the form, displaying the text "RequiredFieldValidator".

9. Click the *Source* button to display the HTML source code for the form. Locate the code for the *RequiredFieldValidator* control toward the bottom of the file, and set the *Style* property to position it underneath the *firstName* text box, as shown here in bold type. (The position of a validation control determines where the error message is displayed.)

   ```
   <asp:RequiredFieldValidator ID="RequiredFieldValidator1"...
    Style="position: absolute; left: 166px; top: 128px"></asp:RequiredFieldValidator>
   ```

10. Click the *Design* button, and then select the *RequiredFieldValidator* control. In the *Properties* window, use the drop-down list to set the *ControlToValidate* property to *firstName*. Setting the *ControlToValidate* property links the validation control to the item it will validate. Enter **You must specify a first name for the employee** in the *ErrorMessage* property. This is the message that will be displayed if the control to be validated (the *First Name* text box) is left blank. Notice that this message replaces the default red text error message ("RequiredFieldValidator") on the form.

11. Add two more *RequiredFieldValidator* controls to the form.

12. Click the *Source* button, and add the *Style* properties shown here in bold type to position these controls under the *lastName* and *employeeID* text boxes.

    ```
    <asp:RequiredFieldValidator ID="RequiredFieldValidator2"...
     Style="position: absolute; left: 508px; top: 128px"></asp:RequiredFieldValidator>
    <asp:RequiredFieldValidator ID="RequiredFieldValidator3"...
     Style="position: absolute; left: 166px; top: 194px"></asp:RequiredFieldValidator>
    ```

13. Click the *Design* button, and then select the *RequiredFieldValidator* control under the *Last Name* text box. Using the *Properties* window, set its *ControlToValidate* property to *lastName*, and enter **You must specify a last name for the employee** in its *ErrorMessage* property. Notice that the *RequiredFieldValidator* control automatically resizes itself to display the complete error message.

14. Select the *RequiredFieldValidator* control under the *Employee Id* text box; set its *ControlToValidate* property to *employeeID*, and enter **You must specify an employee ID** in its *ErrorMessage* property.

15. On the *Debug* menu, click *Start Without Debugging* to run the form in Windows Internet Explorer.

16. When the form first appears, all the required text boxes will be empty. Click *Save*. The error messages belonging to all three *RequiredFieldValidator* controls are displayed.

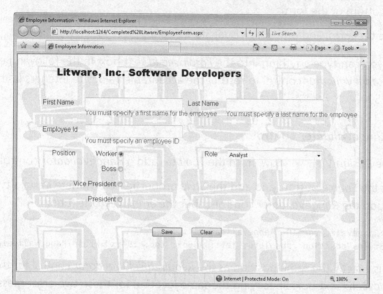

Notice that the *Click* event for the *Save* button did not run, and the label at the bottom of the form did not display the data summary (and the screen did not even flicker). This behavior is because the validation controls prevented the postback to the server; they generate code that can be executed by the browser, and they will continue to block posts back to the server until all the errors have been corrected.

> **Note** If you click the *Clear* button while an error message is displayed, it will not clear the form because the error blocks the postback to the Web server. ASP.NET provides support for client-side scripting so that you can add JavaScript code to clear the Web form. This code is not blocked by postbacks because it runs in the user's Web browser (assuming the browser supports JavaScript). The validation controls actually generate JavaScript code that runs in the user's browser rather than being posted back to the Web server. The details of writing your own client-side JavaScript code in an ASP.NET Web form are outside the scope of this book, but for more information, search for the article "How to Add Client Script Events to ASP.NET Web Server Controls" in the documentation provided with Visual Studio 2008.

17. Type a name in the *First Name* text box.

As soon as you move away from the text box, the corresponding error message disappears. If you return to the *First Name* text box, erase the contents, and then move to the next text box, the error message is displayed again. All this functionality is being performed in the browser with no data being sent to the server over the network.

18. Enter values in the *First Name*, *Last Name*, and *Employee Id* text boxes, and then click *Save*.

    This time the *Click* event runs and the summary is displayed in the *InfoLabel* control at the bottom of the form.

19. Close the form, and return to Visual Studio 2008.

Currently, you can type anything into the *Employee Id* text box. In the following exercise, you will use a *RangeValidator* control to restrict the acceptable values to integers in the range of 1 through 5000.

Add a *RangeValidator* control

1. In the *Design View* window, from the *Toolbox*, add a *RangeValidator* control to the form.

2. Click the *Source* button, and add the *Style* properties shown here in bold type to position the *RangeValidator* control under the *employeeID* text box.

```
<asp:RangeValidator ID="RangeValidator1"...
    Style="position: absolute; left: 166px; top: 194px"></asp:RangeValidator>
```

   This is exactly the same position as the *RequiredFieldValidator* control for the *employeeID* text box. Specifying the same location for these two error messages is not a problem because the validations performed by these controls are mutually exclusive (if the employee ID is blank, the *RangeValidator* control cannot test the value entered by the user), so only one of the error messages can be displayed.

3. Click anywhere in the HTML code for the *RangeValidator1* control. In the *Properties* window, set the *ControlToValidate* property to *employeeID*. Enter **The employee ID must be between 1 and 5000** in the *ErrorMessage* property. Set the *MaximumValue* property to 5000, the *MinimumValue* property to 1, and the *Type* property to *Integer*.

> **Note** You can use the *RangeValidator* control to restrict the range of non-numeric data by setting the *Type* property. The types you can specify are *String*, *Integer*, *Double*, *Date*, and *Currency*. You should specify values of the appropriate type for the *MaximumValue* and *MinimumValue* properties. The *RangeValidator* control uses the collation sequence of the character set used by the current locale when performing range checking for strings, and when checking *Date* ranges, an earlier date is considered to be lower than a later date.

4. Run the form again. Enter a first name and a last name, but leave the employee ID blank. Click *Save*.

    An error message telling you that you must supply an employee ID appears.

5. Type **–1** in the *Employee Id* text box, and then click *Save*.

   An error message telling you that the employee ID must be between 1 and 5000 appears.

6. Type **101** in the *Employee Id* text box, and then click *Save*.

   This time the data is valid. The form is posted back to the server, the *Click* event of the *Save* button runs, and a summary of the information entered in the *InfoLabel* label appears at the bottom of the form.

7. Experiment with other values that are out of range or of the wrong type. Try 5001 and the text "AAA" to check that the *RangeValidator* control works as expected.

8. On the Internet Explorer toolbar, click the *Page* drop-down list, and then click *View Source* to display the source of the HTML page being displayed in the browser.

> **Note** If the *Internet Explorer Security* message box appears, click *Allow* so that you can view the source file for the page.

   Notepad starts and displays the HTML source for the page. Scroll through the file and examine its contents. Near the end, you will find some JavaScript code that performs the validations. This code was generated by using the properties of the validation controls. Close Notepad when you have finished browsing the HTML source code.

9. Close Internet Explorer, and return to Visual Studio 2008.

Disabling Client-Side Validation

In the preceding exercise, you saw that the validations were performed by using JavaScript code running in the browser. The ASP.NET runtime generates this code automatically, depending on the capabilities of the Web browser being used to view the page. If the browser does not support JavaScript, all validation checks will be performed by using code running on the Web server instead. The validation will be performed only when the form is posted back to the server.

If you want, you can suppress client-side validation and force all checks to be performed at the server. To do this, set the *EnableClientScript* property of the validation control to *False*. You might find it useful to do this under certain circumstances, such as those involving custom validations (by using the *CustomValidator* control) that are complex or require access to data that is available only on the server. The *CustomValidator* control also has a *ServerValidate* event that can be used to perform additional validation explicitly on the server, even if *EnableClientScript* is set to *True*.

You have seen how validation controls can validate the data that the user enters, but the error message display is not very pretty. In the following exercise, you will use a *ValidationSummary* control to change the way that the error information is presented to the user.

Add a *ValidationSummary* control

1. In the *Code and Text Editor* window, click anywhere in the HTML code for the *RequiredFieldValidator1* control. In the *Properties* window, set the *Text* property to *****.

   If you set the *Text* property of a validation control, the corresponding text value is displayed on the form rather than the error message. (If no value is specified for the *Text* property, the value of the *ErrorMessage* property is displayed.)

2. Modify the *Style* property of the *RequiredFieldValidator1* control to position it to the right of the *First Name* text box, as shown in bold type here:

   ```
   <asp:RequiredFieldValidator ID="RequiredFieldValidator1" ...
       Style="position: absolute; left: 400px; top: 106px"></asp:RequiredFieldValidator>
   ```

   Now, if a validation error occurs, the user will see a red asterisk appear next to the text box with the error.

3. Click anywhere in the HTML code for the *RequiredFieldValidator2* control, set its *Text* property to *, and then change the *Style* to move it to the right of the *Last Name* text box.

   ```
   <asp:RequiredFieldValidator ID="RequiredFieldValidator2" ...
       Style="position: absolute; left: 744px; top: 106px"></asp:RequiredFieldValidator>
   ```

4. Click anywhere in the HTML code for the *RequiredFieldValidator3* control, set its *Text* property to *, and then change the *Style* property to move it to the right of the *Employee Id* text box.

   ```
   <asp:RequiredFieldValidator ID="RequiredFieldValidator3" ...
       Style="position: absolute; left: 400px; top: 172px"></asp:RequiredFieldValidator>
   ```

5. Click anywhere in the HTML code for the *RangeValidator1* control, set its *Text* property to *, and then change the *Style* property to move it to the right of the *Employee Id* text box.

   ```
   <asp:RangeValidator ID="RangeValidator1" ...
       Style="position: absolute; left: 400px; top: 172px"></asp:RangeValidator>
   ```

6. Click the *Design* button. From the *Toolbox*, add a *ValidationSummary* control to the form.

7. Click the *Source* button, locate the *ValidationSummary* control toward the end of the file, and add the following *Style* property to place it in the space above the button controls and to the right of the radio buttons.

```
<asp:ValidationSummary ID="ValidationSummary1" ...
    Style="position: absolute; left: 300px; top: 260px" />
```

A *ValidationSummary* control displays the *ErrorMessage* values for all of the validation controls on the Web form.

8. In the *Properties* window, verify that the *ShowSummary* property for the *ValidationSummary1* control is set to *True*.

9. Run the Web form. When the form appears in Internet Explorer, leave the *First Name*, *Last Name*, and *Employee Id* text boxes blank, and then click *Save*.

Red asterisks appear next to each of the text boxes, and the corresponding error messages are displayed in the *ValidationSummary* control at the bottom of the form.

10. Enter a first name and a last name, and then type **AAA** in the *Employee Id* text box.

As you move from text box to text box, the asterisks disappear from the *First Name* and *Last Name* text boxes, but an asterisk remains next to the *Employee Id* text box.

11. Click *Save*.

The error message displayed by the *ValidationSummary* control changes.

12. Type **101** in the *Employee Id* text box, and then click *Save*.

All error messages and asterisks disappear, and a summary of the data you entered appears in the *InfoLabel* control as before.

13. Close the form, and return to Visual Studio 2008.

Dynamic HTML and Error Messages

If you are viewing the page with a browser that supports dynamic HTML, you can display the validation summary data in a message box in addition to or rather than on the Web form. To do this, set the *ShowMessageBox* property of the *ValidationSummary* control to *True*. At run time, if any validation errors occur, the error messages will be displayed in a message box. If the Web browser does not support dynamic HTML, the value of the *ShowMessageBox* property is ignored (it defaults to *False*).

- If you want to continue to the next chapter

  Keep Visual Studio 2008 running, and turn to Chapter 29.

- If you want to exit Visual Studio 2008 now

  On the *File* menu, click *Exit*. If you see a *Save* dialog box, click *Yes* and save the project.

Chapter 28 Quick Reference

To	Do this
Perform server-side validation of user input	Use events belonging to server controls, for example, the *TextChanged* event of the *TextBox* control.
Perform client-side validation of user input	Use a validation control. Set the *ControlToValidate* property to the control to be validated, and set the *ErrorMessage* property to an error message to be displayed. Verify that the *EnableClientScript* property is set to *True*.
Force the user to enter a value in a text box	Use a *RequiredFieldValidator* control.
Check the type and range of data values entered into a text box	Use a *RangeValidator* control. Set the *Type*, *MaximumValue*, and *MinimumValue* properties as required.
Display a summary of validation error messages	Use a *ValidationSummary* control. Verify that the *ShowSummary* property is set to *True*. Set the *ShowMessageBox* property to *True* if you want browsers that support dynamic HTML to display the error messages in a message box.

Chapter 29
Protecting a Web Site and Accessing Data with Web Forms

After completing this chapter, you will be able to:

- Restrict access to a Web site by using Microsoft ASP.NET *Login* controls and Forms-based authentication.

- Create Web forms that present data from a database using a *GridView* control.

- Build Web applications that need to display potentially large volumes of data while minimizing resource use.

- Update a database from a Web form.

- Build applications that can pass data between Web forms.

In the previous two chapters, you have seen how to build a Web site that enables the user to enter information and validate the data that was entered. You've also seen in earlier chapters how to build a non-Web-based application that displays and updates data from a database. In this chapter, you'll learn about creating Web applications that display data from a database and that can update the database with any changes made by the user. You will see how to do this in an efficient manner that minimizes use of shared resources, such as the network and the database.

Security is always an important issue, especially when building applications that can be accessed over the Internet, when a Web application accesses sensitive resources such as your company's databases. Therefore, you will start by learning how to configure a Web forms application to use Forms-based security to verify the identity of the user.

Managing Security

Applications built by using the Microsoft .NET Framework have a range of mechanisms available for ensuring that the users who run those applications have the appropriate user rights. Some of the techniques rely on authenticating users based on some form of identifier and password, whereas others are based on the integrated security features of the Microsoft Windows operating system. If you are creating a Web application that will be accessed over the Internet, using Windows security is probably not an option—users are unlikely to be members of any Windows domain recognized by the Web application and might be running an operating system other than Windows, such as UNIX. Therefore, the best option to use in this environment is Forms-based security.

Understanding Forms-Based Security

With Forms-based security, you can verify the identity of a user by displaying a login form that prompts the user for an ID and a password. After the user has been authenticated, the various Web forms that make up the application can be accessed, and the user's security credentials can be examined by the code running in any page if additional authorization is needed. (A user might be able to log in to the system but might not have access to every part of the application.)

To use ASP.NET Forms-based security, you must configure the Web application by making some changes to the web.config file, and you must also supply a login form to validate the user. This login form will be displayed whenever the user tries to gain access to any page in the application if the user has not already been validated. The user will be able to proceed to the requested page only if the logic in the login form successfully verifies the user's identity.

Important To the uninitiated, it might seem that ASP.NET Forms-based security is excessive. It's not. Don't be tempted to simply create a login form that acts as an entry point to your application and assume that users will always access your application through it. Browsers can cache forms and URLs locally on users' computers. Another user might be able to gain access to the browser cache depending on how the computer itself is configured, find the URLs of the sensitive parts of your application, and navigate directly to them, bypassing your login form. You have control over your Web server (hopefully), but you have almost no control over the user's computer. The ASP.NET Forms-based mechanism is robust, and assuming that your Web server is well protected, it should be adequate for most of your applications.

Implementing Forms-Based Security

In the first set of exercises in this chapter, you will create and configure a Web application that implements Forms-based security. The application will ultimately enable a user to view and modify customer information in the Northwind database.

Create the Northwind Web site

1. Start Microsoft Visual Studio 2008 if it is not already running.

2. If you are using Visual Studio 2008 Professional Edition or Enterprise Edition, on the *File* menu, point to *New*, and then click *Web Site*.

3. If you are using Microsoft Visual Web Developer 2008 Express Edition, on the *File* menu, click *New Web Site*.

4. In the *New Web Site* dialog box, click the *ASP.NET Web Site* template. Select *File System* in the *Location* drop-down list box, and specify the \Microsoft Press\Visual CSharp Step By Step\Chapter 29\Northwind folder under your Documents folder. Set the *Language* to *Visual C#*, and then click *OK*.

5. In *Solution Explorer*, right-click *Default.aspx*, click *Rename*, and rename the form to *CustomerData.aspx*.

6. Right-click *CustomerData.aspx*, and click *Set As Start Page*.

7. In the *Code and Text Editor* window displaying the HTML source code for the Web form, click the *Design* button.

8. Using the *Toolbox*, add a *Label* control from the *Standard* category to the Web form. Set the *Text* property of the label to *This form will be implemented later*.

In the next exercises, you will build a login form to authenticate the user and configure Forms-based security for the Web application. When configured to use Forms-based security, the ASP.NET runtime will redirect to the login form attempts made by an unauthenticated user to access the application.

Implementing a login form for Forms-based security is such a common task that Microsoft has implemented a set of Login controls to simplify matters. You will use one of these controls now.

Build the login form

1. On the *Website* menu, click *Add New Item*.

2. In the *Add New Item* dialog box, ensure that the *Web Form* template is selected, and type **LoginForm.aspx** for the name. Verify that the *Language* drop-down list box is set to *Visual C#*, the *Place code in separate file* check box is selected, and the *Select master page* check box is cleared, and then click *Add* to create the form.

    The new Web form is created, and the HTML code for the form is displayed in the *Code and Text Editor* window.

3. Click the *Design* button to display LoginForm.aspx in the *Design View* window.

4. In the *Properties* window, set the *Title* property of the *DOCUMENT* object to *Northwind Traders – Log In*.

5. In the *Toolbox*, expand the *Login* category. Add a *Login* control to the Web form.

    The *Login* control is a composite control that is composed of several labels, two text boxes for the user to type a name and a password, the *Remember me next time* check box, and a button to click to log in. You can configure most of these items by using the *Properties* window for this control, and you can also modify the style of the control.

6. In the *Common Login Tasks* menu displayed by the *Login* control, click *Auto Format* on the *Login Tasks* menu that appears.

> **Tip** If the *Common Login Tasks* menu is not displayed, click the *Login* control, and then click the smart tag icon on the top edge of the control, near the right-hand corner.

The *Auto Format* dialog box appears. You can use this dialog box to change the look and feel of the *Login* control by selecting a predefined scheme. You can also define your own layout by creating a template using the *Convert to Template* command on the *Common Login Tasks* menu for the *Login* control.

7. In the *Auto Format* dialog box, click the *Classic* scheme, and then click *OK*. Click the smart tag icon on the *Login* control to hide the *Login Tasks* menu.

8. In the *Properties* window, change the properties of the *Login* control by using the values in the following table.

Property	Value
DisplayRememberMe	False
FailureText	Invalid User Name or Password. Please enter a valid User Name and Password.
TitleText	Northwind Traders – Log In
DestinationPageUrl	~/CustomerData.aspx

The *Login* control should look like this:

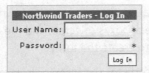

When the user clicks the *Log In* button, the user must be authenticated. If the user name and password are valid, the user should be allowed to proceed to the form specified by the *DestinationPageUrl* property; otherwise, the error message stored in the *FailureText* property of the *Login* control should be displayed and the user prompted to log in again. How do you perform these tasks? You have at least two options:

- Write code that handles the *Authenticate* event for the *Login* control. This event is raised whenever the user clicks the *Log In* button. You can examine the values in the *UserName* and *Password* properties, and if they are valid, allow the user to proceed to the page identified by the *DestinationPageUrl* property. This strategy is highly customizable but requires that you maintain your own secure list of user names and passwords to authenticate against.

- Use the built-in features of Visual Studio 2008 with the ASP.NET Web Site Administration Tool to manage user names and passwords, and let the *Login* control perform its default processing to validate users when the user clicks the *Log In* button. The ASP.NET Web Site Administration Tool maintains its own database of user names and passwords, and it provides a wizard to help you add users to your Web site.

You will use the second option in the following exercise. (You can investigate the first option on your own time.)

Configure Web site security, and activate Forms-based security

1. On the *Website* menu, click *ASP.NET Configuration*.

   The *ASP.NET Configuration* command opens Windows Internet Explorer and starts a Web application called the ASP.NET Web Site Administration Tool, which uses its own instance of the ASP.NET Development Server, independent from your Web application.

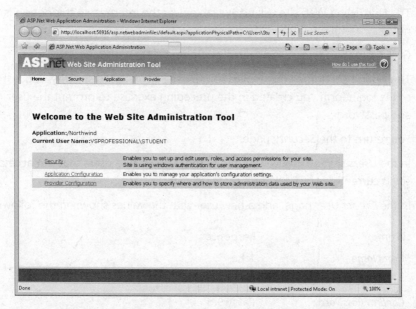

   By using this tool, you can add and manage users for your Web site, specify application settings that you want to be stored in the application configuration file, and specify how security information such as user names and passwords are stored. By default, the ASP.NET Web Site Administration Tool stores security information in a local Microsoft SQL Server database called ASPNETDB.MDF that it creates in the App_Data folder of your Web site. You can configure the ASP.NET Web Site Administration Tool to store security information elsewhere, but that is beyond the scope of this book.

2. In the ASP.NET Web Site Administration Tool, click the *Security* tab.

The *Security* page appears. You can use this page to manage users, specify the authentication mechanism that the Web site uses, define roles for users (roles are a convenient mechanism for assigning rights to groups of users), and specify access rules for controlling access to the Web site.

Note The first time you click the *Security* tab, the ASP.NET Web Site Administrator Tool creates the ASPNETDB.MDF database, so it might take a little time for Internet Explorer to display the next page.

3. In the *Users* section, click the *Select authentication type* link.

   A new page appears, asking how users will access your Web site. You have two options available: *From the internet* and *From a local network*. The *From a local network* option is selected by default. This option configures the Web site to use Windows authentication; all users must be members of a Windows domain that your Web site can access. The Northwind Web site will be available over the Internet, so this option is probably not very useful.

4. Click *From the internet*, and then click *Done*.

   This option configures the application to use Forms-based security. You will make use of the login form you created in the preceding exercise to prompt the user for a name and password.

   You return to the *Security* page.

5. In the *Users* section, notice that the number of existing users that can access your Web site is currently zero. Click the *Create User* link.

6. In the *Create User* page, add a new user with the values shown in the following table.

Prompt	Response
User Name	John
Password	Pa$$w9rd
Confirm Password	Pa$$w9rd
E-mail	john@northwindtraders.com
Security Question	What was the name of your first pet
Security Answer	Thomas

Note You must supply values for all fields in this screen. The E-mail, Security Question, and Security Answer fields are used by the *PasswordRecovery* control to recover or reset a user's password. The *PasswordRecovery* control is available in the *Login* category of the *Toolbar*, and you can add it to a login page to provide assistance to a user who has forgotten his or her password.

7. Ensure that the *Active User* box is selected, and then click *Create User*.

The message "Complete. Your account has been successfully created" appears on a new page.

8. Click *Continue*.

The *Create User* page reappears so that you can add more users.

9. Click *Back* to return to the *Security* page. Verify that the number of existing users is now set to 1.

> **Note** You can use the *Manage users* link on this page to change the e-mail addresses of users and add descriptions, and remove existing users. You can let users change their passwords and recover their passwords if they forget them by adding the *ChangePassword* and *PasswordRecovery* controls to the login page of the Web site. For more information, see the topic "Walkthough: Creating a Web Site with Membership and User Login" in the Microsoft Visual Studio 2008 documentation.

10. In the *Access Rules* section, click *Create access rules*.

The *Add New Access Rule* page appears. You use this page to specify which users can access which folders in the Web site.

11. Under *Select a directory for this rule*, ensure that the Northwind folder is selected by clicking it.

12. Under *Rule applies to*, ensure that *user* is selected, and type **John**.

13. Under *Permission*, click *Allow*, and then click *OK*.

This rule grants John access to the Web site. The *Security* screen reappears.

14. In the *Access Rules* section, click *Create access rules* again.

15. On the *Add New Access Rule* page, under *Select a directory for this rule*, ensure that the Northwind folder is selected. Under *Rule applies to*, click *Anonymous users*. Under *Permission*, ensure that *Deny* is selected, and then click *OK*.

This rule ensures that users who have not logged in will not be able to access the Web site. The *Security* screen reappears.

16. Close the Internet Explorer window displaying the ASP.NET Web Site Administration Tool, and return to Visual Studio 2008.

17. Click the *Refresh* button on the *Solution Explorer* toolbar.

The database file ASPNETDB.MDF appears in the App_Data folder.

18. Double-click the web.config file in the project folder to display it in the *Code and Text Editor* window.

This file was updated by the ASP.NET Web Site Administration Tool and should contain an *<authorization>* and an *<authentication>* element in the *<web.config>* section that look like this:

```
<system.web>
    ...
    <authorization>
        <allow users="John" />
        <deny users="?" />
    </authorization>
    ...
    <authentication mode="Forms" />
    ...
</system.web>
```

The *<authorization>* element specifies the users who are granted and denied access to the Web site ("?" indicates anonymous users). The *mode* attribute of the *<authentication>* element indicates that the Web site uses Forms-based authentication.

19. Modify the *<authentication>* element, replace the terminating delimiter (/>) with an ordinary closing delimiter (>), and add a *<forms>* child element, as shown here in bold type. Make sure you add a closing *</authentication>* element:

```
<authentication mode="Forms">
    <forms loginUrl="LoginForm.aspx" timeout="5"
           cookieless="AutoDetect" protection="All" />
</authentication>
```

The *<forms>* element configures the parameters for Forms-based authentication. The attributes shown here specify that if an unauthenticated user attempts to gain access to any page in the Web site, the user will be redirected to the login page, LoginForm.aspx. If the user is inactive for 5 minutes, she will have to log in again when next accessing a page in the Web site.

In many Web sites that use Forms-based authentication, information about the user is stored in a cookie on the user's computer. However, most browsers allow users to specify that they don't want to use cookies. (Cookies can be abused by malicious Web sites and are frequently considered a security risk.) By inserting *cookieless="AutoDetect"*, you can specify that the Web site can use cookies if the user's browser has not disabled them; otherwise, the user information is passed back and forth between the Web site and the user's computer as part of each request. The user information includes the user name and the password. Obviously, you don't want this to be clearly visible to everyone. You can use the *protection* attribute to encrypt this information, which is what this example does.

20. On the *Debug* menu, click *Start Without Debugging*.

Internet Explorer opens. The start page for the application is CustomerData.aspx, but because you have not yet logged in you are directed to LoginForm.aspx instead.

21. Type a random user name and password, and then click *Log In*.

The *Login* page reappears, displaying the error message "Invalid User Name or Password. Please enter a valid User Name and Password."

22. In the *User Name* box, type *John*; in the *Password* box, type *Pa$$w9rd*; and then click *Log In*.

The *CustomerData* page appears, displaying the message "This form will be implemented later."

23. Close Internet Explorer, and return to Visual Studio 2008.

Querying and Displaying Data

Now that you can control access to your application, you can turn your attention to querying and maintaining data. You will use Web Server data controls to connect to the database, query data, and update data.

Understanding the Web Forms *GridView* Control

When you looked at presenting data from a database in a WPF application in Chapter 26, "Displaying and Editing Data by Using Data Binding," you learned how to display data in a tabular manner by using a *ListView* control. ASP.NET provides a different set of controls from those available with WPF, and one control that is very useful for displaying and managing data in a Web form is the *GridView* control. This control is specifically designed to operate in a network bandwidth–constrained environment. In a Web forms application, it is very likely that the client application (or the browser) will be remote from the server holding the database. It is imperative that you use network bandwidth wisely (this has been stated several times already, but it is very important and worth repeating), and you should not waste resources retrieving vast amounts of data that the user does not actually want to see. The *GridView* control supports *paging*, which you can employ to fetch data on demand as the user scrolls up and down through the data.

 Note Do not confuse the ASP.NET Web Forms *GridView* control used for displaying data retrieved from a database with the WPF *GridView* control that you use for defining the layout of controls in a WPF window. They are different controls that just happen to have the same name.

The information in a Web forms *GridView* control is presented in a grid of read-only labels, rendered as an HTML table in the browser. The properties of the *GridView* control enable the user to enter edit mode, which changes a selected row into a set of text boxes that the user can use to modify the data that is presented.

To save database connection resources, the *GridView* control is designed to operate while it is disconnected from the database. You can create a data source to connect to a database, fetch data and display it in a *GridView* control, and then disconnect from the database. When the user wants to save any changes, the application can reconnect to the database and submit the changes. You will use this technique in the exercises in this chapter.

Displaying Customer and Order History Information

In the following exercises, you will build a Web application that displays in a *GridView* control on a Web form the details of the customers recorded in the Northwind database. You will provide functionality enabling the user to select a customer and display the order history for that customer. To do this, you will make use of data binding by using a LINQ data source.

> **Note** This exercise assumes that you have completed the exercises in Chapter 25, "Querying Information in a Database," and Chapter 26, "Displaying and Editing Data by Using Data Binding," on your computer.

Create a data source for retrieving customer information

1. On the *Website* menu, click *Add New Item*.

2. In the *Add New Item* dialog box, click the *LINQ to SQL Classes* template, type *Customer. dbml* in the *Name* text box, select *Visual C#* in the *Language* drop-down list, and then click *Add*.

3. In the *Microsoft Visual Studio* message box, click *Yes* to place the Linq to SQL file in the App_Code folder.

4. If you are using Visual Studio 2008 Professional Edition or Enterprise Edition, on the *View* menu, click *Server Explorer*.

5. If you are using Visual Web Developer 2008 Express Edition, perform the following tasks:

    5.1. On the *View* menu, click *Database Explorer*.

    5.2. In the *Database Explorer* window, right-click *Data Connections*, and then click *Add Connection*.

    5.3. In the *Add Connection* dialog box, click *Change*.

    5.4. In the *Choose Data Source* dialog box, click the *Microsoft SQL Server Database File* data source, make sure the *.NET Framework Data Provider for SQL Server* is selected as the data provider, and then click *OK*.

Note In contrast with Visual C# 2008 Express Edition, you do not have to connect directly to a database file when creating a data source with Visual Web Developer 2008 Express Edition. If you prefer, you can reattach the Northwind database to SQL Server and then connect by using the *Microsoft SQL Server* data source. For more information about attaching a database, see the *sp_attach_db* command in the MSDN Library for Visual Studio 2008.

5.5. In the *Add Connection* dialog box, in the *Database file name* box, click *Browse*.

5.6. In the *Select SQL Server Database File* dialog box, move to the folder C:\Program Files\Microsoft SQL Server\MSSQL.1\MSSQL\Data, click the Northwind database file, and then click *Open*.

5.7. Select the *Use Windows Authentication* option to log on to the server, and then click *OK*.

6. In *Server Explorer* or *Database Explorer*, expand the new data connection (*YourComputer*\sqlexpress.Northwind.dbo or Northwind.mdf), and then expand *Tables*.

7. Click the *Customers* table, and drag it onto the *Object Relational Designer* window.

Note If you are using Visual Web Developer 2008 Express Edition, a message box appears, asking you whether you want to add the data file for the Northwind database to your project. Click *No*.

8. On the *File* menu, click *Save All*.

Lay out the CustomerData Web form

1. Display the CustomerData.aspx Web form in the *Design View* window. Delete the label displaying the text "This form will be implemented later."

2. In the *Properties* window, set the *Title* property of the *DOCUMENT* object to *Northwind Traders – Customers*.

3. In the *Toolbox*, expand the *Data* category. Add a *LinqDataSource* control to the Web form.

A control called *LinqDataSource1* is added to the Web form.

Note Although the *LinqDataSource* control appears on the Web form at design time, it will not be visible when the Web form runs.

4. Using the *Properties* window, change the *(ID)* property of *LinqDataSource1* to *CustomerInfoSource*.

5. Select the *CustomerInfoSource* control on the Web form. Click the smart tag icon to display the *Common LinqDataSource Tasks* menu, and then click the *Configure Data Source* link.

   The Configure Data Source Wizard appears.

6. On the *Choose a Context Object* page, ensure that *CustomerDataContext* is selected in the *Choose your context object* drop-down list box, and then click *Next*.

7. On the *Configure Data Selection* page, in the *Table* drop-down list box, select the *Customers* table. In the *Select* list box, select the * box, and then click *Finish*.

8. On the *Common LinqDataSource Tasks* menu, select the *Enable Update* box, but leave the *Enable Insert* and *Enable Delete* boxes clear.

   The *Enable Update* check box enables the data source to generate the appropriate SQL UPDATE statements for modifying the data in the *Customers* table. For reasons of referential integrity, the Web form in this application will not allow the user to create or delete customers.

    Note If you don't select any of these options, the data retrieved through the data source is effectively read-only.

9. In the *Toolbox*, click the *GridView* control and drag it onto the form.

   A *GridView* is added to the form and displays placeholder data.

10. Using the *Properties* window, change the *(ID)* property of the *GridView* control to *CustomerGrid*, and set the *Caption* property to *Northwind Traders Customers*.

11. Click the smart tag icon on the top edge of the *GridView* control, near the right-hand corner. On the *Common GridView Tasks* menu, click the *Auto Format* link.

12. In the *AutoFormat* dialog box, select the *Classic* scheme, and then click *OK*.

     Tip If you don't like any of the predefined formats available in the *AutoFormat* dialog box, you can change the styles of the elements of a *GridView* control manually by using the properties in the *Styles* section in the *Properties* window.

13. In the *Properties* window, set the *DataSourceID* property of the *GridView* control to *CustomerInfoSource*.

The column headings for the *Customers* table appear in the *GridView* control on the screen.

14. Click the *Source* button at the bottom of the *Design View* window to display the HTML source code for the CustomerData.aspx page.

Notice that the HTML code for the *GridView* control sets the *DataSourceID* property of the control to *CustomerInfoSource*. The control also contains a *<columns>* element with *boundfield* controls defining the properties of each column displayed. The *DataField* property of each *boundfield* object specifies the name of the property the object is bound to in the data source. The *HeaderText* property is the string displayed in the column header for the column. Currently, the *HeaderText* and *DataField* values for each column are the same.

15. Change the values of the *HeaderText* property for the *boundfield* objects using the information in the following table.

DataField Value	*HeaderText* Value
CustomerID	Customer ID
CompanyName	Company
ContactName	Contact
ContactTitle	Title
Address	Address
City	City
Region	Region
PostalCode	Postal Code
Country	Country
Phone	Phone
Fax	Fax

Test the CustomerData form

1. On the *Debug* menu, click *Start Without Debugging*.

Internet Explorer starts and displays the *Northwind Traders - Login* page.

2. Log in as *John* using the password *Pa$$w9rd*.

The CustomerData Web form appears, displaying the details of every customer in the database:

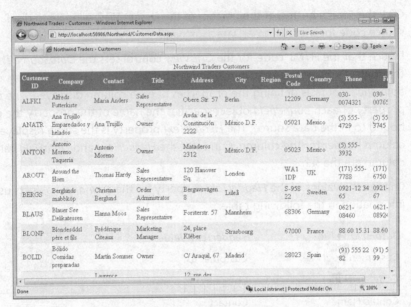

Notice that the page is currently read-only; you cannot modify any of the details displayed. You will enhance the Web form later in this chapter to enable the user to make changes.

3. Close Internet Explorer when you have finished browsing the data, and return to Visual Studio 2008.

Web Site Security and SQL Server

When you use the ASP.NET Development Server to run an application that uses Forms-based security, it executes in the context of the account you are using to run Visual Studio 2008. Assuming you used the same account to create the Northwind database, the Web application should have no problems accessing the database.

However, if you deploy the Web site to a Microsoft Internet Information Services (IIS) server, the situation changes. IIS runs applications that use Forms-based security by using the NETWORK SERVICE account under the Windows Vista operating system or the ASPNET account under Windows XP. This account has very few user rights by default, for security purposes. In particular, it will not be able to connect to SQL Server Express and query the Northwind database. Therefore, you will need to grant the NETWORK SERVICE account (or the ASPNET account) login access to SQL Server Express and add it as a user to the Northwind database. For more details, see the *sp_grantlogin* and *sp_grantdbaccess* commands in the MSDN Library for Visual Studio 2008.

Paging Data

Fetching the details of every customer is very useful, but suppose there are a large number of rows in the *Customers* table. It is highly unlikely that a user would actively want to browse thousands of rows, so generating a long page displaying them all would be a waste of time and network bandwidth. Instead, it would be far better to display data in chunks and enable the user to page through that data. This is what you will do in the following set of exercises.

Modify the *GridView* control to use paging

1. Click the *Design* button to display the CustomerData.aspx Web form in the *Design View* window, and then click the *CustomerGrid* control.

2. In the *Properties* window, set the *AllowPaging* property to *True*.

   A footer is added to the *CustomerGrid* control containing a pair of page numbers. This footer is referred to as the *pager*. The style shown for the footer is the default format, composed of page numbers that the user can click.

3. Expand the *PagerSettings* composite property. You can use the values in this property to customize the format of the page navigation links. You can specify page navigation links in two ways: as page numbers or as next/previous page arrows. Set the *Mode* property to *NumericFirstLast* to display page numbers with the first and last page arrows displayed to enable the user to move quickly to the start or end of the data. Set the *PageButtonCount* subproperty to *5*; this will cause page links to be displayed in groups of five. (You will see what this does when you run the Web application in a moment.)

   If you want to use next/previous page arrows, you can change the default text displayed (">" and "<") by modifying the values of the *NextPageText* and *PreviousPageText* properties. Similarly, you can change the text displayed for the first and last page links by editing the *FirstPageText* and *LastPageText* properties. Notice that the values in these properties require encoding as HTML characters; otherwise, they will not be displayed properly (for example, the ">" symbol must be specified as ">"). If you prefer, you can also specify the name of an image file in the *FirstPageImageUrl*, *LastPageImageUrl*, *PreviousPageImageUrl*, and *NextPageImageUrl* properties. The page navigation links will appear as buttons containing these images if supported by the browser.

4. In the *Properties* window, set the *PageSize* property to *8*.

   This setting causes the *CustomerGrid* to fetch and display data in eight-row chunks.

5. Expand the *PagerStyle* composite property. You can use this property to specify how the pager should be formatted. Set the *HorizontalAlign* subproperty to *Left*.

   The numbers in the pager move to the left margin in the *CustomerGrid* control.

6. Run the Web application, and log in as *John* using the password *Pa$$w9rd*.

   After you log in, the first eight rows of data and a set of page links are displayed on the CustomerData Web form. Page numbers 1, 2, 3, 4, and 5 are displayed, together with ">>" to move directly to the last page. Clicking the ellipsis (...) link displays the next five page numbers together with a "<<" link for moving directly back to the first page. An additional ellipsis (...) link provides access to the previous five pages.

7. Click the links at the bottom of the grid to move from page to page.

8. Close Internet Explorer, and return to Visual Studio 2008 when you have finished browsing the data.

> **Note** The *GridView* control provides the *AllowSorting* property. This property is set to *False* by default. If you set this property to *True*, the user can sort the data by the values in any column by clicking the column header. Whenever the user clicks a column header, the *LinqDataSource* control submits a SQL SELECT statement that fetches the first block of rows in ascending order. If the user clicks the same column header again, the data for the final block is retrieved and displayed in descending order. If the user repeatedly clicks column headers, the Web form will send a SQL SELECT statement to the database for each click.

Editing Data

You have seen how to use a *GridView* control to fetch and browse data. The following set of exercises shows you how to modify data and create new rows.

Updating Rows Through a *GridView* Control

By using the *GridView* control, you can add hyperlinks to the grid to indicate that a command should be performed. You can add your own custom hyperlinks and commands, but Visual Studio 2008 supplies some predefined hyperlinks for inserting, updating, and deleting data. In the following exercise, you will add update functionality to the *GridView* control by adding an *Edit* hyperlink to the grid. When the user clicks the *Edit* hyperlink, the row changes into a set of *TextBox* controls. The user can save the changes or discard them. This is achieved by using two additional automatically created hyperlinks labeled *Update* and *Cancel*.

Create the *Edit*, *Update*, and *Cancel* buttons

1. Display the CustomerData.aspx form in the *Design View* window. Click the smart tag for the *CustomerGrid* control to display the *Common GridView Tasks* menu, and then select *Enable Editing*.

   An *Edit* hyperlink is added to each row in the *GridView* control.

2. Click the *Source* button to display the HTML source code for the Web form. Locate the *<Columns>* collection for the *CustomerGrid* control, and notice that Visual Studio has added a *commandfield* object. The *ShowEditButton* property is set to *True*, like this:

```
<asp:GridView ID="CustomerGrid" runat="server" ...>
    ...
    <Columns>
        <asp:commandfield ShowEditButton="True" ></asp:commandfield>
        ...
    </Columns>
    ...
</asp:GridView>
```

The *ShowEditButton* property determines whether the *commandfield* object displays the *Edit* hyperlink. You can also activate delete and insert functionality by setting the *ShowDeleteButton* and *ShowInsertButton* properties to *True*, which cause further hyperlinks to be displayed.

3. Set the *EditText* property and the *ButtonType* property for the *commandfield* object as shown here in bold type:

```
<asp:commandfield ShowEditButton="True" EditText="Modify" ButtonType="Button">
</asp:commandfield>
```

These properties change the appearance of the *Edit* hyperlink. The *EditText* property specifies the text displayed by the hyperlink, and the *ButtonType* property changes the hyperlink to be displayed as a button instead of as a hyperlink. If you activate the insert and delete hyperlinks for the *commandfield* object, you can change the *InsertText* and *DeleteText* properties to customize the text displayed by these links. However, all links share the same *ButtonType* value—either they all appear as hyperlinks or they all appear as buttons.

4. Run the application. Log in, and then click the *Modify* button on the first row displayed on the CustomerData form.

   The first row changes into a collection of *TextBox* controls, and the *Modify* button is replaced with an *Update* button and a *Cancel* button.

> **Note** The *CustomerID* column remains as a label. This is because this column is the primary key in the *Customers* table. You should not be able to modify primary key values in a database; otherwise, you risk breaking the referential integrity between tables.

5. Change the data in the *Contact* and *Title* columns, and then click *Update*.

   The database is updated, the row reverts to a set of labels, the *Modify* button reappears, and the new data is displayed in the row. Behind the scenes, the *GridView* control changes the data in the LINQ data source and then calls the *SubmitChanges* method of the *CustomerDataContext* object to send the changes to the database.

6. Close Internet Explorer, and return to Visual Studio 2008.

The form currently performs no validation. If you blank out the data in the *Company* column for a customer and then click *Update*, the LINQ data source generates a SQL exception because this column does not allow null values in the database. The message that is displayed is not very user-friendly (although a developer will find it very useful). If a Web form generates an exception, you can arrange for a more friendly message to be displayed by redirecting the user to another page. Set the *ErrorPage* attribute to the *@Page* directive in the form's source definition to redirect the user when errors occur:

```
<%@ Page ... ErrorPage="ErrorPage.aspx" %>
```

You can display a more comforting message to the user on this page. Additionally, you can validate the data before sending it to the database by handling the *Updating* event in the LINQ data source object. The event handler for this method takes a *LinqDataSourceUpdateEventArgs* parameter that contains the original values and the new values for the row. Your code can scrutinize the new values, and if they are invalid, your code can set the *Cancel* property of the *LinqDataSourceUpdateEventArgs* parameter to *false* to indicate that the data source should not attempt to update the database.

Also, notice that the database is updated as soon as the user clicks the *Update* button. This is the default functionality implemented by a *GridView* control that is bound to a LINQ data source and is probably the most suitable mechanism for building interactive Web forms. If you want to modify the update behavior (for example, so that the *GridView* control will store multiple updates locally and then submit them as a single batch), you can implement your own custom mechanism. However, the details for doing this are outside the scope of this book.

Navigating Between Forms

A key aspect of many Web Forms applications is the ability to navigate from one form to another by clicking a hyperlink or button. In addition, you often need to pass information between forms. In the *CustomerData* Web form, it would be useful to be able to click a customer and display another form showing the order history for that customer. This is what you will do in the exercises in this section.

In this section, you will create a new Web form for displaying order history information. You will use a *GridView* control to display the data, but you will populate the data by executing a SQL Server stored procedure rather than querying a table. The Northwind database contains a stored procedure called *CustOrderHist*. This stored procedure takes a customer ID as a parameter and returns a result set containing the name and quantity of each product the customer has ordered. When the user selects a customer in the CustomerData Web form, you must pass the value in the *CustomerID* column to this new form.

The first step, therefore, is to modify the *CustomerData* Web form to enable the user to select a customer.

Modify the *CustomerData* Web form

1. Return to the *Code and Text Editor* window displaying the HTML source code for the CustomerData Web form.

2. Change the definition of the *boundfield* element displaying the customer ID to a *HyperLinkField*, as shown here in bold type:

   ```
   <asp:HyperLinkField DataField="CustomerID" ...></asp:HyperLinkField>
   ```

   The data in this column in the *GridView* control will be displayed as a hyperlink rather than as a label. The user can click this hyperlink. The following steps set properties that specify the actions that occur when this happens.

3. Change the *DataField* property of the control to a *DataTextField* property, as shown in bold type here:

   ```
   <asp:HyperLinkField DataTextField="CustomerID" ...></asp:HyperLinkField>
   ```

   The *HyperLinkField* control does not have a *DataField* property. The *DataTextField* specifies the property from the data source to which the hyperlink binds.

4. Remove the *ReadOnly* property of the control, and add the *Target*, *DataNavigateUrlFields*, and *DataNavigateUrlFormatString* properties shown in bold type here:

   ```
   <asp:HyperLinkField DataTextField="CustomerID" HeaderText="Customer ID"
       Target="_self" DataNavigateUrlFields="CustomerID"
       DataNavigateUrlFormatString="~\OrderHistory.aspx?CustomerID={0}"
       SortExpression="CustomerID">
   </asp:HyperLinkField>
   ```

   The *DataNavigateUrlFormatString* property specifies the address to which the Web application should move when the user clicks the hyperlink. In this example, the application navigates to the OrderHistory.aspx form (which you will create in the next exercise) and includes a query string parameter containing a customer ID. This query string currently contains a placeholder. The *DataNavigateUrlFields* property determines the value that should be used for this placeholder—the data in the *CustomerID* field for the current row in the *GridView* control. The *Target* property specifies where the OrderHistory.aspx Web form should be displayed. The value *_self* causes ASP.NET to reuse the same Internet Explorer window that is currently displaying the *CustomerData* form.

 Note ASP.NET also provides the *HyperLink* control in the *Standard* category in the *Toolbox*. When using this control, you can specify a URL to move to in its *NavigateUrl* property. In addition, you can execute the *Transfer* method of the *Server* property of a Web form if you want to transfer control from one Web form to another programmatically.

The next task is to create a data source that executes the *CustOrderHist* stored procedure in the database. In the final exercise in this chapter, you will see how to invoke the stored procedure by using the data source and pass the *CustomerID* parameter required by this stored procedure.

Create a data source for retrieving customer order history information

1. On the *Website* menu, click *Add New Item*.

2. In the *Add New Item* dialog box, click the *LINQ to SQL Classes* template, type *OrderHistory.dbml* in the *Name* text box, select *Visual C#* in the *Language* drop-down list, and then click *Add*.

3. In the *Microsoft Visual Studio* message box, click *Yes* to place the Linq to SQL file in the App_Code folder.

4. In *Server Explorer* (Visual Studio 2008) or *Database Explorer* (Visual Web Developer 2008 Express Edition), expand the data connection for the Northwind database (*YourComputer*\sqlexpress.Northwind.dbo or Northwind.mdf), and then expand *Stored Procedures*.

5. Click the *CustOrderHist* stored procedure, and drag it onto the *Object Relational Designer* window.

   The stored procedure is added at the top of the right-hand pane of the *Object Relational Designer* window.

6. On the *File* menu, click *Save All*.

You can now construct the OrderHistory Web form that displays the order history for a customer using this data source.

Create the OrderHistory Web form

1. Display the CustomerData.aspx form in the *Design View* window. On the *Website* menu, click *Add New Item*.

2. In the *Add New Item* dialog box , ensure that the *Web Form* template is selected, and type **OrderHistory.aspx** for the name. Verify that the *Language* drop-down list box is set to *Visual C#*, the *Place code in separate file* box is selected, and the *Select master page* box is cleared, and then click *Add* to create the form.

> **Note** If the *Add New Item* dialog box does not display the *Web Form* template, make sure that you have displayed the CustomerData.aspx form rather than the *Object Relational Designer* in the *Design View* window.

3. Click the *Design* button to display OrderHistory.aspx in the *Design View* window.

4. In the *Properties* window, set the *Title* property of the *DOCUMENT* object to *Northwind Traders – Orders for:*.

5. From the *Standard* category in the *Toolbox*, add a *Label* control and a *HyperLink* control to the Web form.

6. Using the *Properties* window, set the properties for the *Label* and *HyperLink* controls to the values shown in the following table.

Control	Property	Value
*Label*1	ID	OrderLabel
	Font-Name (expand the *Font* property and select *Name*)	Arial Black
	Font-Size (in the *Font* property, select *Size*)	X-Large
	Text	Order History for:
*HyperLink*1	*Text*	Return to Customers
	NavigateUrl	~/CustomerData.aspx

7. In the *Data* category of the *Toolbox*, click the *GridView* control and drag it onto the form. A *GridView* is added to the form and displays placeholder data.

8. Using the *Properties* window, set the *(ID)* property for the *GridView* control to *OrderGrid*.

9. In the *Design View* window, click the *OrderGrid* control, and then click the smart tag to display the *Common GridView Tasks* menu.

10. On the *Common GridView Tasks* menu, click the *Auto Format* link.

11. In the *AutoFormat* dialog box, select the *Classic* scheme, and then click *OK*.

   In this form, you are going to write code to bind the *GridView* control to the data source. You will define the columns in the *GridView* control manually.

12. On the *Common GridView Tasks* menu, click the *Edit Columns* link.

13. In the *Fields* dialog box, in the *Available Fields* list box, click *BoundField*, and then click *Add*.

14. In the *BoundField properties* list box, set the *HeaderText* property to *Product Name*.

15. In the *Available Fields* list box, click *BoundField*, and then click *Add* again.

16. In the *BoundField properties* list box, set the *HeaderText* property of the new column to *Total*, and set the *DataFormatString* property to *{0:N0}*. (Both the 0 characters are zeros—this format displays the data as a number with no decimal places.) Expand the

ItemStyle property, and then set the *HorizontalAlign* property to *Right* (by convention, numeric data is displayed right-justified).

17. Clear the *Auto-generate fields* check box, and then click *OK*.

18. Click the *Source* button, and modify the *body* element to lay out the controls on the form underneath one another, with some blank lines between them, as shown in bold type here:

```
<body>
    <form id="form1" runat="server">
    <div>
      <asp:Label ID="OrderLabel" ...></asp:Label>
      <br />
      <br />
      <asp:HyperLink ID="HyperLink1" ...></asp:HyperLink>
      <br />
      <br />
      <asp:GridView ID="OrderGrid" ... >
          ...
      </asp:GridView>
    </div>
    </form>
</body>
```

The final task is to write some code that displays the customer ID on the form, bind the *GridView* control and its columns to the *OrderHistory* data source, and then display the data.

Write code to bind the *GridView* control to the data source

1. In *Solution Explorer*, expand OrderHistory.aspx, and then double-click OrderHistory. aspx.cs to display the C# code for the *OrderHistory* form in the *Code and Text Editor* window.

2. In the *Page_Load* method, add the statement shown here in bold type:

```
protected void Page_Load(object sender, EventArgs e)
{
    string customerId = Request.QueryString["CustomerID"];
}
```

Remember that the *OrderHistory* form is invoked from the *CustomerData* form when the user clicks the *hyperlink* control for one of the customers displayed in the *GridView* control on that form. The hyperlink control specifies a URL with a query string that contains the selected customer ID. For example, if the user clicks the customer with the ID "ALFKI," the hyperlink opens the *OrderHistory* form with the query string value pair "CustomerID=ALFKI". The *Request* object of a Web form is a collection of the query string value pairs passed in to the form. You can access the values either by number or by name. The code you have just written retrieves the value of the pair with the name *CustomerID* from the *Request* object and stores it in a local string variable.

3. Add the code shown here in bold type to the *Page_Load* method:

```
protected void Page_Load(object sender, EventArgs e)
{
    string customerId = Request.QueryString["CustomerID"];
    this.OrderLabel.Text += " " + customerId;
    this.Title += " " + customerId;
}
```

These statements append the customer ID to the text displayed on the form and in the title of the form.

4. Add the following statements to the *Page_Load* method:

```
protected void Page_Load(object sender, EventArgs e)
{
    ...
    OrderHistoryDataContext context = new OrderHistoryDataContext();
    var orderDetails = context.CustOrderHist(customerId);
    this.OrderGrid.DataSource = orderDetails;
}
```

This code creates a new *DataContext* object using the *OrderHistoryDataContext* class. The *OrderHistoryDataContext* class was generated by the *Object Relational Designer* when you created a new data source based on the *CustOrderHist* stored procedure. When you add a stored procedure to a *DataContext* type, the code generated by the *Object Relational Designer* exposes the stored procedure by providing a method with the same name. The second statement in the preceding code example calls the *CustOrderHist* method, which in turn invokes the *CustOrderHist* stored procedure in the Northwind database. The *customerId* variable is passed in as the parameter. The result set generated by this stored procedure is used as the data source for the *OrderGrid* control.

5. Append the code shown here to the end of the *Page_Load* method:

```
protected void Page_Load(object sender, EventArgs e)
{
    ...
    BoundField productName = this.OrderGrid.Columns[0] as BoundField;
    productName.DataField = "ProductName";
    BoundField total = this.OrderGrid.Columns[1] as BoundField;
    total.DataField = "Total";
    this.OrderGrid.DataBind();
}
```

This block of code binds the two columns in the *OrderGrid* control to the corresponding properties in the data source. Notice that you specify the properties by name, as a string. The *DataBind* method of the *OrderGrid* control causes the data source to run the stored procedure and generate the result set, displaying the results in the columns in the grid.

Test the completed application

1. Run the application, and log in as *John*.

2. On the CustomersData Web form, notice that the values in the *Customer ID* column are now displayed as hyperlinks:

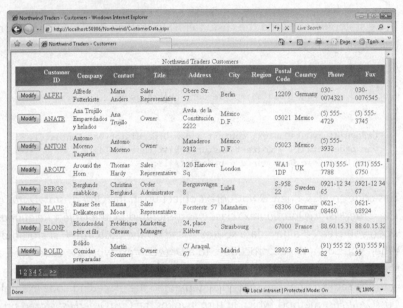

3. Click the hyperlink for the first customer, *ALFKI*. The OrderHistory form should appear, displaying the order history for ALFKI.

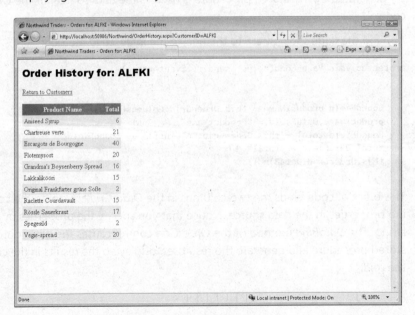

4. Click the *Return to Customers* hyperlink.

5. On the *CustomersData* Web form, click the hyperlink in the *Customer ID* column for another customer. The OrderHistory Web form should display the order history for the appropriate customer.

6. When you have finished browsing the data, close Internet Explorer and return to Visual Studio 2008.

This chapter has shown you the fundamentals of building and protecting a Web application that maintains data in a database. If you are interested in creating highly interactive Web applications that incorporate multimedia capabilities, you should take a look at Microsoft Silverlight. You can find more information about Silverlight at *http://silverlight.net*.

■ If you want to continue to the next chapter

Keep Visual Studio 2008 running, and turn to Chapter 30.

■ If you want to exit Visual Studio 2008 now

On the *File* menu, click *Exit*. If you see a *Save* dialog box, click *Yes* and save the project.

Chapter 29 Quick Reference

To	Do this
Create a login Web form	Create a new Web form. Add a *Login* control for authenticating users.
Configure security for an ASP. NET Web site	Use the ASP.NET Web Site Administration Tool to add and maintain users, define roles, and create access rules. (On the *Website* menu, click *ASP.NET Configuration* to start this tool.)

Implement Forms-based security	Edit the web.config file. Set the *<authentication mode>* attribute to *Forms*, provide the URL of the login form, and specify any authentication parameters required. For example:
	
	``
	<authentication mode="Forms">
	    <forms loginUrl="LoginForm.aspx"
	           timeout="5"
	           cookieless="AutoDetect"
	           protection="All" />
	</authentication>
	``

Create a Web form for displaying data from a database	Add a data source control to the Web form, and configure it to connect to the appropriate database. Add a *GridView* control to the Web form, and set its *DataSourceID* property to the data source control.

Fetch and display data in manageable chunks in a Web form	Set the *AllowPaging* property of the *GridView* control to *True*. Set the *PagerSize* property to the number of rows to be displayed on each page. Modify the *PagerSettings* and *PagerStyle* properties to match the style of the Web form.
Modify rows in a database using a *GridView* control	Ensure that the data source enables updating data. Using the *Common GridView Tasks* smart tag menu, select *Enable Updating*.
Navigate from one Web form to another by selecting a row in a *GridView* control	Define a column as a *HyperLinkField* control. Specify the URL and optional query string for the destination form in the *DataNavigateUrlFormatString* property, and specify any data to pass to the form as query string parameters in the *DataNavigateUrlFields* property. In the destination form, retrieve any query string parameters by accessing the *QueryString* collection of the *Request* property of the Web form.
Bind a *GridView* control to a data source at run time	Set the *DataSource* property of the *GridView* control to the data source. Set the *DataField* property of any *boundfield* columns in the *GridView* to the name of the property holding the data to be displayed in the data source.

Chapter 30
Creating and Using a Web Service

After completing this chapter, you will be able to:

- Create a Web service that exposes simple Web methods.

- Display the description of a Web service by using Windows Internet Explorer.

- Design classes that can be passed as parameters to a Web method and returned from a Web method.

- Create a reference to a Web service in a client application.

- Invoke a Web method.

The previous chapters showed you how to create Web forms and build Web applications by using Microsoft ASP.NET. Although this approach is appropriate for applications where the client is a Web browser, you will increasingly encounter situations where the client is some other type of application. As mentioned in previous chapters, the Internet is just a big network. By using Web services, it is possible to build distributed systems from elements that are spread across the Internet—databases, business services, and so on. The aim of this chapter is to show you how to design, build, and test Web services that can be accessed over the Internet and integrated into distributed applications. You'll also learn how to construct a client application that uses the methods exposed by a Web service.

 Note The purpose of this chapter is to provide a very basic introduction to Web services and Microsoft Windows Communication Foundation (WCF). If you want detailed information about how WCF works and how to build secure services by using WCF, you should consult a book such as *Microsoft Windows Communication Foundation Step by Step*, published by Microsoft Press, 2007.

What Is a Web Service?

A Web service is a business component that provides some useful function to clients, or consumers. A Web service can be thought of as a component with truly global accessibility— if you have the appropriate access rights, you can make use of a Web service from anywhere in the world as long as your computer is connected to the Internet. Web services use a standard, accepted, and well-understood protocol, Hypertext Transfer Protocol (HTTP), to transmit data and a portable data format that is based on XML. HTTP and XML are both standardized technologies that can be used by other programming environments outside the Microsoft .NET Framework. You can build Web services by using Microsoft Visual Studio 2008. Client applications running in a totally different environment, such as Java, can use them. The reverse is also true: you can build Web services by using Java and write client applications in C#.

With Visual Studio 2008, you can build Web services by using Microsoft Visual C++, Microsoft Visual C#, or Microsoft Visual Basic. However, as far as a client application is concerned, the language used to create the Web service, and even how the Web service performs its tasks, is not important. The client application's view of a Web service is of an interface that exposes a number of well-defined methods, known as Web methods. All the client application needs to do is call these Web methods by using the standard Internet protocols, passing parameters in an XML format and receiving responses also in an XML format.

One of the driving forces behind the recent releases of the Windows operating system, the .NET Framework, and its associated development tools is the concept of the "programmable Web." The idea is that you can construct systems by using the data and services supplied by multiple Web services. Web services provide the basic elements for systems, the Web provides the means to access them, and developers glue them together in meaningful ways to add functionality to their applications. Web services are a key integration technology for combining disparate systems, and they are the basis for many business-to-business (B2B) and business-to-consumer (B2C) applications.

The Role of SOAP

Simple Object Access Protocol (SOAP) is the protocol used by client applications for sending requests to and receiving responses from Web services. SOAP is a lightweight protocol built on top of HTTP—the protocol used by the Web to send and receive HTML pages. SOAP defines an XML grammar for specifying the names of Web methods that a consumer can invoke on a Web service, for defining the parameters and return values, and for describing the types of parameters and return values. When a client calls a Web service, it must specify the method and parameters by using this XML grammar.

SOAP is an industry standard. Its function is to improve cross-platform interoperability. The strength of SOAP is its simplicity and also the fact that it is based on other industry-standard technologies, such as HTTP and XML. The SOAP specification defines a number of things. The most important are the following:

- The format of a SOAP message
- How data should be encoded
- How to send messages (method calls)
- How to process replies

Descriptions of the exact details of how SOAP works and the internal format of a SOAP message are beyond the scope of this book. It is highly unlikely that you will ever need to create and format SOAP messages manually because many development tools, including Visual Studio 2008, automate this process, presenting a programmer-friendly API to developers building Web services and client applications.

What Is the Web Services Description Language?

The body of a SOAP message is an XML document. When a client application invokes a Web method, the Web server expects the client to use a particular set of tags for encoding the parameters for the method. How does a client know which tags, or XML schema, to use? The answer is that, when asked, a Web service is expected to supply a description of itself. The Web service response is another XML document that describes the Web service. Unsurprisingly, this document is known as the Web Service Description. The XML schema used for this document has been standardized and is called Web Services Description Language (WSDL). This description provides enough information so that a client application can construct a SOAP request in a format that the Web server should understand. Again, the details of WSDL are beyond the scope of this book, but Visual Studio 2008 contains tools that parse the WSDL for a Web service in a mechanical manner. Visual Studio 2008 then uses the information to define a proxy class that a client application can use to convert ordinary method calls on this proxy class to SOAP requests that the proxy sends over the Web. This is the approach you will use in the exercises in this chapter.

Nonfunctional Requirements of Web Services

The initial efforts to define Web services and their associated standards concentrated on the functional aspects for sending and receiving SOAP messages. Not long after Web services became a mainstream technology for integrating distributed services, it became apparent that there were issues that SOAP and HTTP alone could not address. These issues concern many nonfunctional requirements that are important in any distributed environment, but much more so when using the Internet as the basis for a distributed solution. They include the following items:

- **Security**   How do you ensure that SOAP messages that flow between a Web service and a consumer have not been intercepted and changed on their way across the Internet? How can you be sure that a SOAP message has actually been sent by the consumer or Web service that claims to have sent it, and not some "spoof" site that is trying to obtain information fraudulently? How can you restrict access to a Web service to specific users? These are matters of message integrity, confidentiality, and authentication and are fundamental concerns if you are building distributed applications that make use of the Internet.

In the early 1990s, a number of vendors supplying tools for building distributed systems formed an organization that later became known as the Organization for the Advancement of Structured Information Standards, or OASIS. As the shortcomings of the early Web services infrastructure became apparent, members of OASIS pondered these problems (and other Web services issues) and produced what became known as the WS-Security specification. The WS-Security specification describes how to protect the messages sent by Web services. Vendors that subscribe to WS-Security provide their own implementations that meet this specification, typically by using technologies such as encryption and certificates.

- **Policy**   Although the WS-Security specification defines how to provide enhanced security, developers still need to write code to implement it. Web services created by different developers can often vary in how stringent the security mechanism they have elected to implement is. For example, a Web service might use only a relatively weak form of encryption that can easily be broken. A consumer sending highly confidential information to this Web service would probably insist on a higher level of security. This is one example of policy. Other examples include the quality of service and reliability of the Web service. A Web service could implement varying degrees of security, quality of service, and reliability and charge the client application accordingly. The client application and the Web service can negotiate which level of service to use based on the requirements and cost. However, this negotiation requires that the client and the Web service have a common understanding of the policies available. The WS-Policy specification provides a general-purpose model and corresponding syntax to describe and communicate the policies that a Web service implements.

- **Routing and addressing**   It is useful for a Web server to be able to reroute a Web service request to one of a number of computers hosting instances of the service. For example, many scalable systems make use of load balancing, in which requests sent to a Web server are actually redirected by that server to other computers to spread the load across those computers. The server can use any number of algorithms to try to balance the load. The important point is that this redirection is transparent to the client making the Web service request, and the server that ultimately handles the request must know where to send any responses that it generates. Redirecting Web service requests is also useful if an administrator needs to shut down a computer to perform maintenance. Requests that would otherwise have been sent to this computer can be rerouted to one of its peers. The WS-Addressing specification describes a framework for routing Web service requests.

Note   Developers refer to the WS-Security, WS-Policy, WS-Addressing, and other WS-specifications collectively as the WS-* specifications.

The Role of Windows Communication Foundation

As standardization of Web services security, policy, and addressing became more important, Microsoft provided its own implementation of the WS-Security, WS-Policy, and WS-Addressing specifications in its Web Services Enhancements (WSE) package, available as a free download from the Microsoft Web site. What does all this mean if you are developing Web services using Visual Studio 2008? Well, with Visual Studio 2008, you can build Web services by using two technologies—ASP.NET and WCF. Ordinary ASP.NET Web services do not directly support the various WS-* specifications. Instead, you can use Microsoft's WSE package to provide features such as security.

So, by using Visual Studio, the .NET Framework, and WSE, you can quickly build Web services and client applications that can communicate and interoperate with Web services and client applications running on any operating system. Why then do you need WCF? Well, first, WCF is a more recent technology that emerged as part of version 3.0 of the .NET Framework. It provides its own fully integrated implementation of the common WS-* specifications without requiring you to download, install, and configure additional packages. Second, Web services are just one technology that you can use to create distributed applications for the Windows operating systems. Others include Enterprise Services, .NET Framework Remoting, and Microsoft Message Queue (MSMQ). If you are building a distributed application for Windows, which technology should you use, and how difficult will it be to switch later if you need to? The purpose of WCF is to provide a unified programming model for many of these technologies so that you can build applications that are as independent as possible from the underlying mechanism being used to connect services and applications. (Note that WCF applies as much to services operating in non-Web environments as it does to the World Wide Web.) It is actually very difficult, if not impossible, to completely divorce the programmatic structure of an application or service from its communications infrastructure, but WCF lets you come very close to achieving this aim much of the time. Additionally, by using WCF, you can maintain backward compatibility with many of the earlier technologies. For example, a WCF client application can easily communicate with a Web service that you created by using WSE.

To summarize, if you are considering building distributed applications and services for Windows, you should use WCF. The exercises in this chapter will show you how.

Building a Web Service

In this section, you will create the ProductsService Web service. This Web service exposes two Web methods. The first method enables the user to calculate the cost of buying a specified quantity of a particular product in the Northwind database, and the second method takes the name of a product and returns all the details for that product.

Creating the ProductsService Web Service

In the first exercise, you will create the ProductsService Web service and examine the example code generated by Visual Studio 2008 whenever you create a new WCF service project. In subsequent exercises, you will define and implement the *HowMuchWillItCost* Web method and then test the Web method to ensure that it works as expected.

Create the Web service, and examine the example code

1. Start Visual Studio 2008 if it is not already running.

2. If you are using Visual Studio 2008 Professional Edition or Enterprise Edition, on the *File* menu, point to *New*, and then click *Web Site*.

3. If you are using Microsoft Visual Web Developer 2008 Express Edition, on the *File* menu, click *New Web Site*.

4. In the *New Web Site* dialog box, click the *WCF Service* template. Select *File System* in the *Location* drop-down list box, and specify the \Microsoft Press\Visual CSharp Step By Step\Chapter 30\NorthwindServices folder under your Documents folder. Set the *Language* to *Visual C#*, and then click *OK*.

   Visual Studio 2008 generates a Web site containing folders called App_Code and App_Data, a file called Service.svc, and a configuration file called Web.config. The code for an example Web service is defined in the *Service* class, stored in the file Service.cs in the App_Code folder, and displayed in the *Code and Text Editor* window. The *Service* class implements an example interface called *IService*, stored in the file IService.cs in the App_Code folder.

5. Click the C:\...\NorthwindServices\ project. In the *Properties* window, set the *Use dynamic ports* property to *False*, and set the *Port number* property to *4500*.

   By default, the Development Web server provided with Visual Studio 2008 picks a port at random to reduce the chances of clashing with any other ports used by other network services running on your computer. This feature is useful if you are building and testing ASP.NET Web sites in a development environment prior to copying them to a production server such as Microsoft Internet Information Services (IIS). However, when building a Web service, it is more useful to use a fixed port number because client applications need to be able to connect to it.

6. Expand the App_Code folder, right-click the Service.cs file, and then click *Rename*. Change the name of the file to ProductsService.cs.

7. Using the same technique, change the name of the IService.cs file to IProductsService.cs.

8. Double-click the IProductsService.cs file to display it in the *Code and Text Editor* window.

This file contains the definition of an interface called *IService*. At the top of the IProductsService.cs file, you will find *using* statements referencing the *System*, *System. Collections.Generic*, and *System.Text* namespaces (which you have met before), followed by two additional statements referencing the *System.ServiceModel* and *System.Runtime. Serialization* namespaces.

The *System.ServiceModel* namespace contains the classes used by WCF for defining services and their operations. WCF uses the classes in the *System.Runtime.Serialization* namespace to convert objects to a stream of data for transmission over the network (a process known as *serialization*) and to convert a stream of data received from the network back to objects (*deserialization*). You will learn a little about how WCF serializes and deserializes objects later in this chapter.

The primary contents of the IProductsService file are an interface called *IService* and a class called *CompositeType*. The *IService* interface is prefixed with the *ServiceContract* attribute, and the *CompositeType* class is tagged with the *DataContract* attribute. Because of the structure of a WCF service, you can adopt a "contract-first" approach to development. When performing contract-first development, you define the interfaces, or *contracts*, that the service will implement, and then you build a service that conforms to these contracts. This is not a new technique, and you have seen examples of this strategy throughout this book. The point behind using contract-first development is that you can concentrate on the design of your service. If necessary, it can quickly be reviewed to ensure that your design does not introduce any dependencies on specific hardware or software before you perform too much development; remember that in many cases client applications might not be built using WCF and might not even be running on Windows.

The *ServiceContract* attribute marks an interface as defining methods that the class implementing the Web service will expose as Web methods. The methods themselves are tagged with the *OperationContract* attribute. The tools provided with Visual Studio 2008 use these attributes to help generate the appropriate WSDL document for the service. Any methods in the interface not marked with the *OperationContract* attribute will not be included in the WSDL document and therefore will not be accessible to client applications using the Web service.

If a Web method takes parameters or returns a value, the data for these parameters and value must be converted to a format that can be transmitted over the network and then converted back again to objects—this is the process known as serialization and deserialization mentioned earlier. The various Web services standards define mechanisms for specifying the serialized format of simple data types, such as numbers and strings, as part of the WSDL description for a Web service. However, you can also define your own complex data types based on classes and structures. If you make use of these types in a Web service, you must provide information on how to serialize and deserialize them. If you look at the definition of the *GetDataUsingDataContract* method in the

IService interface, you can see that it expects a parameter of the type *CompositeType*. The *CompositeType* class is marked with the *DataContract* attribute, which specifies that the class must define a type that can be serialized and deserialized as an XML stream as part of a SOAP request or response message. Each member that you want to include in the serialized stream sent over the network must be tagged with the *DataMember* attribute.

9. Double-click the ProductsService.cs file to display it in the *Code and Text Editor* window.

   This file contains a class called *Service* that implements the *IService* interface and provides the *GetData* and *GetDataUsingDataContract* methods defined by this interface. This class is the Web service. When a client application invokes a Web method in this Web service, it generates a SOAP request message and sends it to the Web server hosting the Web service. The Web server creates an instance of this class and runs the corresponding method. When the method completes, the Web server constructs a SOAP response message, which it sends back to the client application.

10. Double-click the Service.svc file to display it in the *Code and Text Editor* window.

    This is the service file for the Web service; it is used by the host environment (IIS, in this case) to determine which class to load when it receives a request from a client application.

 Note If the *Error List* window is open, you will notice that the Service.svc file appears to contain two errors: "Keyword, identifier, or string expected after verbatim specifier: @" and "A namespace does not directly contain members such as fields or methods." When you rebuild the solution later, these errors will disappear and you can safely ignore them.

The *Service* property of the @ *ServiceHost* directive specifies the name of the Web service class, and the *CodeBehind* property specifies the location of the source code for this class.

 Tip If you don't want to deploy the source code for your WCF service to the Web server, you can provide a compiled assembly instead. You can then specify the name and location of this assembly by using the @ *Assembly* directive. For more information, search for "@ *Assembly* directive" in the documentation provided with Visual Studio 2008.

Now that you have seen the structure of a WCF service, you can define the interface and class that specifies the service and data contracts for the ProductsService Web service and then create a class that implements the service contract.

Define the contracts for the ProductsService Web service

1. Display the IProductsService.cs file in the *Code and Text Editor* window. Change the name of the *IService* interface to *IProductsService*, as shown here in bold type:

```
[ServiceContract]
public interface IProductsService
{
    ...
}
```

2. In the *IProductsService* interface, remove the definitions of the *GetData* and *GetDataUsingDataContract* methods, and replace them with the *HowMuchWillItCost* and *GetProductInfo* methods, shown in the following code. Make sure you retain the *OperationContract* attribute for each Web method.

```
[ServiceContract]
public interface IProductsService
{
    [OperationContract]
    decimal HowMuchWillItCost(int productID, int howMany);

    [OperationContract]
    ProductInfo GetProductInfo(int productID);
}
```

The *HowMuchWillItCost* method takes a product ID and a quantity and returns a *decimal* value specifying the amount this quantity will cost.

The *GetProductInfo* method takes a product ID and returns a *ProductInfo* object containing the details of the specified product. You will define the *ProductInfo* class in the next step.

3. Remove the *CompositeType* class from the IProductsService.cs file, and add the *ProductInfo* class, including the *DataContract* attribute, like this:

```
[DataContract]
public class ProductInfo
{
}
```

4. Add the following public properties to the *ProductInfo* class. There is one property for each of the columns in the *Products* table in the database. Mark each property with the *DataMember* attribute:

```
[DataContract]
public class ProductInfo
{
    [DataMember]
    public int ProductID {get; set;}

    [DataMember]
    public string ProductName {get; set;}
```

```
        [DataMember]
        public int? SupplierID {get; set;}

        [DataMember]
        public int? CategoryID {get; set;}

        [DataMember]
        public string QuantityPerUnit {get; set;}

        [DataMember]
        public decimal? UnitPrice {get; set;}

        [DataMember]
        public short? UnitsInStock {get; set;}

        [DataMember]
        public short? UnitsOnOrder {get; set;}

        [DataMember]
        public short? ReorderLevel {get; set;}

        [DataMember]
        public bool? Discontinued {get; set;}
    }
```

Notice that the properties that correspond to columns that allow null values in the database are defined by using nullable types (apart from *QuantityPerUnit*, which is a reference type that allows null values automatically because it is a *string*). Also, you should ensure that all properties support read and write access. The serialization mechanism used by WCF is automatic and largely transparent as long as you follow a few simple rules when defining the class. In particular, serialization can be used only when the runtime transmits objects that contain public fields and properties; private members will not be serialized. Also note that all properties must have both *get* and *set* accessors. This is because the XML serialization process must be able to write this data back to the object after it has been transferred. Additionally, the class must provide a default (with no parameters) constructor.

It is common to design classes used for SOAP purely as containers for transmitting data. If necessary, you can define additional functional classes that act as façades, providing the business logic for these data structures. Users and applications can gain access to the data by using these business façades.

 Note You can customize the serialization mechanism using the various SOAP attribute classes of the *System.Xml.Serialization* namespace or define your own XML serialization mechanism by implementing the *ISerializable* interface of the *System.Runtime.Serialization* namespace.

The next stage is to define the *ProductsService* class that implements the *IProductsService* interface. The methods in this class will retrieve product information from the Northwind database, so you will start by adding an entity class and data context for retrieving this information.

Implement the *IProductsService* interface

1. On the *Website* menu, click *Add New Item*.

2. In the *Add New Item* dialog box, click the *LINQ to SQL Classes* template, type *Product. dbml* in the *Name* text box, select *Visual C#* in the *Language* drop-down list, and then click *Add*.

3. If you are using Visual Studio 2008 Professional Edition or Enterprise Edition, on the *View* menu, click *Server Explorer*.

4. If you are using Visual Web Developer 2008 Express Edition, on the *View* menu, click *Database Explorer*.

5. In *Server Explorer* (if you are using Visual Studio 2008) or *Database Explorer* (if you are using Visual Web Developer 2008 Express Edition), expand the data connection for the Northwind database (*YourComputer\sqlexpress.Northwind.dbo* or *Northwind.mdf*), and then expand *Tables*.

6. Click the *Products* table, and drag it onto the *Object Relational Designer* window.

7. On the *File* menu, click *Save All*.

8. Display the ProductsService.cs file in the *Code and Text Editor* window. Remove the *Service* class from this file.

9. Add the *ProductsService* class to the file, and specify that it should implement the *IProductsService* interface, as shown here:

```
public class ProductsService : IProductsService
{
}
```

10. Add the *HowMuchWillItCost* method to the *ProductsService* class, as follows:

```
public class ProductsService : IProductsService
{
    public decimal HowMuchWillItCost(int productID, int howMany)
    {
        ProductDataContext pdc = new ProductDataContext();
        decimal? cost = pdc.Products.Single(
            p => p.ProductID == productID).UnitPrice;

        decimal totalCost = 0;
        if (cost.HasValue)
```

```
        {
            totalCost = cost.Value * howMany;
        }

        return totalCost;
    }
}
```

This method connects to the database and executes a DLINQ query to retrieve the cost of the product matching the supplied product ID from the Northwind database. If the cost returned is not null, the method calculates the total cost of the request and returns it; otherwise, the method returns the value 0.

> **Note** This method performs no validation of the input parameters. For example, you can specify a negative value for the *howMany* parameter. In a production Web service, you would trap errors such as this, log them, and return an exception. However, transmitting meaningful reasons for an exception back to a client application has security implications in a WCF service. The details are beyond the scope of this book. For more information, see *Microsoft Windows Communication Foundation Step by Step*.

11. Add the *GetProductInfo* method shown below in bold type to the *ProductService* class:

```
public class ProductsService : IProductsService

    ...

    public ProductInfo GetProductInfo(int productID)
    {
        ProductDataContext pdc = new ProductDataContext();
        Product product = pdc.Products.Single(p => p.ProductID == productID);

        ProductInfo prodInfo = null;
        if (product != null)
        {
            prodInfo = new ProductInfo();
            prodInfo.CategoryID = product.CategoryID;
            prodInfo.Discontinued = product.Discontinued;
            prodInfo.ProductID = product.ProductID;
            prodInfo.ProductName = product.ProductName;
            prodInfo.QuantityPerUnit = product.QuantityPerUnit;
            prodInfo.ReorderLevel = product.ReorderLevel;
            prodInfo.SupplierID = product.SupplierID;
            prodInfo.UnitPrice = product.UnitPrice;
            prodInfo.UnitsInStock = product.UnitsInStock;
            prodInfo.UnitsOnOrder = product.UnitsOnOrder;
        }

        return prodInfo;
    }
}
```

These statements use DLINQ to connect to the Northwind Traders database and retrieve the details for the specified product from the database. Note that like the *HowMuchWillItCost* method, this method does not handle exceptions.

Before you can use the Web service, you must update the configuration in the Service.svc file to refer to the *ProductsService* class in the ProductsService.cs file. You must also modify the Web.config file to reflect the new name of the Web service.

Configure the Web service

1. In *Solution Explorer*, double-click the Service.svc file to display it in the *Code and Text Editor* window. Update the *Service* and *CodeBehind* attributes of the *ServiceHost* directive, as shown here in bold type:

```
<%@ ServiceHost Language="C#" Debug="true" Service="ProductsService"
     CodeBehind="~/App_Code/ProductsService.cs" %>
```

 Note The *Error List* window for the Service.svc file might display the same errors as before. Again, these errors should disappear when you rebuild the application, so you can safely ignore them.

2. In *Solution Explorer*, double-click the Web.config file. In the *Code and Text Editor* window, locate the *<system.serviceModel>* element. This element contains the following *<services>* element, specifying the endpoint binding information for the Web service implemented by this solution. (You can ignore the service that implements the *IMetadataExchange* contract in this chapter.)

```
<system.serviceModel>
  <services>
    <service name="Service" behaviorConfiguration="ServiceBehavior">
      <!-- Service Endpoints -->
      <endpoint address="" binding="wsHttpBinding" contract="IService"/>
      ...
    </service>
  </services>
  <behaviors>
    ...
  </behaviors>
</system.serviceModel>
```

WCF uses the notion of endpoints to associate a network address with a specific Web service. If you are hosting a Web service by using IIS or the ASP.NET Development Server, you should leave the *address* property of your endpoint blank because IIS listens for incoming requests on an address specified by its own configuration information.

 Note You can build your own custom host applications if you don't want to use IIS or the ASP.NET Development Server. In these situations, you must specify an address for the service as part of the endpoint definition. For more information about endpoints and custom hosts, see *Microsoft Windows Communication Foundation Step by Step*.

3. In the Web.config file, change the *name* attribute of the *Service* element and the *contract* attribute of the endpoint element to refer to the *ProductsService* service and the *IProductsService* contract, as shown here in bold type:

```
<system.serviceModel>
  <services>
    <service name="ProductsService" behaviorConfiguration="ServiceBehavior">
      <!-- Service Endpoints -->
      <endpoint address="" binding="wsHttpBinding" contract="IProductsService"/>
      ...
    </service>
  </services>
  <behaviors>
    ...
  </behaviors>
</system.serviceModel>
```

4. On the *File* menu, click *Save All*.

5. In *Solution Explorer*, right-click *Service.svc*, and then click *View in Browser*.

Internet Explorer starts and displays the following page, confirming that you have successfully created and deployed the Web service and providing helpful information describing how to create a simple client application that can access the Web service.

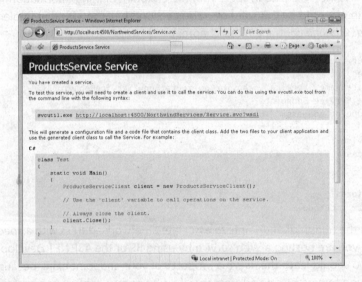

 Note If you click the link shown on the Web page (*http://localhost:4500/ NorthwindServices/Service.svc?wsdl*), Internet Explorer displays a page containing the WSDL description of the Web service. This is a long and complicated piece of XML, but Visual Studio 2008 can take the information in this description and use it to generate a class that a client application can use to communicate with the Web service.

6. Close Internet Explorer, and return to Visual Studio 2008.

Web Services, Clients, and Proxies

You have seen that a Web service uses SOAP to provide a mechanism for receiving requests and sending back results. SOAP uses XML to format the data being transmitted, which rides on top of the HTTP protocol used by Web servers and browsers. This is what makes Web services so powerful—SOAP, HTTP, and XML are well understood (in theory anyway) and are the subjects of several standards committees. Any client application that "talks" SOAP can communicate with a Web service. So how does a client "talk" SOAP? There are two ways: the difficult way and the easy way.

Talking SOAP: The Difficult Way

In the difficult way, the client application performs a number of steps. It must do the following:

1. Determine the URL of the Web service running the Web method.

2. Perform a Web Services Description Language (WSDL) inquiry using the URL to obtain a description of the Web methods available, the parameters used, and the values returned. You saw how to do this by using Internet Explorer in the preceding exercise.

3. Parse the WSDL document, convert each operation to a Web request, and serialize each parameter into the format described by the WSDL document.

4. Submit the request, along with the serialized data, to the URL by using HTTP.

5. Wait for the Web service to reply.

6. Using the formats specified by the WSDL document, deserialize the data returned by the Web service into meaningful values that your application can then process.

This is a lot of work just to invoke a method, and it is potentially error-prone.

Talking SOAP: The Easy Way

The bad news is that the easy way to use SOAP is not much different from the difficult way. The good news is that the process can be automated because it is largely mechanical. As mentioned earlier, many vendors, including Microsoft, supply tools that can generate a proxy class based on a WSDL description. The proxy hides the complexity of using SOAP and exposes a simple programmatic interface based on the methods published by the Web service. The client application calls Web methods by invoking methods with the same name in the proxy. The proxy converts these local method calls to SOAP requests and sends them to the Web service. The proxy waits for the reply, deserializes the data, and then passes it back to the client just like the return from any simple method call. This is the approach you will take in the exercises in this section.

Consuming the ProductsService Web Service

You have created a Web service call that exposes two Web methods: *GetProductInfo* to return the details of a specified product and *HowMuchWillItCost* to determine the cost of buying *n* items of product *x* from Northwind Traders. In the following exercises, you will use this Web service and create an application that consumes these methods. You'll start with the *GetProductInfo* method.

Open a Web service client application

1. Start another instance of Visual Studio 2008. This is important. The ASP.NET Development Server stops if you close the NorthwindServices Web service project, meaning that you won't be able to access it from the client. (An alternative approach you can use if you are running Visual Studio 2008 and not Visual Web Developer 2008 Express Edition is to create the client application as a project in the same solution as the Web service.) When you host a Web service in a production environment by using IIS, this problem does not arise because IIS runs independently of Visual Studio 2008.

> **Important** If you have been using Visual Web Developer 2008 Express Edition for the exercises in this part of the book, start Visual C# 2008 Express Edition rather than a second instance of Visual Web Developer 2008 Express Edition (leave Visual Web Developer 2008 Express Edition running).

2. In the second instance of Microsoft Visual Studio 2008, open the ProductClient solution in the \Microsoft Press\Visual CSharp Step By Step\Chapter 30\ProductClient folder in your Documents folder.

3. In *Solution Explorer*, double-click the file ProductClient.xaml to display the form in the *Design View* window. The form looks like this:

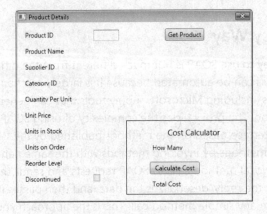

The form enables the user to specify a product ID and retrieve the details of the product from the Northwind database. The user can also provide a quantity and retrieve a price for buying that quantity of the product. Currently, the buttons on the form do nothing. In the following steps, you will add the necessary code to invoke the methods from the ProductsService Web service to obtain the data and then display it.

Add code to call the Web service in the client application

1. On the *Project* menu, click *Add Service Reference*.

   The *Add Service Reference* dialog box opens. In this dialog box, you can browse for Web services and examine the Web methods that they provide.

2. In the *Address* text box, type **http://localhost:4500/NorthwindServices/Service.svc**, and then click *Go*.

   The ProductsService service appears in the *Services* box.

3. Expand the ProductsService service, and then click the *IProductsService* interface that appears. In the *Operations* list box, verify that the two operations, *GetProductInfo* and *HowMuchWillItCost*, appear, as shown in the following image.

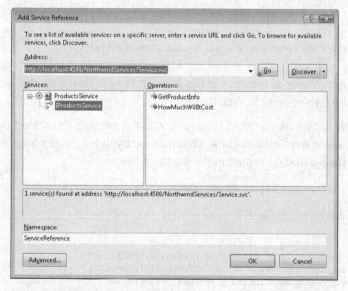

4. Change the value in the *Namespace* text box to *NorthwindServices*, and then click *OK*.

   A new folder called Service References appears in *Solution Explorer*. This folder contains an item called NorthwindServices.

5. Click the *Show All Files* button on the *Solution Explorer* toolbar. Expand the NorthwindServices folder, and then expand the Reference.svcmap folder. Double-click the Reference.cs file and examine its contents in the *Code and Text Editor* window.

This file contains several classes and interfaces, including a class called *ProductsServiceClient* in a namespace called *ProductClient.NorthwindServices*. The *ProductsServiceClient* is the proxy class generated by Visual Studio 2008 from the WSDL description of the *ProductsService* Web service. It contains a number of constructors, as well as methods called *HowMuchWillItCost* and *GetProductInfo*. The client application can instantiate the *ProductsServiceClient* class and call these methods. When this happens, these methods invoke code that packages up the information supplied as parameters into a SOAP message that they transmit to the Web service. When the Web service replies, the information returned is unpacked from the SOAP response and passed back to the client application. In this way, the client application can call a method in a Web service in exactly the same way as it would call a local method.

6. Display the ProductClient.xaml form in the *Design View* window. Double-click the *Get Product* button to generate the *getProduct_Click* event handler method for this button.

7. In the *Code and Text Editor* window, add the following *using* statement to the list at the top of the ProductClient.xaml.cs file:

```
using ProductClient.NorthwindServices;
```

8. In the *getProduct_Click* method, create the variable shown here in bold type:

```
private void getProduct_Click(object sender, RoutedEventArgs e)
{
    ProductsServiceClient proxy = new ProductsServiceClient();
}
```

This statement creates an instance of the *ProductsServiceClient* class that your code will use to call the *GetProductInfo* Web method.

9. Add the code shown here in bold type to extract the product ID entered by the user on the form, execute the *GetProductInfo* Web method by using the *proxy* object, and then display the details of the product in the labels on the form.

```
private void getProduct_Click(object sender, RoutedEventArgs e)
{
    ProductsServiceClient proxy = new ProductsServiceClient();
    try
    {
        int prodID = Int32.Parse(this.productID.Text);
        ProductInfo product = proxy.GetProductInfo(prodID);
        this.productName.Content = product.ProductName;
        this.supplierID.Content = product.SupplierID;
        this.categoryID.Content = product.CategoryID;
        this.quantityPerUnit.Content = product.QuantityPerUnit;
        this.unitPrice.Content = String.Format("{0:C}", product.UnitPrice);
        this.unitsInStock.Content = product.UnitsInStock;
        this.unitsOnOrder.Content = product.UnitsOnOrder;
        this.reorderLevel.Content = product.ReorderLevel;
        this.discontinued.IsChecked = product.Discontinued;
    }
```

```
    catch (Exception ex)
    {
        MessageBox.Show("Error fetching product details: " +
            ex.Message, "Error", MessageBoxButton.OK,
            MessageBoxImage.Error);
    }
    finally
    {
        if (proxy.State == System.ServiceModel.CommunicationState.Faulted)
            proxy.Abort();
        else
            proxy.Close();
    }
}
```

You are probably aware of how unpredictable networks are, and this applies doubly to the Internet. The *try/catch* block ensures that the client application catches any network exceptions that might occur. It is also possible that the user might not enter a valid integer into the *ProductID* text box on the form. The *try/catch* block also handles this exception.

The *finally* block examines the state of the proxy object. If an exception occurred in the Web service (which could be caused by the user supplying a nonexistent product ID, for example), the proxy will be in the *Faulted* state. In this case, the *finally* block calls the *Abort* method of the proxy to acknowledge the exception and close the connection; otherwise, it calls the *Close* method. The *Abort* and *Close* methods both close the communications channel with the Web service and release the resources associated with this instance of the *ProductsServiceClient* object.

10. Display the ProductClient.xaml form in the *Design View* window again. Double-click the *Calculate Cost* button to generate the *calcCost_Click* event handler method for this button.

11. In the *calcCost_Click* method, add the code shown here in bold type:

```
private void calcCost_Click(object sender, RoutedEventArgs e)
{
    ProductsServiceClient proxy = new ProductsServiceClient();
    try
    {
        int prodID = Int32.Parse(this.productID.Text);
        int number = Int32.Parse(this.howMany.Text);
        decimal cost = proxy.HowMuchWillItCost(prodID, number);
        this.totalCost.Content = String.Format("{0:C}", cost);
    }
    catch (Exception ex)
    {
        MessageBox.Show("Error obtaining cost: " +
            ex.Message, "Error", MessageBoxButton.OK,
            MessageBoxImage.Error);
    }
    finally
```

```
    {
        if (proxy.State == System.ServiceModel.CommunicationState.Faulted)
            proxy.Abort();
        else
            proxy.Close();
    }
}
```

This code follows a similar pattern to the *getProduct_Click* method. It creates an instance of the *ProductsServiceClient* class and calls the *HowMuchWillItCost* method using this instance, passing the product ID and the quantity required as parameters. The return value is displayed on the form. The exception handler traps any errors, and the *finally* block ensures that the network connection is closed when the method finishes.

Test the application

1. Build and run the project. When the Product Details form appears, type **3** in the *Product ID* text box, and then click *Get Product*.

   After a short delay while the client instantiates the proxy and builds a SOAP request containing the product ID, the proxy sends the request to the Web service. The Web service deserializes the SOAP request to extract the product ID, reads the database, creates a *Product* object, serializes it as XML, and then sends it back to the proxy. The proxy deserializes the XML data and creates a *ProductInfo* object and then passes this object to your code in the *getButton_Click* method. The details for Aniseed Syrup then appear in the form as shown in the following graphic:

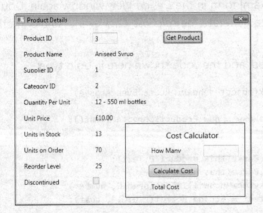

Tip If you get an exception with the message "Could not connect to http://localhost:4500/ NorthwindServices/Service.svc. TCP error code 10061: No connection could be made because the target machine actively refused it," the ASP.NET Development Server has probably stopped running. (It shuts down if it is inactive for a time.) To restart it, switch to the Visual Studio 2008 instance for the ProductsService Web service, right-click *Service.svc* in *Solution Explorer*, and then click *View in Browser*. Close Internet Explorer when it appears.

2. Type **24** in the *Product ID* text box, and then click *Get Product*.

    For reasons that are outside the scope of this book to explain, you will probably find that the details are displayed more quickly this time.

3. Type **10** in the *How Many* text box, and then click *Calculate Cost*. Verify that the value displayed in the *Total Cost* field on the form is 10 times the value shown in the *Unit Price* field.

4. Experiment by typing the IDs of other products. Notice that if you enter an ID for a product that does not exist, the Web service returns an exception. As described earlier, the error message returned does not contain any information that could be useful to an attacker, although it does describe how you can enable meaningful error messages and log exceptions. When you have finished, close the form and return to Visual Studio 2008.

Congratulations. You have now built your first WCF service together with a client application that calls its methods.

You have also completed all the exercises in this book. You should now be thoroughly conversant with the C# language and understand how to use Visual Studio 2008 to build professional applications. However, this is not the end of the story. You have jumped the first hurdle, but the best C# programmers learn from continued experience, and you can gain this experience only by building C# applications. As you do so, you will discover new ways to use the C# language and the many features available in Visual Studio 2008 that I have not had space to cover in this book. Also, remember that C# is an evolving language. Back in 2001, when we wrote the first edition of this book, C# introduced the syntax and semantics necessary for you to build applications that made use of .NET Framework 1.0. Some enhancements were added to Visual Studio and .NET Framework 1.1 in 2003, and then in 2005, C# 2.0 emerged with support for generics and .NET Framework 2.0. As you have seen in this book, C# 3.0, the latest release of the language aligned with Visual Studio 2008 and .NET Framework 3.5, has added numerous features such as anonymous types, lambda expressions, and most significantly, LINQ. What will the next version of C# bring? Watch this space!

Chapter 30 Quick Reference

To	Do this
Create a Web service	Use the WCF Service template. Define a service contract that specifies the Web methods exposed by the Web service by creating an interface with the *ServiceContract* attribute. Tag each method with the *OperationContract* attribute. Create a class that implements this interface.
Display the description of a Web service	Right-click the .svc file in *Solution Explorer*, and click *View in Browser*. Internet Explorer runs, moves to the Web service URL, and displays a page describing how to create a client application that can access the Web service. Click the WSDL link to display the WSDL description of the Web service.
Pass complex data as Web method parameters and return values	Define a class to hold the data and tag it with the *DataContract* attribute. Ensure that each item of data is accessible either as a public field or through a public property that provides *get* and *set* access. Ensure that the class has a default constructor (which might be empty).
Add a service reference to a client application and create a proxy class	On the *Project* menu, click *Add Service Reference*. Type the URL of the Web service in the *Address* text box at the top of the dialog box, and then click *Go*. Specify the namespace for the proxy class, and then click *OK*.
Invoke a Web method	Create an instance of the proxy class. Call the Web method using the proxy class.

Index

John Sharp

John Sharp is a Principal Technologist at Content Master (*www.contentmaster.com*), part of CM Group, a technical authoring company in the United Kingdom. He researches and develops technical content for training courses, seminars, and white papers. John is deeply involved with Microsoft .NET Framework application development and interoperability. He has written papers and courses, built tutorials, and delivered conference presentations covering distributed systems and Web services, application migration and interoperability between Microsoft Windows/.NET Framework and UNIX/Linux/Java, as well as development using the C# and J# languages. John has also authored *Microsoft Windows Communication Foundation Step by Step* and *Microsoft Visual J# Core Reference*, both published by Microsoft Press.